SNOW

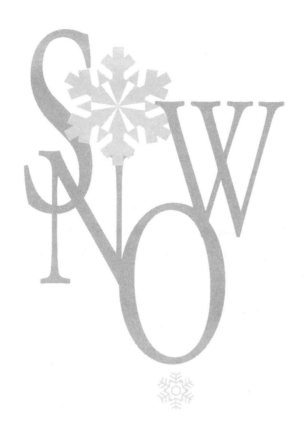

SNOW

IN AMERICA

BERNARD MERGEN

SMITHSONIAN INSTITUTION PRESS
WASHINGTON AND LONDON

Copyright acknowledgments appear on pp. 311–12

Copy editor: Susan M. S. Brown
Production editor: Duke Johns
Designer: Linda McKnight

Library of Congress Cataloging-in-Publication Data
Mergen, Bernard.
Snow in America / Bernard Mergen.
p. cm.
Includes bibliographical references and index.
ISBN 1-56098-780-4 (alk. paper)
1. Snow—North America—History. 2. Snow—Environmental aspects—North America—History. I. Title.
QC929.S7M44 1997
551.57'84'097—dc21 97-12247

British Library Cataloguing-in-Publication Data is available

Manufactured in the United States of America
04 03 02 01 00 99 98 97 5 4 3 2 1

∞ The paper used in this publication meets the minimum requirements of the American National Standard for Information Sciences—Permanence of Paper for Printed Library Materials ANSI Z39.48-1984.

For Claudia

Maybe falling in love, the piercing knowledge that we ourselves will someday die, and the love of snow are in reality not some sudden events; maybe they are always present. PETER HØEG, *SMILLA'S SENSE OF SNOW*

Contents

ACKNOWLEDGMENTS
ix

INTRODUCTION
xiii

ACCUMULATION
xix

1. EARLY PHILOSOPHIES OF SNOW
1

Snows of Yesteryear 1
Frozen Women 5
Transcendental Weathermen 8
"Beautiful Snow" 13
Snow in War and Peace 22
Discovery of the Snow Frontier 33

2. STALLED MAGNIFICENCE
41

Trains Follow the Plow 41
In Advance of the Broken Arm: Shoveling Snow 50
Urban Snow: Refuse or Art? 53
Shoveling out of the Depression and into War 62
The Flurry Flurry and Political Snows 69
"Even As We Speak": The Livable Winter City 75

3. SNOWMEN AND SNOWMANSHIP: RECREATION IN THE SNOW
81

Snowballs and Snowmen 81
Ice Palaces, Winter Carnivals, and Snow Sculptures 87
"Seaming a Virgin Face": Skiing and the Redefinition of Snow 94
Designer Snow 108
Avalanche Hunting 114

4. OPENING THE SNOW FRONTIER
121

Snow into Water 121
The Mount Rose Snow Sampler and the Origins of Snow Science 126
Interception: The Forest versus Snow Controversy 133
A New Deal for Snow 139
Niphometrology and the War Years 145
Snow Tubes, SNOTEL, and Airborne Snow Surveys 148

5. THE NAMES OF THE SNOWS
159

Qanik to Slush: From Two Kinds of Snow to Twelve 159
From Slush to Crud: Skiers' Contributions to Snow Terminology 166
Snowflakes to Snow Grains: From a Dozen to 101 Kinds of Snow 171

6. ECOLOGY OF THE SNOW COMMONS
183

Nivean Frontiers 183
Niveculture 190
Eating Yellow Snow: The Nivean Food Chain and Pollution 198

7. THE MODERN MINDS OF WINTER
207

"Nothing that is not there . . ." 207
Rhyme on Rime 225
". . . and the nothing that is" 239

ABLATION
247

NOTES
251

SOURCE CREDITS
311

INDEX
313

Acknowledgments

In the decade that it has taken to research, write, and publish this book, I have acquired many intellectual debts that can be only partially repaid here. Earliest encouragement came from Wilcomb Washburn and Nathan Reingold of the Smithsonian Institution and James Fleming, who was a Fellow at that time. No aspect of this project has been more rewarding than my association with the members of the professional associations of snow scientists—Eastern Snow Conference, Western Snow Conference, and International Snow Science Workshop. David McClung encouraged me to present a paper in 1988 at the ISSW meeting in Whistler, British Columbia, where I met Sam Colbeck of the Cold Regions Research and Engineering Laboratory; Richard Armstrong of the World Data Center, CIRES, University of Colorado; Ed Adams of Montana State University, who was at that time at the Institute of Snow Research at Michigan Technological University; and many others who were indispensable sources of information. Ed introduced me to Al Wuori of the Keweenaw Research Center, and to Brad Baltensperger of Michigan Technological University, from whom I learned much about the "yooper scooper" and other important snow management technologies.

Later, on research trips and at meetings of the ESC and WSC, I benefited from conversations with Chris Pacheco, of the Soil Conservation Service; Hal Klieforth, of the Desert Research Institute, University of Nevada, Reno; and Tom Carroll, National Operational Hydrological Remote Sensing Center, National Weather Service, National Oceanic and Atmospheric Administration. Hal sent me to three of the historians of western snow—Doug Powell, Charles Bradley, and Bill Berry—from whom I learned many things unavailable in archives. At meetings of the ESC and WSC I have benefited from conversations with Skip Walker, University of Colorado; Peter Adams, Trent University, Peterborough, Ontario; Ron Hoham, Colgate University; and Don Wiesnet, U.S. Weather Bureau, retired.

Thanks to Bob Rydell and Bob Brown of Montana State University, I was able to visit that institution and learn from them and Steve Custer, Ted Lange, and Glen Liston. Special thanks to Bob Brown and his wife, Maryanne, for taking me cross-country skiing near Bozeman. Doug Helms, historian at the Soil Conservation Service, U.S. Department of Agriculture, a longtime friend and fellow historian of snow surveying, put me in touch with Dave Johnson of SCS in Portland, Oregon, who supported my application to the West-Wide Snow Survey School held in January 1991 near Bend, Oregon. A week in the snow with the instructors, especially Doug Fesler, Stan Fox, Don Huffman, Ron Jones, Peter Palmer, and Garry Schaefer, provided much needed hands-on experience. In the snow of the bivouac area, I learned some survival skills from Scott Guenther of the Bureau of Reclamation in Billings and Martin Barnaby of the Conference of Kootenai and Salish Tribes in Pablo, Montana.

I thank the editors of the *New York Times Book Review* for publishing my author's query and the good readers who responded, many of whom are acknowledged in the notes. Even if I did not use all your information, I learned much from your responses.

Airfare to Vancouver, which allowed me to attend a meeting of the International Snow Science Workshop, was provided by Dr. Norman London of the Canadian Embassy in Washington, D.C.

George Washington University contributed to this book by granting me sabbatical leaves at the beginning and end of this project. My colleagues in the American Studies Department, especially Howard Gillette and John Vlach, have been supportive. Joel Kuipers of the Anthropology Department told me about Laura Martin, who then shared her Eskimo material with me. Quadir Amiryar, interlibrary loan librarian of the university, obtained many rare and important books for me.

Scholars at other institutions in Washington, D.C., were equally generous with their time; John Sherrod gave me invaluable assistance at the NOAA library. At the Smithsonian Institution, Jack White's knowledge of railroad history saved me time in searching for materials; Pam Henson provided information on both the history of the institution and the history of biology; Reggie Blazechezk found interesting illustrations of snow in the course of her own research; Liza Kirwin of the Archives of American Art helped me find Rockwell Kent material; Lois Fink of the National Museum of American Art invited me to participate in her colloquia; and Jeff Stine made me feel welcome in environmental history.

Many former and current students helped me, none more than Pete Rothenhofer and Peggy Hermann, but contributions from Margie Baer, Sarah Brown, James Deutsch, Meredith Fisher, Martin Gordon, Noriko Ishii, Juan Liu, Jane Loeffler, Jeanne Schinto, Carol Slatik, and Phil Terrie were also significant. Jim Deutsch's assistance in securing permissions for quotations and illustrations was crucial. Colleagues in other institutions shared their knowledge of snow in history, folklore, photography, and literature with me: Julie Brown, Miriam Fromanek-Brunnell, M. Thomas Inge, Alan Jabbour, and Jean Spraker. Encouragement from abroad came from Orm Øverlund in Norway, Jean Malaurie in France, Maya Koroneva and Louisa Bashmakova in Russia, and Mick Gidley, Robert Lawson-Peebles, and Robert Lewis in England. As I neared the end of the project I met Cullen Murphy, managing editor of the *Atlantic,* who shares my enthusiasm for snow and provided needed encouragement.

In the editing process I benefited from the interest and good judgment of Mark Hirsch, Bob Lockhart, and Duke Johns of the Smithsonian Institution Press. Freelance editor Susan M. S. Brown was particularly helpful and thorough.

Some of the final revisions were made while I was Fulbright professor at the National University of Mongolia in Ulaanbaatar, an experience that yielded yet another perspective on the role of snow in human affairs.

Within my small fort besieged with snow, good friends Bob and Johanna Humphrey, my daughter, Alexa, and son-in-law, Matt Weiser, my son, Andrew, and daughter-in-law, Kate, my mother, Katharine, and my wife, Claudia, who read and edited many drafts, provided an avalanche of information, a blizzard of insights, and the firm crust of love.

Introduction

To write a snow poem you must ignore the snow
falling outside your window.
RICHARD HUGO, "SNOW POEM"

I decided to write a history of snow in America knowing full well that many of my friends and colleagues considered the topic too "flaky." "Are you trying to snow us?" they asked, laughing. "You must be snowed under," they said, and howled. You get the drift.

The fact that snow metaphors constitute an important part of our daily speech proves that the subject has linguistic relevance. My task is to demonstrate its importance to other issues central to the study of history, culture, and nature. This book is an attempt to tell the story of snow and responses to it.[1]

My argument is that neither the environmental nor the symbolic importance of snow is fully appreciated. I believe that snow requires the same attention as its liquid collateral, water, and that snow has had and continues to have great metaphorical power to illuminate paradoxes in nature and culture. Snow may, in fact, be the perfect metaphor for science itself, the search for unifying theories in the face of innumerable bits of information. The various attempts to name and define snow illuminate the ways nature and culture interact. By examining how we know what we know about snow, we may begin to understand the truth about it.

For me this book is a way of combining interests in ecology, history, literature, and art. Doing what I think American studies should be. My childhood was spent on the eastern slope of the Sierra Nevada, and the debt I owe to that place was also a significant stimulus to write, as was my personal acquaintance with James Church, inventor of the Mount Rose Snow Sampler.

Snow falls on more than half of the North American continent every year. Annually, a majority of Americans experience some snow. Snow is an important influence on global climate and therefore on ecology. Snow alters the landscape and challenges us physically and mentally. The crucial variables in discussing snowcover and its effects are the four D's: dates of occurrence and disappearance, depth, density (water content), and duration. The fact that these variables are constantly changing is one of the reasons it is so difficult to measure snow. Buffalo, Rochester, and Syracuse, New York, vie for the claim of snowiest major city in the contiguous forty-eight states, averaging 1.83 to 2.74 meters (6 to 9 feet) each winter. Denver, Salt Lake City, and Minneapolis–St. Paul get 1.22 to 1.83 meters (4 to 6 feet) of snow a year; and Chicago, Boston, and Pittsburgh are usually not far behind. Among smaller cities, Marquette, Michigan, averages over 3.3 meters (11 feet) in the long white months between October and May. T-shirts in Houghton, on Michigan's Keweenaw Peninsula, proudly proclaim: "Spend four seasons in the Keweenaw—Early Winter, Mid Winter, Late Winter, and Next Winter."[2]

Snowcover begins, according to one standard definition, when 2 centimeters (0.75 inch) of snow lies on the ground. This is misleading, however, since such a light snow may melt soon after being recorded. Another definition, used by city governments and those responsible for snow removal, is "snow day," a period that may exceed twenty-four hours but during which at least 2.54 centimeters (1.00 inch) of snow falls. For biologists concerned with the effects of snow on plant and animal life, 20 centimeters (8 inches) of snowcover is the minimum necessary to sustain life under the snow, a subnivean environment, for creatures such as voles, shrews, and white-footed mice.[3] Like art, snowcover is difficult to define but recognizable when seen.

The anticipation of snow, followed by preparation for it, the actual experience of it, recovery from its effects, and finally the memory of snow after the spring melt, make snow consciousness a part of everyday life. It is this quotidian aspect that makes snow a good topic for the historian.

As one way of finding out what people think about snow, I published an author's query in the *New York Times Book Review,* asking for "interesting or unusual experiences with snow" and memories of first encounters with it. Initially I was surprised that so many of the nearly one hundred letters I received were written by southerners. I soon realized that by asking for first experience of snow, I probably excluded many whose earliest experience predated their conscious memories. Those who did not see snow until they were adults, by contrast, vividly recalled being surprised that snow did not cover the ground like a big blanket, as it does in cartoons. They remembered thinking the first snowflakes they saw were soap, rice, sugar, flowers, even nuclear fallout. Above all, they believed that once they had experienced snow they were "real" Americans. Three of the hundred letters I received mentioned that Mayor Felisa Rincón de Gautier of San Juan had snow flown from New York to pile in Muñoz Rivera Park so Puerto Rican children could experience an "American" Christmas. Snow's geopolitical dimensions are even clearer in the poetry of immigrants from the tropics. The Philippine-American Carlos Bulosan put it simply and beautifully: "I walked homeward to America, across the snow."[4]

These and similar stories illustrate another dimension of snow as a subject for serious inquiry. That is, that while our responses to snow are conditioned in part by who we are, snow itself is a constant. To understand responses to snow it is necessary to ask: What is snow? How do we know an element of nature except through cultural responses to it? The answer, I think, is that while we can never fully define snow, we can know it in the sense that we feel something we call cold, white, and wet falling from the sky and see it covering the ground. "To see nature as active," Carolyn Merchant writes, "is to recognize its formative role over geologic and historical time. Only by according ecology a place in the narrative of history can nature and culture be seen as truly interactive."[5]

So I begin with the premise that what we call snow exists before our human invention of it. I ask: What names have people bestowed on snow? What questions have they asked about snow? What answers have they given? What symbolic meanings have they found? How have they managed snow? These questions lead to all kinds of historical inquiry—environmental, cultural, social, legal, philosophical, literary, artistic, architectural, scientific, and technological.

In the following pages I look at what men and women have written about snow, the ways they have used snow, and the ways they have depicted it visually. There is a rich heritage of cultural artifacts associated with snow. Curiously, historians have been silent on this realm. Two earlier books on snow—Corydon Bell's *Wonder of Snow* (1957) and Ruth Kirk's *Snow* (1977)—provide some history, but both are intended to introduce snow science rather than to analyze cultural responses. Douglas Helms, historian for the Soil Conservation Service, has begun to write about government snow surveys, and the physicist Samuel Colbeck has written on the history of snowcover research for the *Journal of Glaciology*, but you would never know from recent histories of the arid West and California that snow is the chief source of water in these regions.[6]

Historians Stephen Pyne and Donald Worster offer models of the kind of history attempted here. Pyne uses fire as a means of historical understanding. "By studying fire," he writes, "one can extract information from the historical record that is otherwise inaccessible or overlooked, much as burning often flushes infertile biotas with nutrients and cooking renders palatable many otherwise inedible foodstuffs. Fire can reconfigure historical landscapes and remake raw materials into humanly usable history. . . . In the end, it describes as few phenomena can the interplay between humans and landscape, which is to say it illuminates the character of each." I think that the commonplaceness of snow makes it an even better euhemerist of the myths of history and nature. Like Donald Worster, who seeks ways to describe ecosystems in the context of technological change, I have tried to do three things: discover past natural environments; see how technology has restructured human ecological relations; and show how perceptions, ideologies, ethics, laws, and myths have become part of an individual's or a group's dialogue with the natural world.[7]

The first chapter of this book surveys literary and artistic responses to snow from the beginning of the nation through the nineteenth century. I find several "philosophies of snow," which come together in a nationalist ideology marked by a belief in the benefits of winter weather in shaping character and the discovery of the economic and psychological importance of the snow frontier. Chapter 2 examines the impact of snow on railroads and on cities, how Americans learned to manage snow as refuse. Chapter 3 looks at play in the snow, from snowball fights to skiing, and how the demand for winter recreation has led to further snow management. Chapters 4, 5, and 6 focus on the scientific study of snow in the twentieth century—snow as a resource. The final chapter returns to literature and art, reviewing the ways snow has been used as a symbol of the modern age. The organization is thus roughly chronological but partly topical. Readers with special interests in or knowledge of one of these areas might begin with the chapter or chapters that focus on their specialities, then go to other chapters. I want both specialists and general readers to think about snow in new ways.

I regret that I do not have the space to discuss Native American concepts of snow. Although I use Canadian writers and artists for comparison, my treatment of them is also too brief. The Arctic and Antarctic have rich literatures of their own, so I have left out polar snows and ice. There are other gaps to be filled, material to be plowed through by future historians of snow. The poet John Haines has imagined "the coldest scholar on earth, who followed each clue in the snow, writing a book as he went . . . the history of snow, the book of winter."[8] I have followed some clues, others remain. There are many books of winter.

I hypothesize that Americans have developed five relatively distinct attitudes toward snow. The first emerged in the century before 1830. During this time Americans endured snow and adapted to it. They began to make it a symbol of moral and physical fitness and ultimately a part of national identity.

A second set of attitudes appeared in the 1840s, when creative writers and creative scientists started to look at snow in new ways. Snow was celebrated for its symbolic paradoxes, which seemed to mirror Yankee character. But snow was more than metaphor, it was part of weather, which was measurable and therefore partially predictable. Following the great blizzards of the 1880s, this belief developed into a third cluster of opinions. Demands to settle and irrigate the West and open rail transportation to all regions in all seasons came to dominate. Snow had to be studied, described, and named. The Weather Bureau grew in importance within the federal bureaucracy. An era of polar exploration contributed to snow consciousness. Naturalism replaced Romanticism in the arts, and snow became a symbol more of illusion than of creativity and revelation.

By the 1920s the automobile had so revolutionized transportation that snow was a major problem for city and highway administrators. This fourth attitude defined snow as refuse as well as resource. At the same time, western development and the reclamation movement demanded more accurate data on snow density for water management. Winter recreation, especially skiing, became a major commercial activity. Vilhjalmur Stefansson extolled the northward course of empire in his book by that title, and Wallace Stevens expressed a modernist creed that demanded "a mind of winter."[9]

A fifth set of beliefs emerged in the metaphorical winter of the Cold War and continues to the present. Space-age weather satellites and fear of global climate changes—floods, drought, and pollution—have reshaped snow consciousness. A cry has gone up for "Livable Winter Cities," and snow has become a symbol of lost innocence, what might have been.

As human responses to snow were growing more complex, snow fell. And snow will continue to fall each winter for as long as any of us will care. This is the story of snow. Is it a good story? Does it make us care about the world we inhabit? How do we use it? Does it explain why and when snow is considered refuse and why and when it is seen as a resource? Do our cultural responses to snow help us get true knowledge of the real world? The answers to these questions are important for understanding the history of culture and nature and their interrelationships.[10]

Accumulation

Not slowly wrought, nor treasured for their form
In heaven, but by the blind self of the storm
Spun off, each driven individual
Perfected in the moment of his fall.

HOWARD NEMEROV, "SNOWFLAKES"

Snow, like you and me, has a history. Like ours, its ancestry can be traced. It begins in the sky, in the swirling atmosphere of oxygen, nitrogen, and other elements surrounding the earth. At 9 to 12 kilometers (30,000 to 40,000 feet) above us, temperatures cool to –60°C (–75°F), a level rarely reached even at the poles. At this altitude, the boundary between the troposphere and the stratosphere, virtually all weather begins. We fly along this boundary on most long-distance commercial flights.

Looking into the sky you may see clouds, masses of water vapor dense enough to be visible. Clouds have shapes determined by the amount of water in vapor or solid form, air movement, temperature, and, I think it is fair to say, imagination. The same cloud may be a person's fist, a duck, the bust of Caesar, the Beast of the Apocalypse, or a fair-weather cumulus. The naming, the classification, of clouds, or anything else, depends on purpose. Moreover, clouds are constantly changing.

Living in these ephemeral cloud lands are thousands of species of bacteria, fungi, and protozoa; the pollen of more than 10,000 species of flowering plants; and an incalculable number of specks of dust. Some of these particles will form the nuclei of snow crystals. Water vapor condenses and becomes liquid in cold, saturated air. It freezes around some nuclei at just below 0°C, whereas a pure water droplet will remain unfrozen until the temperature falls to between –18° and –31°C (0° and 24°F), depending on its size. As they descend toward earth, these ice crystals go through a process called sublimation—transforming from solid to gaseous to solid without liquefying and growing from invisible particles to aggregations of crystals we call snowflakes.

Water gets into the atmosphere when the heat of the sun evaporates the liquid from a river, ocean, lake, or puddle. The transpiration of plants and trees, exuding tiny droplets of moisture, also contributes. Sun-warmed moist air rises. Cold air sinks. All around us air is warming and cooling, rising and falling, spinning with the rotation of the globe. The winds from this carousel of microorganisms and water vapor bounce crazily off mountains; the earth tilts back and forth on its axis. For a few splendid moments of each day, somewhere all the ingredients for making snow—temperature, water vapor, movement, latitude, and season of the year—are in place.

The colder the air temperature, the greater the number of particles capable of acting as nuclei. More nuclei, potentially more snow crystals. Temperature affects the shape of snow crystals. At just below 0°C (32°F), single hexagonal plates begin to form. At –3° to –5°C (27° to 23°F), the crystals are needlelike. From –5° to –8°C (23° to 18°F), the crystals become hollow prismatic columns. Plates again predominate between –8° and –12°C (18° and 10°F). From –12° to –16°C (10° to 3°F), star-shaped stellar dendrites are formed; below –16°C the plates appear again, and below –22°C (–8°F) hollow columns reappear. As the crystals fall they evolve from one form to another. Currents of air may carry them from 0° to –30°C (–22°F) and back again. They may melt completely, leaving the nuclei to attract other water droplets.

Nature is wasteful. The chance of the right humidity, temperature, and nuclei occurring at just the right moment to create snow crystals is slight. The chance of any one snow crystal growing from a speck smaller than the width of a human hair to a collection of crystals as large as a dime is even slighter. The chance that the collection of crystals we call a snowflake will survive buffeting winds and fluctuating temperatures to reach the ground is hardly better than the proverbial snowball's chance in hell. Yet some snow crystals get the chance to become snowballs.

Descending at the rate of 30 centimeters (about 1 foot) a second—faster for needles, slower for stellar dendrites—a snow crystal can spend from a few minutes to five or six hours in the air. Each snowflake is the product of a separate moment of nucleation and falls uniquely to "perfection." This is why there are no two alike but they can be classified by shape and process of formation. As to all questions of quiddity, the answer to the riddle of identical snowflakes lies in defining the degree of similarity. The more interesting question is why we continue to seek twin snowflakes.

For many people falling snow is more beautiful and fascinating than fallen, the accumulation more intriguing than the ablation. I have listened to meteorologists who were eloquent on the subject of snow formation but indifferent to snow on the ground. For them the life of the snowflake ended when it reached the earth. By contrast, for glaciologists and hydrologists the snowflake is prenatal. They seldom look up. Their snow is born in the crust of the fallen snow and dies when the last crystal melts back into water or evaporates into the air. From air to solid to water is the life cycle of snow, and its worship requires a true trinitarian creed. Solid, liquid, and vaporous forms of H_2O—each deserves its exegesis.[1]

The falling snow accumulates like letters on this page, but it makes a page itself. In the following pages, and in the layers of snow, is the story of snow. As you turn this page, the process begins.

Early Philosophies of Snow

I

Announced by all the trumpets of the sky,
Arrives the snow, and, driving o'er the fields,
Seems nowhere to alight: the whited air
Hides hills and woods, the river, and the heaven,
And veils the farm-house at the garden's end.

RALPH WALDO EMERSON, "THE SNOW-STORM"

SNOWS OF YESTERYEAR

Thomas Jefferson was happy to report in 1782 that winters in Virginia were getting warmer. "Snows are less frequent and less deep," he wrote in *Notes on the State of Virginia.* "They do not often lie, below the mountains, more than one, two, or three days, and very rarely a week. They are remembered to have been formerly frequent, deep, and of long continuance. The elderly inform me the earth used to be covered with snow about three months every year."[1]

Jefferson's belief in a warming trend might be dismissed as promotion for his native state, or naive faith in oral testimony, but it did not go unchallenged. In a paper read to the Connecticut Academy of Arts and Sciences in 1799, Noah Webster offered an explanation for the disappearing snow:

> It appears that all the alterations in a country, in consequence of clearing and cultivation, result only in making a different distribution of heat and cold, moisture and dry weather, among the several seasons. The clearing of lands opens them to the sun, their moisture is exhaled, they are more heated in summer, but more cold in winter near the surface; the temperature becomes unsteady, and the seasons irregular. This

> is the fact. A smaller degree of cold, if steady, will longer preserve snow and ice, than a greater degree, under frequent changes. Hence we solve the phenomenon, of more constant ice and snow in the early ages; which I believe to have been the case. It is not the *degree,* but the *steadiness* of the cold which produced this effect. Every forest in America exhibits this phenomenon. We have, in the cultivated districts, deep snow to-day, and none to-morrow; but the same quantity of snow falling in the woods, lies there till spring. . . . This will explain all the appearances of the seasons, in ancient and modern times, without resorting to the unphilosophical hypothesis of a general increase of heat.[2]

With that parting blast, Webster turned a discussion of climate into a political debate—hard winter Federalist against warming trend Republican. Corollary to Jefferson's and Webster's positions on weather are their implicit attitudes toward memory as a source of information on winters past. Whereas Jefferson trusted the statements of the elderly, Webster was skeptical and urged better record keeping.

Webster, loyal son of New England, expressed his love of snow in the 1828 edition of his *American Dictionary of the English Language,* which concluded the definition of the word *snow* with the observation "When there is no wind these crystals fall in flakes or unbroken collections, sometimes extremely beautiful."[3] Americans in general may have preferred Jefferson's optimism concerning the warming climate, but New Englanders would not let them forget the importance of snow in the formation of their national identity. In scientific papers, stories, and poems, New Englanders created an image of the United States as a nation of sharply defined seasons, in which winter, with its white blanket of snow and its mysteriously complex and beautiful snowflake, was the most inspiring.

The attempts by Jefferson and Webster to define, explain, and interpret snow in America are part of the process by which we all create our identities and struggle to understand our natural environment. What is this stuff we call snow? Frozen water. Chemical compounds. The geometry of crystals. A coded message that reveals the reality of nature and ourselves. All these answers seem incomplete. We believe that there is more to discover about snow, and we invent new meanings, bestow more names. We often confuse knowledge with truth, that chimera that haunts our dreams of reason. Snow is the stuff of which such dreams are made.

The numerous paradoxes we find in snow—that it falls in soft crystals of infinite variety yet lies in a heavy sameness on the land, that it obscures the familiar yet reveals new shapes, that it comes in the season of darkness yet makes both day and night more brilliant with reflected light, that it arrives with the killing frost and the disappearance of many plants and animals yet preserves seedlings and tiny creatures under its warming blanket, that snow is pure and beautiful yet volatile and transitory, that it confines the body yet releases the imagination—had already been explored by Greek and Roman writers and were further developed by French, Spanish, German, and English poets in the Renaissance, as the colonization of the Americas began. If the ordinary farmer in New England or New France found snow mundane, something to be endured, poets saw it as a symbol for love and loss.

Thanks to the research of Quentin Hope, we can trace the comparison of snow to anything white, especially the face of a beautiful woman, from Homer through the Latin poets and the Renaissance to the present:

> Ivory, alabaster, marble, milk, and lilies are also emblems of whiteness and purity, but love poetry shows a preference for the snow. The mineral similes yield the paradox beautifully white but hard and unyielding. The snow simile suggests cold and undefiled, hence unattainable. Since it falls from heaven it has a greater purity than anything earthbound. The snow is often described as untrodden, untouched, and is at its purest when the wind has sifted it. Newfallen, wind-sifted snow reflecting the sunlight, dazzling and blinding the beholder, furnishes a hyperbole that poets describing woman's beauty find hard to resist.[4]

Obviously, poets do not mean that a woman's complexion is literally as white as snow, since this kind of whiteness is associated with lepers and ghosts, Hope continues. Rather the image is meant to evoke related qualities of purity and softness. The whiteness of snow is often used metaphorically in contrast to the red of lips or nipples, or the gold or black of hair. Snow in seventeenth-century love poetry often creates a white playground for lovers.[5]

The impermanence of snow in the temperate zone caused it to be used as a metaphor for change, the inevitable passage of time, and loss. This, according to Hope, is the meaning of François Villon's frequently quoted line "Mais où sont les neiges d'antan?" usually translated as "Where are the snows of yesteryear?" Villon's lament specifically referred to the disappearance of beautiful and seductive women known from myth and history. The ephemeral nature of snow may be tragic when emblematic of the brevity of life and the loss of beauty, but it may also indicate triviality and worthlessness.[6]

It was in this sense that the great German astronomer and mathematician Johannes Kepler wrote in the dedication of his study of the snowflake, *De Nive Sexangula* (1610):

> . . . embarrassed by my discourtesy in having appeared before you without a New Year's present, except in so far as I harp ceaselessly on the same chord and repeatedly bring forward Nothing: vexed too at not

> finding what is next to Nothing, yet lends itself to sharpness of wit. Just then by a happy chance water-vapour was condensed by the cold into snow, and specks of down fell here and there on my coat, all with six corners and feathered radii. 'Pon my word, here was something smaller than any drop, yet with a pattern; here was the ideal New Year's gift for the devotee of Nothing, the very thing for a mathematician to give, who has Nothing and receives Nothing, since it comes down from heaven and looks like a star.
>
> . . . Ask a German what Nix means and he will answer "nothing" (if he knows Latin).
>
> So accept with unclouded brow this enrichment by Nothing, and (if you have the sense) hold your breath for fear of once again receiving nothing.

The playfulness of Kepler's introduction obscures the importance of his contribution to an understanding of snowflakes. His little "gift" was translated into English in 1611 and is one of the first scientific studies of snow.[7]

Proverbs, love poems, and incipient scientific experiments were parts of the philosophy of snow that colonists brought to New England, but the winters of 1637–38, 1640–41, 1641–42, and 1645–46 were severe and forced the Puritans to ask why New England winters were colder and snowier than those in England. Moreover, hotter summers made the extremes of temperature greater, and some travelers and colonial writers questioned those who promoted the superiority of the American climate. As Karen Ordahl Kupperman has shown, New England settlers began to see the weather as an element of the "howling wilderness" they must subdue. Whatever they accomplished toward this goal—mostly in the improvement of their wooden houses and stone fireplaces—was threatened by the extremely hard winters of 1680–81, 1697–98, 1704–5, 1705–6, and 1716–17.[8]

While they offered many ingenious explanations as to why New England was colder than Britain despite the fact that it lay farther south on the globe, colonial Americans shifted the focus of discussion from the climate to self-examination. The deeply religious listened to the jeremiads of their ministers and blamed themselves for moral weakness. The more nationalistic proclaimed that the winters were a test of character and that those who survived demonstrated their physical and moral superiority. Kupperman cites several examples of colonial farmers experimenting with new kinds of hardier grain and merchants promoting trade in woolen cloth. By the middle of the seventeenth century, American colonists, especially New Englanders, had made snowy winters the primary characteristic of their identity.[9]

Their identification with snow was codified by St. John de Crèvecoeur. His essay "A Snow Storm as It Affects the American Farmer," written while he was living in Herkimer County, New York, during the winter of 1778, develops the themes of winter as a test and snow as a liberating force. "Of all the scenes which this climate offers," it observes, "none has struck me with a greater degree of admiration than the ushering in of our winters, and the vehemence with which their first rigour seizes and covers the earth; a rigour which, when once descended, becomes one of the principal favors and blessings this climate has to boast of." Crèvecoeur emphasizes the hardships of the farmer in winter. He must be sure that he has stored enough feed for his livestock; repaired his sheds, stables, barnyards, partitions, racks, and mangers; and he and his wife must prepare "raiment, fuel, and victuals" for their family and tenants. Writing in English, the French settler and diplomat uses words such as *prudent, vigilant,* and *assiduous* to describe the successful Yankee farmer. The need to provide for a long and snowy winter teaches foresight and makes the American farmer superior to "the wretched of Europe."[10]

Moral superiority is not the only reward of northeastern winters. Deep snow and frozen lakes bring improved communication among neighbors. The need to clear drifting snow from roads inspires communities to work together. Sleighs move quickly. Winter is "the season of merriment and mutual visiting. All the labors of the farm are now reduced to those of the barn; to fetching fuel and to cleaning their own flax. The fatigues of the preceding summer require now some relaxation."[11] This idyllic life continues until spring thaws mire the roads, a condition Crèvecoeur believed crippled the South throughout the winter.

The moral and recreational superiority of the North established, writers turned to variations on the theme. The era of Romanticism made suffering and death from the cold and snow a popular coda on winter's symphony.

FROZEN WOMEN

In 1834 *The People's Almanac* printed what it called an "Affecting Narrative of a Woman's Sufferings in a Snow Bank," the story of Elizabeth Woodcock, who, in February 1799, was riding from Cambridge to Impington, Massachusetts, when she was forced to dismount and became trapped in a snowbank for eight days. After her rescue her feet had to be amputated, and she lived only five months. The incident is described in detail and illustrated with a woodcut, captioned with two stanzas of a ballad that begins:

She was in a prison as you see,
  All in a cave of snow;
And she could not relieved be
  Though she was frozen so.

The story is just one of bad luck. Winter was a test that Woodcock failed. Life, like snow, is fleeting.[12]

In 1841 the same almanac continued the pessimistic mood and frozen traveler theme in the entry for January:

> Cold as a coy damsel on her first introduction to the man she is destined to marry, commences the year. On upland plain, in the valley and dell, the bleached bones of the deceased year are found. The white pith of winter piled upon the ground proclaims that we have arrived at its centre. Nature sits trembling on her throne like a bereaved mother, wrinkled with age, and her desolate moan is heard among the naked branches. . . .
>
> The lonely traveller, weary with floundering through the drifted snow, and circumnavigating the treacherous bogs and half hidden spring of water, sits down to rest in the rustling woods, screened from the cold wind by the thick but naked branches of many trees; and here he ponders on his far off home, the quiet fire-side, the hissing tea-urn, and the busy wife, and sighs as he recollects the distance which still lies between him and the centre of his hopes and fears.[13]

The mixture of metaphors, the bleakness of the description, and a woodcut that shows a man covering his eyes with his hand all suggest a depth of despair beyond previous winter scenes. That the January weather is "a coy damsel" and Nature an aged mother, while the traveler "ponders" his "busy wife," expresses some ambiguity toward women and their power to destroy as well as to create and nurture.

This ambiguity is apparent in two popular songs thought to have been composed in the 1840s. "Young Charlotte," the best known of these, may have been written in 1843 by Seba Smith, a New York journalist and humorist. Versions have been collected by folklorists from the Appalachians to the Ozarks and as far west as Utah. The song tells of Charlotte, a beautiful young woman who lives with her parents fifteen miles from the nearest village. On New Year's Eve she is picked up by her escort in a sleigh, but she refuses to cover herself with a lap robe because she wants everyone to see her ball gown. When they reach the dance, he discovers Charlotte has frozen to death. He returns her body to her parents, then dies of grief. Clearly Charlotte was killed by vanity as much as by the cold, but there is a sense in which she is a symbol of lost innocence, broken promises, and dashed hopes.

Another ballad of the period, also attributed to Smith, is variously titled "The Mother Perishing in a Snowstorm," "She Perished in the Snow," and "The Mother's Sacrifice" and describes a mother in a storm who wraps her child in her mantle before she freezes to death. The child is rescued in some versions; in others it dies. William Morris Hunt painted the frozen

Figure 1.1. *The Snow Storm* (1859), William Morris Hunt. Oil on chipboard, 47.0 × 31.1 cm (18½ × 12¼ in.), Courtesy, Museum of Fine Arts, Boston, bequest of Elizabeth S. Gregerson. Hunt's painting depicts a mother and infant frozen in the snow, a common theme in nineteenth-century songs and stories in which women who lacked the protection of men were especially vulnerable to the power of nature.

mother in 1859, possibly as an illustration for a publication of the poem (Figure 1.1). His version is bleak, the mother's body and upturned face thrusting rigidly from the snow. The persistence of the frozen woman motif suggests its power to affect the emotions of its audiences, whether the victim was to be scorned or pitied.[14]

Of all the snow disaster tales of the nineteenth century, the best known, that of the Donner party, is also a frozen woman narrative. An account published in 1879 and based on survivors' memories retains much of the sentiment that circulated first in oral form, then in newspaper and magazine stories. In this version the members of the emigrant wagon train that left Illinois in April 1846 who died in the snow of the Sierra Nevada that winter are martyrs to the development of a new land. This is also a story of gender roles. The two principal families were those of George and Jacob Donner. George and his wife, Tamsen, had three children of their own and two from his first marriage. His brother, Jacob, had seven children, two from his wife Elizabeth's first marriage. The other seventy-four members of the party were parts of eight families or were unattached males.

When the party was unable to cross the summit of the mountains because of early winter snow, some members went ahead for help while others tried to build shelters in the snow. "Lonely, desolate, forsaken apparently by God and man, their situation was painfully, distressingly terrible. . . . The snow was wrapped about cliffs and forest and gorge. It varied in depth from twelve to sixty feet." Starving, at least one of the survivors ate those who died, thereby fixing the story in the public mind. The mothers, however, chose death rather than cannibalism and "were actively administering to the wants of the dying, and striving to cheer and comfort the living." Tamsen Donner, it is implied, could have saved herself but refused to leave her dying husband, then died trying to return to her children.[15]

The frozen women of American mythology range from the unlucky Elizabeth Woodcock to the vain Charlotte, from the pathetic mother who "perished in the snow" to the heroic Tamsen Donner. They share little except their vulnerability. Joy Kasson has made an argument that

> a fascination with female victimization dominated American idea sculpture in the mid-nineteenth century, most notably in such captivity subjects as *The Greek Slave,* but also in a variety of works that represented women in peril. Artworks and the narratives that elaborated on them told an overt story of female passivity, resignation, and self-abnegation, yet these same tales throbbed with anxious uncertainty as they insisted that victimized women could triumph over their enslavers, wielding a spiritual power that belied their temporal powerlessness. Tales of female vulnerability reflected a widespread uneasiness about the stability of women's identity in a period of rapid and disorienting change.

In addition to Hiram Powers's *The Greek Slave* (1844), Kasson mentions Edward A. Brackett's *Shipwrecked Mother and Child* (1850) and Winslow Homer's painting *The Wreck of the Atlantic—Cast Up by the Sea* (1873) as examples of male artists depicting women facing death with heroic calm, thereby assuaging society's anxiety over the threat to the family caused by changing roles. This is plausible but hard to prove. Frozen women present a variation on Kasson's interpretation. Because they are victims of a more specific threat than enslavement by Turks or death by shipwreck, their stories serve as both practical warnings about the dangers of winter travel and lessons in charity. Although victims of cultural expectations, Young Charlotte and Tamsen Donner exhibit some individuality; they do what they want, even when it costs them their lives.[16]

TRANSCENDENTAL WEATHERMEN

The serious young philosophers of Concord, no less than the journalists and almanac writers, were topographers of snow. Nathaniel Hawthorne,

Henry David Thoreau, and Ralph Waldo Emerson discovered in snow tropes of creativity, memory, and illusion. One of Hawthorne's early essays, "Snow-flakes," published in *Twice-Told Tales* in 1837, is about the creative process. Watching a gathering storm, he first remarks on "the two or three people visible on the side-walks, [who] have an aspect of endurance, a blue-nosed, frosty fortitude, which is evidently assumed in anticipation of a comfortless and blustering day." After this reminder of earlier interpretations of snowstorms as tests, he comments on the total transformation snow will bring, covering the earth and preventing it from seeing "her sister sky." Losing sight of mother earth, humans will look heavenward more often.

This scene set, Hawthorne becomes a meteorologist of Romantic Weather, to borrow Arden Reed's phrase, forecasting hours of inspiration. "Gloomy as it may seem," Hawthorne writes, "there is an influence productive of cheerfulness and favorable to imaginative thought, in the atmosphere of a snowy day. . . . Blessed, therefore, and reverently welcomed by me, her trueborn son, be New England's winter, which makes us, one and all, the nurslings of the storm and sings a familiar lullaby even in the wildest shriek of the December blast." Looking out his study window into the storm, he sees schoolboys having a snowball fight; then he imagines a traveler frozen to death in the snow, and finally a flock of snowbirds, "playmates of the storm," which cheer his spirit. After presenting this anthology of snow images, Hawthorne offers some interpretations. The snowball fight, "a pretty satire on war and military glory," should, he thinks, end in building a monument of snow, of which future observers will ask, "How came it here?"[17] For Hawthorne, it might be said, children's snow forts are the American equivalent of ruined castles of Europe.

Despite the fact that the cold winters he spent at Walden Pond presaged the possible extinction of life, Thoreau also saw winter as a time when nature is simplified, allowing close examination. In the essay "A Winter Walk," published in 1843, he writes: "In Winter nature is a cabinet of curiosities, full of dried specimens, in their natural order and position." In *Walden* he dwells on the symbol of the ice pond. "Why is it," he asks, "that a bucket of water soon becomes putrid, but frozen remains sweet forever? It is commonly said that this is the difference between the affections and the intellect." Thoreau needs to isolate nature, freeze its parts in a museum of which he is curator. His approach to winter, to nature, is cerebral, not emotional. His philosophy of snow serves as a bridge between the Romantics and the scientists of his day. For Thoreau the snowscape is a code to be broken. "This plain sheet of snow which covers the ice of the pond," he writes, "is not a blancness [*sic*] as yet unwritten, but such as is unread."[18]

Thoreau read many messages in the winter landscape, some of them seemingly contradictory. In his journal for the winter of 1840–41, he describes a scene that anticipated the paintings of George Henry Durrie: "The snow hangs on the trees as the fruit of the season. . . . The whole tree exhibits a kind of interior and household comfort—a sheltered and covert aspect—It has the snug inviting look of a cottage on the Moors, buried in snow." A month later he pursues part of this image and discovers a visual pun in "the snow [that] collects upon the plumes of the pitch pine in the form of a pineapple, which if you divide in the middle will expose three red kernels like the tamarind stone. So does winter with his mock harvest jeer at the sincerity of summer. The tropical fruits which will not bear the rawness of our summer, are imitated in a thousand fantastic shapes by the whimsical genius of winter." Thoreau amused himself by turning the conventional and sentimental image of the New England farm into a kind of frozen jungle. His fertile imagination needed little help, but snowstorms enforced an isolation that stimulated his memory. He notes, "For human society I was obliged to conjure up the former occupants of these woods."[19]

Conjuring, summoning God from the snow, was the task of Emerson, the most serious of the early philosophers of snow. In 1836, in his essay "Nature," he wrote: "Crossing a bare common, in snow puddles, at twilight, under a clouded sky, without having in my thoughts any occurrence of special good fortune, I have enjoyed a perfect exhilaration. . . . I became a transparent eyeball; I am nothing; I see all." This moment of transcendence by a snow puddle, a glimpse of the order of the universe, became the basis for Emerson's subsequent insights. Two years later, in his "Address to the Harvard Divinity School," he wrote: "I once heard a preacher who sorely tempted me to say I would go to church no more. Men go, thought I, where they are wont to go, else had no soul entered the temple in the afternoon. A snow-storm was falling around us. The snow-storm was real, the preacher merely spectral, and the eye felt the sad contrast in looking at him, and then out of the window behind him into the beautiful meteor of the snow." In 1841 he published "The Snow-Storm," a poem that achieves a complete, if elliptical philosophy of snow.

The first five lines, quoted at the beginning of this chapter, capture with deliberately inverted sentence structure the complex rhythm of wind-blown snow, which immediately screens out familiar landmarks. The next four lines depict the conventional interior scene but also seal off human life from nature:

> The sled and traveller stopped, the courier's feet
> Delayed, all friends shut out, the housemates sit
> Around the radiant fireplace, enclosed
> In a tumultuous privacy of storm.

In the farmhouse, with its radiant fire, the occupants are both protected and helpless. Outside, nature mocks culture. In the remaining lines—an imperative and four increasingly complex sentences—Emerson explores the conundrum of appearance and reality:

> Come see the north wind's masonry.
> Out of an unseen quarry evermore
> Furnished with tile, the fierce artificer
> Curves his white bastions with projected roof
> Round every windward stake, or tree, or door.
> Speeding, the myriad-handed, his wild work
> So fanciful, so savage, naught cares he
> For number or proportion. Mockingly,
> On coop or kennel he hangs Parian wreaths;
> A swan-like form invests the hidden thorn;
> Fills up the farmer's lane from wall to wall,
> Maugre the farmer's sighs; and at the gate
> A tapering turret overtops the work.
> And when his hours are numbered, and the world
> Is all his own, retiring, as he were not,
> Leaves, when the sun appears, astonished Art
> To mimic in slow structures, stone by stone,
> Built in an age, the mad wind's night-work,
> The frolic architecture of the snow.

Combining the images of Hawthorne and Thoreau, Emerson goes well beyond their insights to find in a New England snowstorm a primordial force that is neither rational nor serious. Although he begins conventionally, using the storm as a screen between humanity and nature, order and disorder, the summons "Come see the north wind's masonry" forces the reader to see that behind the apparent order of nature lies chaos, the "nothing that is not there and the nothing that is," as Wallace Stevens would put it eighty years later. Yet as in recent chaos theory, the disorder may be an order we cannot yet see.[20] Emerson's choice of the snowstorm to epitomize ultimate reality was timely as well as apt. Eighteen forty-one was also the year James Pollard Espy published *The Philosophy of Storms,* which helped to popularize the developing science of meteorology.

The period 1800 to 1870 is significant in American meteorology, according to its most recent historian, James Fleming, because of the emergence of organized systems of data collection, the growth of institutions and journals, and the standardization of instruments and measurements. In 1814 James Tilton, surgeon general of the U.S. Army, directed physicians in military hospitals to record daily temperatures, precipitation, and related phenomena. The data provided the basis of Samuel Forry's *The Climate of the United States and Its Endemic Influences* (1842) and Lorin Blodgett's *Climatology of the United States* (1857). Forry cautiously, and Blodgett positively, rejected the theory of climate improvement through settlement.

Meanwhile a generation of scientists began to study weather by organizing volunteers to track specific storms. In 1825 New York authorized funds to equip colleges and academies in the state with instruments for making daily observations. In 1834 the colorful and controversial James Pollard Espy persuaded the American Philosophical Society and the Franklin Institute of Philadelphia to support a network of observers from Pennsylvania to the Mississippi River. By 1838 Espy had fifty stations equipped with barometers, thermometers, and rain gauges, and by 1842 he had 110 observers, although many lacked the necessary instruments. Espy lobbied Congress to establish a national weather service and promoted himself as official meteorologist.[21]

Espy's popular book, *The Philosophy of Storms,* is less a fully developed theory than a series of case studies, but it is written in a lively, combative style that earned him an international reputation and the nickname Storm King. His theories of the origin and behavior of storms, based on the vertical convection of warm air and its cooling and condensation while correct in some details were erroneous in others and brought him into frequent conflict with other meteorologists. William C. Redfield rejected Espy's theory of elliptical wind patterns and correctly deduced that hurricanes circulate around a center and that the movement of winds is caused by the rotation of the earth.

Robert Hare, a University of Pennsylvania chemist, challenged Espy with an electromagnetic theory of storms. The debate raged for thirty years without being completely settled, but it had several important consequences for the study of the weather in America and ultimately for the study of snow. As Fleming points out, these men's obsession with windstorms, influenced partly by the expansion of the commercial sailing fleet, meant that meteorology focused on violent, short-lived phenomena. It also meant that the United States developed a geographically extensive weather reporting system that welcomed the participation of amateur observers. Some of these volunteers, including the founder of the modern snow survey system, James E. Church, made substantial contributions to snow science.[22]

While Espy and others were struggling to make sense of the weather, two events took place that shaped the direction of meteorological research forever. The first occurred in 1844, when Samuel F. B. Morse demonstrated his telegraph. It was soon possible to transmit weather data ahead of even the most rapidly advancing storm. Two years later the Smithsonian Institution was organized, and its director, Joseph Henry, took over and expanded Espy's network of weather observers, increasing their number to 616 just before the Civil War. Weather reports were exhibited daily on a map in the Smithsonian building and published in the *Washington Star.*

Weather data collection was taken over by the Army Signal Corps during the war, and the weather service remained there until it was transferred to the Department of Agriculture in 1891. Henry made an important contribution to the study of snow when he hired Arnold H. Guyot, a Swiss, to prepare instructions for the volunteer observers. In these instructions, first issued in 1850 and elaborated in subsequent editions, Guyot recommended either using the rain gauge to catch snow as it fell or pressing the open end of the gauge into the snow on the ground and obtaining a cylinder of snow. In either case the content of the gauge was to be melted and the amount of water measured. This procedure remained in effect for most of the following century and still forms the basis for many snow measurements.[23]

One of the most original meteorologists of the nineteenth century, Elias Loomis, included a brief but comprehensive section on snow in his influential textbook, *A Treatise on Meteorology,* which went through thirteen American editions between 1868 and 1899. Loomis begins his discussion with the formation of snowflakes. He knows that snow is not simply frozen water; in fact, he suggests that raindrops are melted snow crystals. Lacking a knowledge of the molecular structure of water, he speculates that the crucial variables are moisture and temperature. Loomis also has a sense of the annual accumulations and their water content. He illustrates twenty-five hexagonal snowflakes and comments on the relation of size and shape to temperature and wind. He even mentions snow rollers—windblown hollow cylinders of snow. He discusses the whiteness of snow as a function of its reflective qualities and explains colored snow as the result of foreign matter and microorganisms. Loomis concludes his section on snow with descriptions of glaciers and avalanches, although his examples are drawn only from the Alps and the Himalayas.[24] If Loomis and other scientists were content to provide a modest beginning for a philosophy of snow, poets, preachers, and painters were not. They boldly went beyond science in their pursuit of snow's ultimate secrets.

"BEAUTIFUL SNOW"

The first flaucht in the blizzard of snow poems and stories that struck the popular press in the 1850s appeared in the *Ladies' Repository* in January 1853. "Harmony," the author of a column called "Talks with You," announces that "my subject is *snow,* which I propose treating in the manner following:

First, of the beauties of snow.

Second, of the music of snow.

Third, of the pleasures of snow.

Fourth, of the benefits of snow.

Fifth, of the moral teachings of snow."

Under beauty, the author writes of symmetry and variety in snowflakes and of the dazzling light reflected by these crystals. On music she quotes an authority who advises the reader to experience rather than think about the sound: to go to the south side of a wood in the morning after a soft rain and listen to the frozen drops falling onto the snow. The pleasures of snow are, of course, sleighing and tumbling in soft drifts. On the benefits of snow, Harmony follows tradition, comparing snow to wool in its insulating qualities and calling for charity to the poor because of "the peculiar incentives to cultivate the social feelings in a time of snowing." Finally, snow's moral teachings include purity, the cumulative effects of seemingly unimportant actions, and the existence of life after death. This recitation of conventional wisdom on snow sets the agenda for two decades of visual and verbal depictions of the symbolic snowscape.[25]

A few American artists had painted snowscapes as part of landscape studies before the 1850s, and a few more, including amateurs and "plain painters," had produced scenes of everyday life incorporating snow, but the popularity of winter scenes in paintings and lithographs was confined to the 1850s and 1860s. Fred B. Adelson has made a good case for the Boston painter Alvan Fisher as the pioneer artist of snow scenes in the United States. Fisher's earliest known snow painting is *Farmer in a Pung,* a large oil on wood panel dated 1814. In it a man rides on a pung, a small wooden box on runners, pulled by a single horse. A dog walks beside him, and people skate on a frozen millpond in the background.[26]

No American theorist of weather in art was as influential as the English critic John Ruskin, whose *Modern Painters* codified artistic taste for more than a generation. "Pictures of winter scenery," Ruskin wrote in 1846, "are nearly as common as moonlights, and are usually executed by the same order of artists, that is to say, the most incapable; it being remarkably easy to represent the moon as a white wafer on a black background, or to scratch out white branches on a cloudy sky." Ruskin praises the Flemish painters and rhapsodizes about snow:

> In the range of inorganic nature, I doubt if any object can be found more perfectly beautiful than a fresh, deep snowdrift, seen under warm light. Its curves are of inconceivable perfection and changefulness, its surface and transparency alike exquisite, its light and shade of inexhaustible variety and inimitable finish, the shadows sharp, pale, and of heavenly color, the reflected lights intense and multitudinous. . . .
>
> Snow is modified by the under forms of the hill in some sort, as dress is by the anatomy of the human frame. And as no dress can be well laid on without conceiving the body beneath, so no Alp can be drawn unless its under form is conceived first, and its snow laid on afterwards. . . .

> We have, therefore, every variety of indication of the under mountain form; first, the mere coating, which is soon to be withdrawn, and which shows as a mere sprinkling or powdering after a storm on the higher peaks; then the shallow incrustation on the steep sides glazed by the running down of its frequent meltings, frozen again in the night; then the deep snow more or less cramped or modified by sudden eminences of emergent rock, or hanging in fractured festoons and huge blue irregular cliffs on the mountain flanks, and over the edges and summits of their precipices in nodding drifts, far overhanging, like a cornice (perilous things to approach the edge of from above); finally, the pure accumulation of overwhelming depth, smooth, sweeping, and almost cleftless, and modified only by its lines of drifting. . . .
>
> We want the pure and holy hills, treated as a link between heaven and earth.[27]

The final moralizing line aside, Ruskin was a good observer of snow, and his challenge to artists is to incorporate scientific knowledge into their work. The response by American painters was not immediate, but by the 1860s artists such as Frederic Church, George Inness, Charles Herbert Moore, William Bradford, and Eastman Johnson were producing sophisticated snowscapes. In the interim popular demand for snow scenes was met by Currier & Ives and other companies that mass-produced hand-colored lithographs based on paintings.

The first of these were sporting scenes, sleighs racing along country roads, skaters gliding over millponds, and hunters stalking deer. Arthur Fitzwilliam Tait painted many of the hunting scenes, and his work provides some understanding of the effect of forests on snow in the Adirondacks. About 1855 winter scenes of farms, many of them painted by George Henry Durrie, became popular. "Durrie was unusual," his biographer writes, "in that he chose to be thought of as a winter scene painter." Eighty-four of the 125 paintings attributed to him are snowscapes, more than enough to make him the most prolific snow scene painter of his time. In his diary for January 21, 1845, he observed, "Snowed considerably this afternoon—tolerably good sleighing. This evening the moon shone out beautifully on the new fallen snow. The trees sparkling with icy limbs, made the scene almost enchanting." On March 4 he noted, "There have been a great many avalanches of snow from roofs of buildings today, coming with tremendous force to the ground and endangering the lives of passersby."[28]

Little of the enchantment and less of the dangers come through in his paintings, however. Durrie opted for the picturesque rather than the sublime. Many of his paintings depict ox-drawn sleds loaded with wood, an activity familiar to most Americans of the mid-nineteenth century but quaintly nostalgic for the growing number of coal-burning city dwellers. Durrie worked in New Haven, Connecticut, most of his life, and his snow-

Figure 1.2. *Winter Scene in New Haven, Connecticut* (1858), George Henry Durrie. Oil on canvas, 45.6 × 60.9 cm (18 × 24 in.), National Museum of American Art, Smithsonian Institution. In this idealization of New England life, snow highlights the orderliness of the farm and the joys of winter.

scapes are recognizably New England, with small farms and wooded hills. His snow is reasonably realistic, broken by sleigh and wagon tracks and reflecting some of the spectrum of light from the sun. His chief defect is that snow in his paintings always clings evenly to roofs, as if the wind never blew. In his 1858 *Winter Scene in New Haven,* for example, there is a simple one-story house with an attached lean-to and a small barn behind them (Figure 1.2). Despite the different pitches of the roofs and the smoke indicating a warm chimney, no snow has melted or slid off. This is the result not of Durrie's lack of skill but of his philosophy of snow.

His Canadian contemporary Cornelius Krieghoff was both more talented and more interested in snow as a cause of disorder. Krieghoff, who emigrated from Holland, has been described as "a romantic adventurer," a good musician, storyteller, actor, dancer, woodsman, hunter, and botanist. In *The Habitant Farm,* painted in 1856, the farmhouse and barn of a French Canadian are depicted in a state of disrepair accentuated by snow sliding from the roofs (Figure 1.3). The snow on the road is dirtier than Durrie's, with cast-off objects poking through. There is little doubt that Krieghoff was making fun of the rustic Quebecois, whereas Durrie idealized his Yankee farmers, whose lives are so well ordered that snow would not dare drip down their stiff necks. Durrie's paintings and the lithographs produced from his work in the 1860s were intended for middle-class consumption. Their purchasers were generally unconcerned with Ruskinian aesthetics; for them the snow in the pictures was less a link between heaven and earth than one more obstacle to be domesticated, a symbol of the moral and material superiority of the Northern states.[29]

Durrie and Krieghoff explored the aesthetics and metaphysics of snow on the ground while others rediscovered the snowflake. The British scientist James Glaisher's work was the basis of a two-part essay in 1857 in the *Art-Journal,* an English publication widely circulated in the United States. The anonymous author discusses "crystals of snow as applied to the purposes of design" and includes thirty-five drawings made by Mrs. Glaisher. After describing various shapes of snow crystals, including some that are irregular and "imperfect," the author reviews the principles of "fitness, proportion, and harmony," and concludes with the suggestion that the form of snow crystals is appropriate for tiles, dessert plates, and dress designs. A few years later the ingenious Louis Tiffany designed and marketed ceiling paper in a snow-crystal motif.[30]

Figure 1.3. *The Habitant Farm* (1856), Cornelius Krieghoff. Oil on canvas, 61.0 × 91.5 cm (24 × 36 in.), National Gallery of Canada, Ottawa, gift of Gordon C./v. Edwards, Ottawa, 1923, in memory of Senator and Mrs. W. C./v. Edwards. In contrast to Durrie, the Canadian artist sees disorder in the snow which slides from the roof and only partially hides the rubbish-strewn ground.

As artists were satisfying public demand for snow scenes to hang in the parlor, poets, journalists, and ministers continued to publish their homilies, gathering momentum with the approach of the Civil War. The moods evoked by snow in popular verse range from escape from sorrow to contemplation of it, from the gaiety of a snowstorm to the sufferings of the poor, from madness to redemption. Snow remains an ambiguous and contradictory symbol, useful because it can be made to represent so many human conditions. As the literary scholar Tim Armstrong puts it, "The snows of winter have become part of the ground of the American imagination, a white screen across which questions of origins and cultural differences have played."[31]

Two of the best-known poems of this period also reflect the diversity of themes. James W. Watson's "Beautiful Snow" sets up a contrast between the falling snow and a derelict begging on a street corner. Its concluding stanza, in which the reader is told that "Christ stoopeth low/To rescue the soul that is lost in its sin," is often omitted from later anthologies, leaving the poor beggar "to lie and to die in [his] terrible woe,/With a bed and a shroud of beautiful snow." Remembered now solely for the refrain, "beautiful snow," Watson's poem is surprisingly complex. The snow is described as "fickle," "flirting," a "frolicsome freak"; even the alliteration conveys a sinister undertone, so when the poet says, "it can do no wrong," the reader is suspicious. Nature is licentious. In the second stanza the snowflakes whirl about in "maddening fun," and "even the dogs, with a bark and a bound,/ Snap at the crystals that eddy around." The poem is unusual for its time in having an urban setting. In the city, Watson writes, the pure snow falls only

> To be trampled in mud by the crowd rushing by;
> To be trampled and tracked by the thousands of feet,
> Till it blends with the horrible filth of the street.[32]

Watson's miserable city contrasts with the coziness of the farmhouse in northwestern Massachusetts recollected by John Greenleaf Whittier in his often reprinted poem "Snow-Bound." Subtitled "A Winter Idyl," the poem uses snow to set the scene for a long description of Whittier's childhood, as his family entertained themselves during a snowstorm that kept them isolated by "the chill embargo of the snow" for about thirty-six hours. Prefaced by the first nine lines of Emerson's "Snow-Storm," Whittier's poem reprises the second part of the earlier work, but whereas Emerson's north wind is a "fierce artificer," "wild," "fanciful," "savage," "mocking," and "mad," Whittier's is "gusty" and does a "whirl-dance." There is the same transformation of familiar objects by the drifting snow, but whereas Emerson saw "white Bastions," "Parian wreaths," swans and turrets, "the frolic architecture of the snow," Whittier remembers,

The old familiar sights of ours
Took marvellous shapes; strange domes and towers
Rose up where sty or corn-crib stood,
Or garden-wall, or belt of wood;
A smooth white mound the brush-pile showed,
A fenceless drift that once was road;
The bridle-post an old man sat
With loose-flung coat and high cocked hat;
The well-curb had a Chinese roof;
And even the long sweep, high aloof,
In its slant splendor seemed to tell
Of Pisa's leaning miracle.

Emerson's grotesque and Whittier's idyllic established the extremes of snowscape in the middle years of the nineteenth century. Emerson's snow is an intellectual challenge, Whittier's is emotional pabulum. One deals with survival, the other with sentiment.[33]

One poet, Emily Dickinson, stands outside this spectrum. Her elliptical, dense style places her with the modern snow poets for whom snow is a matter of neither survival nor sentiment but a glimpse of chaos. Her snow imagery is sparse and frequently unique. While I agree with L. Edwin Folsom's assessment that "winter for Dickinson is the season that forces reality, that strips all hope for transcendence," that it is "an endurance test, a problem of persisting through adversity until the ease and warmth of spring return," I think the "solitary singer" of Amherst, Massachusetts, had a more subtle agenda when she thought of snow. Without the drama of snow in winter, seasons are transient and trivial. As the snow performs the New England eye can see truth, the paradox of concealing and revealing:

The Seasons flit—I'm taught—
Without the Snow's Tableau
Winter, were lie—to me—
Because I see—New Englandly—

Having established the close connection between winter and snow in this 1861 poem, Dickinson went on the following year to give her version of the Emerson-Whittier transformations of the snowscape:

It sifts from Leaden Sieves—
It powders all the Wood.
It fills with Alabaster Wool
The Wrinkles of the Road—

It makes an Even Face
Of Mountain, and of Plain—

Unbroken Forehead from the East
Unto the East again—

It reaches to the Fence—
It wraps it Rail by Rail
Till it is lost in Fleeces—
It deals Celestial Vail

To Stump, and Stack—and Stem—
A Summer's empty Room—
Acres of Joints, where Harvests were,
Recordless, but for them—

It Ruffles Wrists of Posts
As Ankles of a Queen—
Then stills its Artisans—like Ghosts—
Denying they have been—

The wonderful wordplays with "Vail" and "Joints," the image of the human figure in the landscape with its not so "veiled" reference to "Stump, and Stack—and Stem," these by themselves set Dickinson apart, but she went on in later poems to see beyond the order in snow, to look into the face of winter and see, like Wallace Stevens, the nothingness that is there as well:

The Snow that never drifts—
The transient, fragrant snow
That comes a single time a Year
Is softly driving now—

So thorough in the Tree
At night beneath the star
That it was February's Foot
Experience would swear—

Like Winter as a Face
We stern and former knew
Repaired of all but Loneliness
By Nature's Alibi—

Were every storm so spice
The Value could not be—
We buy with contrast—Pang is good
As near as memory—

In this cryptic poem Dickinson points out what other poets have missed, that all snowstorms do not symbolize the same thing, that knowledge is gained by comparing one storm with another. The special snow that never drifts contrasts with the snows that offer hidden messages, "Nature's Alibi." Even if we read this poem as Martin Bickman does, as being about something other than snow, blossoms falling from a fruit tree, for example, or "a meditation on the nature of analogy-making itself, a meta-statement about metaphor," we must still see that Dickinson links snow, loneliness, memory, and pain, or "Pang," to use her obscure synonym. The face of winter, even when repaired by snow/blossoms, is still winter—lonely and painful. Anthropomorphizing nature, Dickinson implies, only obscures its reality. She also reminds us that we carry our most significant snowstorms within:

Absent Place—an April Day—
Daffodils a-blow
Homesick curiosity
To the Souls that snow—

Drift may block within it
Deeper than without—
Daffodil delight but
Him it duplicate—[34]

Dickinson is one of the "Souls that snow," one of the philosophers of snow who seeks to understand its physical and emotional meanings. For her, nature, self-knowledge, God are all closer to the ineffable mystery of the snowdrift than they are to the daffodil. This is why, I think, she could write the ultimate poem about arctic exploration without ever leaving Amherst:

I think the Hemlock likes to stand
Upon a Marge of Snow—
It suits his own Austerity—
And satisfies an awe

That men, must slake in Wilderness—
And in the Desert—cloy—
An instinct for the Hoar, the Bald—
Lapland's—necessity—

The Hemlock's nature thrives—on cold—
The Gnash of Northern winds
Is sweetest nutriment—to him—
His best Norwegian Wines—

To satin Races—he is nought—
But Children on the Don,
Beneath his Tabernacles, play,
And Dnieper Wrestlers, run.[35]

By suggesting that the frontier is an aesthetic as well as an ascetic experience, by identifying with the small evergreen tree that seems to be noticed only when its branches are snow-laden, Dickinson laid claim to the discovery of the snow line but left its mapping to the Weather Bureau twenty years later. While she may have been partially inspired by the search for the ill-fated Sir John Franklin expedition by the American explorer Elisha Kent Kane, her perceptions are so sharp, her voice so distinctive, that we must conclude that the snowy winters of New England were her ultimate muse. The storms that stranded travelers freed her spirit.

SNOW IN WAR AND PEACE

Snow in the years of the Civil War and Reconstruction was not overtly politicized, but the publication of two books devoted to snowflakes and to the scientific, religious, and aesthetic discussion of snow is evidence of a growing interest in snow and its symbolic meanings. *Snow-Flakes: A Chapter from the Book of Nature* appeared in 1863, published in Boston by the American Tract Society, a Protestant organization that published and distributed Bibles and religious pamphlets. A preface initialed by I. P. W. indicates that Israel Perkins Warren of the society was the editor and that a brief article on snowflakes with illustrations had been published by the society in the winter of 1862–63.

Citing the work of Glaisher and others, Warren assures the reader that all the reproductions of snowflakes in the volume were "actually observed and sketched with the aid of the microscope." Although the purpose of the book is primarily religious, the twelve chapters are, like the essay in the *Ladies' Repository,* organized around topics that allow the inclusion of a variety of sources commenting on the aesthetics and science of snow as well as moral philosophy. Chapter 1, on snow structure, for example, attempts a classification of snow crystals based on size and shape—prismatic, pyramidal, and lamellar. After briefly discussing the relation of temperature to snowfall and confessing his ignorance of the physical process, Warren reprints poems by Julia H. Scott and Henry Wadsworth Longfellow that carry implicit political messages: Scott's to memories of a time "when hopes were bright" and Longfellow's to the power of snow to cover the country on "one uninterrupted level."[36]

A second, more interesting book of the Civil War period, *Cloud Crystals; A Snow-Flake Album,* "Collected and edited by A Lady," appeared in 1864. Slightly longer and containing a few more drawings of snowflakes than does the earlier volume, *Cloud Crystals* reprints some of the same poems by William Cullen Bryant, Helen F. Gould, and others but distinguishes itself by prefacing the collection with a letter from Louis Agassiz, the Swiss-born naturalist who was at the height of his influence in American science. Although its tone is somewhat condescending, it is interesting

Figure 1.4. Page from Frances E. Knowlton Chickering's *Cloud Crystals; A Snow-Flake Album* (1864). Chickering's drawings of hexagonal snow crystals and collection of essays and verse on snow represent an era in which snow symbolized an America unified by nature despite the Civil War then in progress.

because it reveals how little difference there was between the professional scientist and the amateur in knowledge of snow. Agassiz commends the accuracy of the renderings, then urges the "Lady" to arrange the crystals systematically according to their resemblances, to record the atmospheric conditions that produced each one, and to make measurements of the angles in the more complicated crystals.

The Lady, who was later revealed to be Frances E. Knowlton Chickering, answers Agassiz by describing in some detail the method she used to copy the snowflakes, the conditions under which they were collected, and the process of metamorphosis as the frozen crystals melted to drops of water. Mrs. Chickering was a keen observer and intuitive theorist, if still confined by religious orthodoxy. "Some of these forms would probably be rejected by science," she writes, "as being in a transition state; yet one could hardly say why. For in their transitions they are as much under law as when fixed; and it would seem that the beauty of these evanescent forms should not be lost. The Good Father formed them. . . . The copyist of these figures only claims to show them as they were shown to her" (Figure 1.4).[37]

Chickering answers Agassiz's request for information on the electrical condition of the atmosphere during snowstorms by including an essay by

Charles Smallwood, professor of meteorology at McGill University, that had been presented to the Natural History Society of Montreal. Smallwood, known as the Canadian Merlin, attempted to measure the electricity in the atmosphere using a copper rod attached to a 21-meter (70-foot) pole. The lower part of the rod rested on a glass pillar. Electrometers were used to measure the quantity and intensity of the snowflakes striking the apparatus at the top. On the basis of his observations, Smallwood was convinced that stellar crystals occur with positive electrical charges and plate forms with negative. Electrical theories of precipitation were already out of date, but Smallwood made another contribution to snow study:

> The method I have adopted to obtain enlarged outline copies of the snow crystal, consists in first throwing a magnified image either on photographic paper or by means of the common camera obscura. By this means the different angles may be measured and drawn out on paper. The copies now shown are obtained by the chromotype process, which consists in exposing to the sun for a few minutes paper prepared by washing with a solution of chromate of potash and sulphate of copper, having the outline-drawing superimposed; it is then washed with a weak solution of nitrate of silver, and afterward with water, and then allowed to dry.

By this laborious process Smallwood may have taken the first photographs of snow crystals.[38]

Our glimpse of Mrs. Chickering's methods and motives is tantalizingly brief. She apparently never published again. Her seriousness is not to be doubted, however, as she concludes her preface with this credo:

> While the object of this work is not strictly scientific, aiming to present the results, rather than the processes, of nature's operations, it may furnish hints, or at least serve as a guidepost, directing to scientific investigation.
>
> In its present aspect, the subject is commended to popular taste and attention, and its accessibility and simplicity are much in its favor. No weary journeys need be taken, no expensive machinery employed, no "midnight oil" consumed in abstruse study. A winter's storm, an open window, a bit of fur or velvet, and a common magnifier, will bring any curious inquirer upon his field of observation, with all the necessary apparatus, and he has only to open his eyes to find the grand and beautiful laboratory of nature open to his inspection.

It is tempting to see Mrs. Chickering as a scientific Emily Dickinson, liberated briefly by the falling snow. The other sixty selections in her book are mostly poems by British and American writers on the familiar themes of snow as destroyer and creator, evoker of memories and cause of forgetfulness, weak as a single flake and strong in the mass, beautiful and gloomy. Her selection is broader than Warren's and consequently more interesting. There is a piece by John Tyndall, the British physicist and mountaineer, on snow in the Alps, and another from *Atlantic* magazine echoing Thoreau's idea that "in winter each separate object interests; in summer the mass. Natural beauty in the winter is the poor man's luxury, infinitely enhanced in quality by diminution in quantity."[39]

The politics of snow is apparent in the inclusion of Emerson's "Voluntaries," with its ringing lines on freedom: "The snow-flake is her banner's star,/Her stripes the boreal streamers are," and in the anonymous "Extract from a Letter to a Friend in one of the West Indian Islands," in which the writer describes the beauty of a New England winter in contrast to perpetual summer. The number of selections by women is noteworthy too; at least ten of sixty-six are credited to female authors, and several of the unidentified pieces may also have been written by women. Three of these present snow in the form of a woman. In one anonymous essay titled "The Winter Queen," winter bestows beauty, leisure, and renewal: "she comes and goes a friend, a noble, generous, truthful, faithful, joyous friend—will you not welcome the white-robed queen, listen to her counsels, cooperate in her high purposes, and follow her in her glad course of duty?" In a long poem, "The Spirit of the Snow," D. F. McCarthy imagines mountains, fields, trees, ships, trains, and churches sprinkled with a snow that is feminine but that is also destructive and can temporarily stop wars:

> With her flag of truce unfurled,
> She makes peace o'er all the world—
> Makes bloody Battle cease awhile, and War's unpitying woe;
> Till, its hollow womb within,
> The deep dark-mouth culverin
> Encloses, like a cradled child, the Spirit of the Snow.

Finally, in "The Snow Bride," C. T. Brooks tells the fable of a "Glacier-nymph" who falls in love with a young hunter and "caresses" him in an avalanche so that "in the snow-bride's arms, at rest,/The hunter bridegroom lies."[40]

Chickering's view of snow as revealed in her compilation is intriguing. Unable to achieve what she wants in terms of scientific discourse, she plays with the conventions of snow by arranging her anthology in a way that leaves the reader with more questions than answers. The last selection in the book, "On Seeing the Plates of These Snow-Flakes," by "Mrs. F.E.H.H.," calls attention to Chickering's "patient care" in rendering the snow crystals. However, Frances Chickering made no lasting impact with this book. It was simply part of a final flurry of books that mark the end of one era of snow consciousness.

In Reconstruction America, the poet Oliver Wendell Holmes felt that he was "Nearing the Snow-Line," where, he wrote,

> . . . with unsaddened voice thy verge I hail,
> White realm of peace above the flowering line;
> Welcome thy frozen domes, thy rocky spires!
> O'er thee undimmed the moon-girt planets shine,
> On thy majestic altars fade the fires
> That filled the air with smoke of vain desires,
> And all the unclouded blue of heaven is thine!

The blue-clad Union soldiers, snow forts at their backs, have held the line and extinguished the fires of rebellion.[41]

The creation of the Dominion of Canada in 1867, which provided that country a measure of political independence, was also celebrated with new symbols of nationalism. One of these was a photograph, captioned "Young Canada," of a boy wearing snowshoes, carrying a small bow, and dressed in a heavy coat with a hood (Figure 1.5). Although the boy leans wearily on a rock in a bleak, snow-covered scene shot in a studio, he embodies the spirit of a young nation that will grow with a new identity. The model was the son of a Scottish immigrant, William Notman, who had established a photographic studio in Montreal in 1861 and whose snowy scenes had become enormously popular. As Jana Bara has discovered, "Though the devices developed in Notman's studio to create winter scenes were simple, they were used with artistry and care, and created a convincing illusion. The effect of deep snow was created by the use of Arctic fox fur or cotton in the foreground. Packed snow in camp, or snow on clothes or branches, was actually a sprinkling of coarse-grained salt. . . . Falling snow was created by spraying white paint into the air and passing the glass plate negative through the cloud as the droplets fell." The Canadian government never encouraged the identification of snow with the nation, but by the end of the century artists such as Maurice Cullen, Marc-Aurèle de Foy Suzor-Côté, and James Wilson Morrice had painted snowscapes that established snow as an icon of cultural independence.[42]

Snow in American art and literature in the 1870s and 1880s became a symbol of continuity, eternity, and tranquillity. James Russell Lowell summarized these feelings in a long essay, "A Good Word for Winter," in 1870. Winter shouldn't be a symbol of death, he writes, but of comfort and cheer. He reviews the image of winter in European and American literature, rejects most of the earlier interpretations, and concludes: "Winter, too, is, on the whole, the triumphant season of the moon, a moon devoid of sentiment, if you choose, but with the refreshment of a purer intellectual light,—the

Figure 1.5. *Young Canada* (1867), William Notman. Notman Photographic Archives, McCord Museum of Canadian History, Montreal. Notman's studio photograph of his son dressed in furs and wearing snowshoes makes a political statement by identifying Canada with snow and the region's Native American cultures.

cooler orb of middle life." Exchanging sentiment for fortitude and a sense that life endures after its apparent end, the postbellum generation of snow philosophers sought peace. The poets who had challenged their generation to explore the limits of personal emotion—Whittier, Longfellow, Bryant, Emerson, even Lowell—were safely tucked into anthologies with titles such as *Winter Poems by Favorite American Poets.* Periodic hagiology of the poets who celebrated winter continued well into the twentieth century, in publications often illustrated with soft-focus, impressionistic photographs whose mood belies the message of the printed page.[43]

The outstanding snow icon of the postwar years—celebrated in a photograph, an engraving, a painting, and a poem—is a mountain in Colorado where ravines filled with snow form a natural cross (Figure 1.6). The photographer William Henry Jackson, accompanying the survey expedition of Ferdinand V. Hayden in 1873, was apparently the first to record the phenomenon of the Mountain of the Holy Cross. His photo was reproduced as an engraving in several publications and was among the photographs shown at the Centennial Exhibition in Philadelphia in 1876. Thomas Moran, an artist who was also with the Hayden party, produced a wood engraving based on Jackson's photo for William Cullen Bryant's book *Picturesque America* in 1874, and an oil painting in 1875. Both the photograph and the painting are easily interpreted as symbols of Manifest Destiny, a God-given sign that Americans should settle the West. A few years later Henry Wadsworth Longfellow, inspired by the pictures and grieving for his de-

Figure 1.6. *The Mountain of the Holy Cross* (1873), William Henry Jackson. National Archives and Records Administration. This photograph of a mountain in Colorado was copied in engravings and paintings and interpreted as a sign that nature favored westward expansion and settlement.

ceased wife, turned "The Cross of Snow" into a personal symbol of eternal fidelity and love. In 1880 Moran's painting was purchased by William A. Bell, an English physician and businessman who used it to promote his tuberculosis sanitarium and hot springs resort at the base of Pike's Peak in Colorado. The Denver & Rio Grande Railroad used reproductions of Jackson's photo and Moran's engraving to attract tourists. Thus, as Joni Louise Kinsey observes, the Mountain of the Holy Cross came to be a shrine for pilgrims seeking both spiritual and physical health.[44]

As early as 1864 one American artist began to paint snow in a more realistic, nonidealized way. John La Farge's *Snow Field, Morning, Roxbury* places two small spruce trees in the middle of a snowfield with a dark line of woods in the background, surmounted by a strip of sky and blanketed by the dark undersides of unbroken cumulus clouds. The trees are free from snow and break through the snowcover with a promise of life continuing. A New York painter, Charles Herbert Moore, also completed several realistic snowscapes in the 1860s. One, *Winter Landscape, Valley of the Catskills,* depicts a village surrounded by terraced hills (Figure 1.7). The snowfields are intersected by rail fences that accentuate the horizontal element in the small oil painting, producing a sense of layers of snow over time. The art critic

Figure 1.7. *Winter Landscape, Valley of the Catskill* (1866), Charles Herbert Moore. Oil on canvas, 18.0 × 25.5 cm (7 × 10 in.), The Art Museum, Princeton University, gift of Frank Jewett Mather, Jr. This painting is typical of the growing scientific interest in snow, its drifting and its effects on fields, trees, and human life.

Russell Sturgis caught the spirit of Moore's art in reviewing another of his paintings:

> The snow is deep. It has lain for awhile under the winter sun, for around each little tree it has melted away in a conical pit, the dark trunk having absorbed the heat, just as black coal will sink into the snow three inches a day, while a white pebble will lie upon the surface. The shadows from a cluster of cedars fall right across the foreground upon the snow, almost pure blue. Away across the snowy fields are other cedars, in their reddish-brown winter dress, and bare, leafless maples, and fences and stone wall, and a few houses, all white with snow, and then come the mountains, their anatomy very visible now all the light and shadows brought out in white and grayish blue.[45]

Such attention to small details of snow on the ground, its ablation and metamorphosis, by both painters and writers became standard in the 1870s and 1880s. Whether this is attributable to a psychological desire to confront nature on a scale at which it could be domesticated and manipulated or simply to increased familiarity with snow as the West was settled, there is a noticeable trend toward detail and the interrelationships of vegetation, small animal life, and humans in snow.[46]

The man who probably reached more people with the message that

Figure 1.8. Wilson Alwyn Bentley, a Vermont farmer, made thousands of microphotographs of snow crystals in the years 1884–1931, popularizing the belief that no two snowflakes are alike. National Archives and Records Administration.

snow is part of the unseen order of life was not an artist in the conventional sense. Wilson Alwyn Bentley, the Vermont farmer who devoted much of his winter time to photographing and classifying snow crystals in the years 1884 to 1931, produced more than 6,000 microscopic photographs of snowflakes, many of which were reproduced in popular magazines (Figures 1.8 and 1.9). Although Cleveland Abbe published some of Bentley's work in the *Monthly Weather Review,* he dismissed its scientific value, and Bentley never made more than a halfhearted attempt to correlate his typology with atmospheric conditions. His was a labor of love, collecting and arranging by design. When 2,000 examples of his work were published, the public was easily convinced that no two were alike. Bentley's obsession left a legacy of stereotypes about snowflakes that are still believed. The interesting questions raised by his work and its popular reception are how the camera came to validate reality and why Americans were eager to believe in the infinite variety of snowflakes. It is clear that Bentley's photographs filled a need, both scientific and emotional, that was only partly met by Frances Chickering's drawings. For popular scientific illustration, Bentley's photographs remain in use.[47]

Figure 1.9. Page of hexagonal snow crystals photographed by Bentley. National Archives and Records Administration.

Bentley's microscopic photographs of snowflakes are an extreme example of a pervasive miniaturization of snow scenes in the late nineteenth century. Natural history essays, often illustrated with drawings by artists such as Arthur Burdett Frost, Sr., were another medium for describing snow on a small scale. John Burroughs's essay "The Snow-Walkers" is a prototype. Less mystical than Thoreau, Burroughs pursues an idea introduced by the Transcendentalists that "winter . . . give[s] the bone and sinew to Literature, summer the tissues and blood. The simplicity of winter has a deep moral." In "Snow-Walkers" Burroughs mentions the "beautiful masquerade" of the winter: "All life and action upon the snow have an added emphasis and significance." As he looks at tracks in the snow, he describes the habits of each animal—red fox, hare, partridge, squirrel, skunk, and raccoon.[48]

William Gibson was the Burroughs of plants in winter. In "A Winter Walk," he celebrates "how this clear, purged atmosphere sharpens the sight and opens up the horizon, as the merciful mantle of the snow smooths away all former invidious distinctions, and confounds our arbitrary judgments! In these white fields you shall not know poverty from affluence, worldly distinction from obscure humility. The princely park and the plebeian potato patch are one; their artificial barrier is blotted out in this universal baptism of beneficent whiteness." Such rhetorical excess is appropriate to an era of farmer and labor protests, and Gibson goes on to make trees in winter stand for American individualism in the face of collectivism and conformity. "It is a common error to suppose that winter effaces the distinctions of individuality among the various trees," he writes, naming more than a dozen species and describing their color, bark, and shape. Weeds, too, are enumerated and their virtues listed. In a remarkable passage he invites readers to "lie down upon the snow and shut out the distant trees, divest [themselves] of [their] physical identity, and look up at this beetling range as an ant might do. . . . At this focal range . . . man learns his true status as a constituent of the universe." Gibson plays with this perspective for several paragraphs before returning to a human scale. His intention is to amuse rather than to impart new information about snow, but the dramatic shift in scale from whole fields to a square inch augurs a new sensibility, the closing of old frontiers and the opening of new ones.[49]

This shift is evident, too, in the essays of Frank Bolles, a Harvard administrator who published his thoughts on nature in Boston and its suburbs in 1891. *Land of the Lingering Snow* is one of the first of a genre of natural history writing that would become more common in the urbanized twentieth century. Walking a few miles from his Cambridge home, Bolles follows the tracks of a boy with a sled, his dog, and a domestic cat, as well as an occasional rabbit, yet he is as delighted with his discoveries as Thoreau and Burroughs were with theirs. The microscopic perspective is here too:

> While recrossing pasture and field, swamp and thicket, I noticed countless black specks upon the snow. They moved. They were alive. Wherever a footprint, a sharp edge of drift, or a stone wall broke the monotony of the snow surface, these black specks accumulated, and heaped themselves against the barrier. For miles every inch of snow had from one to a dozen of these specks upon it. What were they? Snowfleas or springtails (*Achorutes nivicola*), one of the mysteries of winter, one of the extravagances of animal life.

The next moment he catches a streetcar and goes home. In another chapter he explores the snowy streets of the Boston waterfront—Crab Alley, Bread and Milk Streets, Batterymarch Street, and Rowe's Wharf—where he sees sparrows nesting in the "iron caps of the electric light lamps," ducks, gulls, doves, and crows who have made the city their permanent abode. Although Bolles finds the empty kiosks and padlocked buildings desolate, he also sees beauty: "Snow clung to decks, masts, yards, furled sails and rigging. It whitened the water-front of the city, purified the docks, and made even Crab Alley seem picturesque as I ploughed through it homeward bound." On another day he even dons snowshoes in his search for birds in Boston. Like Gibson's, Bolles's purpose is primarily entertainment rather than education, but his reference to snow fleas indicates his awareness of early biological research at universities and government laboratories.[50]

The combination of amateur naturalist and scientist may be seen in the work of John Muir. Muir's essays on Yosemite began appearing in eastern magazines in the 1870s. A largely self-taught geologist, Muir correctly understood the role of glaciers in the formation of mountain valleys and the importance of the snowpack for water in the arid West. His chapter on snow in *The Mountains of California,* published in 1894, gives one of the first and most thorough descriptions of the processes of accumulation and ablation of snow by an American. His popularization of winter hiking and camping also helped to inspire more systematic snow study. Muir was one of the first to write about the snow banner, snow crystals blown from the tops of peaks in clouds up to a kilometer and a half long. Symbolically the white banners flying from the tops of the Sierra served as the equivalent of the snow cross of the Rockies, and, as Muir notes, the banners could only be created by the north wind.[51]

DISCOVERY OF THE SNOW FRONTIER

Clark Orton was a boy living on his family's farm near the headwaters of the Minnesota River on the South Dakota border in the winter of 1879–80 when a blizzard struck. As he recalled it almost fifty years later, the storm was memorable not only because it killed a neighbor and his son but because of the rapidity and ferocity with which it struck. Within thirty minutes of the appearance of the first storm cloud, the wind was blowing tiny particles of ice through cracks and keyholes with the velocity of bullets. Orton savors the stories of men who rescued their schoolteacher-sweethearts and of others who froze to death within hailing distance of well-stocked cabins. He is utterly convinced that settlement has improved the climate and that the terrible blizzards of the past are banished forever.[52]

The origins of the term *blizzard* are obscure, but Mitford Matthews in *A Dictionary of Americanisms on Historical Principles* shows a plausible connection between an earlier meaning of violent windstorm and the later one restricted to a windblown snowstorm. *Blizzard* in the first sense appeared in print as early as 1829; the latter definition is traced to the April 23, 1870, issue of the *Estherville* [Iowa] *Vindicator*, where it appeared in quotation marks and spelled with one *z*. Within a year the Estherville baseball team changed its name to the Blizzards, and by 1872 the word was linked with its frequent modifier, *raging*. The term was used even earlier: *A Dictionary of Canadianisms on Historical Principles* cites an advertisement for the "Blizzard Saloon" in a New Westminster, British Columbia, newspaper in 1866, but it is not clear what kind of weather is being memorialized. In Tennessee and Kentucky, *blizzard* refers to a period of very cold weather, even without snow and wind. In Alaska early American settlers adopted the Russian word *purga* for arctic blizzards.[53]

Increased settlement of the prairies and experience with their winter weather demanded new words. Terrible storms struck various regions of the plains in the 1880s, killing hapless farmers and wiping out large parts of the range cattle industry. The image of the killer storm was well established by the press when a heavy, windblown snowstorm struck New York City and the northeastern United States in March 1888, carrying the term *blizzard* to that region. Blizzards are remembered not so much for the amount of snow, or even the intensity of winds and cold, but for the damage they cause. The first in this sense was the blizzard of January 6 through 9, 1886, in which as many as 90 percent of the range cattle of the southeastern plains died from the cold and starvation.

As David L. Wheeler points out in his study of the effects of this storm, the cattle boom of the 1880s was a disaster waiting to happen. By 1885 ranges were overstocked. Fences 50 to 65 kilometers (30 to 40 miles) long had been erected to prevent cattle moving from one range to another. The winter of 1885–86 began mild, and many ranchers laid off cowhands. Then, on New Year's Day, 1886, a storm left 20 to 30 centimeters (8 to 12 inches) of snow in western Kansas, and a week later a blizzard hit the Texas Panhandle, the Oklahoma Indian Territory, and Kansas. Record low temperatures were measured at weather stations from the Rio Grande Valley through the Gulf States. Dead cattle piled up in railroad cuts and along fences for kilometers. According to Wheeler, the blizzard of 1886 changed cattle ranching forever. Afterward fewer but higher quality cattle were raised in enclosed pastures. A similar storm with similar results occurred to the north in the Dakota Badlands a year later. The artist Charles M. Russell's watercolors of starving cattle in Montana are stark reminders of the devastation of these storms. A third blizzard, January 12, 1888, killed hundreds of

Figure 1.10. *Frank Leslie's Illustrated Newspaper* (March 24, 1888). Museum of the City of New York. The artist humorously comments on the blizzard that crippled New York City and much of the eastern United States.

pioneers from Canada to Texas and has become the region's most celebrated in song and story.[54]

When a great snowstorm paralyzed New York City, the concept of blizzard was brought to the consciousness of every American (Figure 1.10). The reasons for this are obvious. New York in 1888 was the center of much of American commerce, industry, and communications—especially publishing. The storm of March 11 through 14, 1888, left snow accumulations of 30 to 122 centimeters (1 to 4 feet) throughout the Northeast. As Paul Kocin reconstructs the storm, it was atypical of late winter storm tracks along the East Coast. A large high-pressure system covered most of the eastern half of the United States, creating clear skies and freezing temperatures on Saturday, March 10. By Sunday a slow-moving cold front from the west brought rain, but Monday's forecast called for clearing. What the weather observers could not know, given the state of instrumentation and number of observations, was that a low-pressure area was developing over the Carolina coast that would bring a moisture-laden storm to the New York area, where, meeting with extremely cold air from the west, it turned to snow. New Yorkers and New Englanders went to bed Sunday night expecting to go about their usual activities on Monday but awakened to almost a meter of drifting snow in streets and yards. A few hearty individuals managed to get to work that morning, only to be trapped away from home because snow fell all day. There was less snowfall on Tuesday, but temperatures fell and winds reached 110 kilometers (70 miles) an hour at times. On Wednesday, March 14, the skies cleared and temperatures rose above freezing; the processes of recovery and mythmaking could begin.[55]

Newspaper coverage from the first day emphasized that the storm was a turning point in urban transportation and communication. "People vexed at the collapse of all the principal means of intercommunication and transportation became reflective, and the result was a general expression of opinion that an immediate and radical improvement was imperative," read the lead story in the *New York Times.* New Yorkers might have continued to endure the nuisance of electrical wires dangling from poles, slow trains running on trestlework, and slower and dangerous horsecars on the streets if the blizzard had not demonstrated their vulnerability, according to the reporter. Moreover, the storm was a direct challenge to belief in the conquest of nature. "In looking back at the events of yesterday," the story continues, "the most amazing thing to the residents of this great city must be the ease with which the elements were able to overcome the boasted triumph of civilization, particularly in those respects which philosophers and statesmen have contended permanently marked our civilization and distinguished it from the civilization of the old world—our superior means of intercommunication."

"Plucky Women, Freezing Sparrows, Heroic People, and Hat Chasers" were profiled. The tone of the stories is a mixture of jocular reassurance and nervousness. New Yorkers, especially the young, were quick to adapt. "Some little girls, who had evidently read the story of the perils through which some of their contemporaries in Dakota had gone, were observed to clasp hands in a row when necessity compelled them to venture into some particularly serious drifts." By contrast, older citizens retained a distinctly Manhattanite identity:

> The simplest and the most common means adopted to protect the legs was to tie the trousers at the ankle with a stout piece of string or fasten them with a rubber band. From this there were frequent deviations and the exceptions to the twine and rubber band forms of protection were often extremely ludicrous. One boy was seen with his legs encased from the ankle up in the straw-covers of two champagne bottles which he fastened securely about him with pieces of string. Others had pieces of paper protecting both trousers and shoes; still others had old rags and

bits of cloth wound around their ankles. A few men wore woolen leggins, and those who first pulled a pair of woolen socks over their shoes and then covered their legs with leggins rejoiced in a complete protection from the wet and cold.[56]

The impact of this storm has been noted many times, and there are good eyewitness accounts. Hugh Flick, writing in 1935, mentioned the impact on the ferries as well as the elevated railroads. When ice blocked the rivers, Manhattan was isolated. Even the Brooklyn Bridge, the engineering marvel of its day, was closed. For people accustomed to technological mastery of nature, the blizzard was a rude awakening. Flick quotes the *Evening Sun:*

> It was as if New York had been a burning candle upon which nature had clapped a snuffer leaving nothing of the city's activities but a struggling ember. . . . The streets were littered with blown down signs, tops of fancy lamps, and all the wreck and debris of projections, ornaments and moveables. Everywhere horse cars were lying on their sides, entrenched in deep snow, lying across the tracks, jammed together and in every conceivable position. The city's surface was like a wreck-strewn battlefield.

Telephone and telegraph lines were pulled down by the weight of ice and snow, disrupting communication. It is clear from the reporter's account and Flick's reminiscences that the storm was seen as an assault on the city's image of itself. With communication cut off, street signs down, post office and city hall empty of workers, the city lost its voice. Flick calls it a "great communal experience" and savors stories of merchants facing bankruptcy because they were overstocked with food and snow shovels who then profited from the storm. He was one of the organizers of the "Blizzard Men of 1888" for the New-York Historical Society, who encouraged all who had lived through the experience to record their memories.[57]

One who did was thirteen years old at the time of the storm. Meta Stern [Lilienthal], the daughter of a doctor who lived near Stuyvesant Square on the Lower East Side of the island, was not allowed to wander far, but her memories are vivid. Here is her description of what she did when she was finally allowed to go out on Wednesday:

> [I] made a beeline for the ten foot snowdrift in the back yard. For two days I had pictured myself going to Elena's and Pauline's house across the fence, and now I was going to do it. The climb was more difficult than I had imagined. The surface of the snow was frozen hard, and each time I had struggled up two or three feet, I found myself sliding back down again. But finally, with considerable effort, digging my fingers and the toes of my boots into the snow, I managed to reach the top of the fence and stood there looking proudly about me, experienc-

> ing a sensation similar to one I had experienced in the Catskills when, for the first time, I had climbed a mountain. For a few moments I reveled in the feeling of freedom and achievement; then I spread out my arms as a young eagle spreads his wings, and leaped into the deep snow in the neighboring yard.

The familiar snow paradox of liberation and confinement was especially poignant for a young upper-class woman. Stern also recalled the spirit of helpfulness and good humor the storm evoked. The signs she remembered seeing stuck in snowdrifts—"Keep off the Grass," "Do not pick the flowers," and "Important notice: this is Twenty-Third Street"—are additional comments on the city's search for identity.[58]

The year 1888 ended an era in the history of snow in America. The blizzards of that year climaxed centuries of snowstorms that tested the physical strength of those who survived them and created a philosophy of snow that delighted in its many paradoxes, providing Americans with metaphors for beauty and gloom, purity and wickedness, weakness and strength, creativity and destruction, variety and unity, mask and revelation, cold and warmth, usefulness and uselessness, order and disorder, the momentary and the eternal. New conditions demanded new metaphors, new attitudes, new solutions to problems of snow management and the understanding of snowstorms.

As 1888 came to a close, Cleveland Abbe, editor of the *Monthly Weather Review,* decided that there were sufficient data to publish a map showing the depth of snow in the United States each month of the winter. The snowcover map for November 30, 1888, the first to appear in the United States, showed only scattered patches in Montana, the southern Rockies, near the Great Lakes, the Appalachians, and parts of New England. By December 31, however, a continuous line running from the state of Washington to Maine, dipping south into Arizona and New Mexico, clearly established a frontier north of which the ground was covered with at least 2.5 centimeters (1 inch) of snow and south of which the earth was bare; snow-covered areas included most of the Great Plains and part of New England[59] (Figure 1.11).

This nivean frontier shifted during the winter, but unlike the imaginary line of frontier settlement that ended in 1890, it was perennial. It was also a frontier with as many social, cultural, economic, political, and environmental consequences as Frederick Jackson Turner's frontier line. North of the snow line all life had to adjust to the transformations, however brief, caused by snowcover. The line was as real as the snow. Unlike the line drawn by demographers to divide East from West, wilderness from civilization, the snow line was laid down by nature and divided two kinds of civilization. It

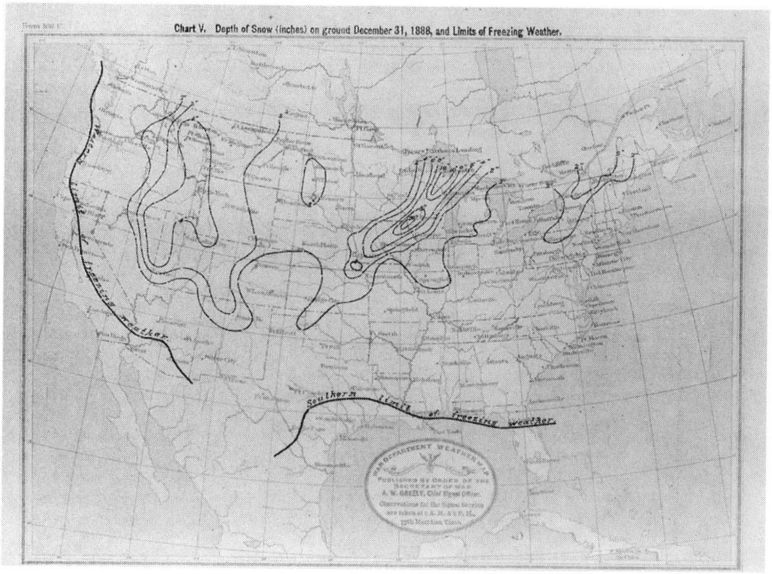

Figure 1.11. Map of depth of snow on the ground, December 31, 1888, published in the *Monthly Weather Review.*
The drawing of a snow line across the United States established a new kind of frontier, one that divided the country by weather and made Americans more conscious of environmental factors in western settlement.

required the coordination of several sciences just to describe it. That this dynamic frontier was opened at almost the same time that the old frontier of settlement was closed is one of history's neglected ironies.

The snows that fell on the United States from 1778 to 1888 evoked two basic responses with a single result. For artists and philosophers from Emerson to Bentley, snow was an enciphered message from the cosmos. For scientists and naturalists from Loomis to Abbe, snow was a substance to be measured and weighed, its meanings deciphered in its numerical aggregates. From both perspectives snow was part of the order of nature, its reality unquestioned. It could be used as a symbol, as it was by many New Englanders, but none doubted that the thing named snow was fully fixed on Bentley's glass plates and Abbe's maps, nor did any of these students of snow doubt its benefits to humanity.

Stalled Magnificence

2

I believe in this stalled magnificence,
this churning chaos of traffic,
a beast with broken spine,
its hoarse voice hooded in feathers
and mist; the baffled eyes
wink amber and slowly darken.

JOHN HAINES, "THE SNOWBOUND CITY"

TRAINS FOLLOW THE PLOW

Doubts about the benefits of snow arose when the streets of the burgeoning cities and the tracks of the expanding railroads were blocked by annual snowfalls. Traffic delays were measured in dollars lost. The cost of snow removal had to be added to budgets or considered a loss (Figure 2.1). Snow was redefined as refuse, to be removed as quickly and efficiently as possible. Snow could, like the rest of nature, be managed by applying the knowledge of science and the power of technology. Once under control, snow could even be made beautiful again.

By 1869, when the transcontinental railroad was nearing completion, the railroads had accumulated considerable knowledge of snowstorms, drifting, and snow removal, but building the Central Pacific across the Sierra Nevada near the route of the Donner party presented special problems. Not only could the route receive nearly a meter of snow in a single storm but the steep grades and narrow ledges on which the track was laid made the line vulnerable to avalanches. The solution was snowsheds, called snow galleries at first, built to cover the track in areas subject to heavy snow, drifting, and avalanches. Thirty-seven kilometers (23 miles) of snowsheds

Figure 2.1. "Utah—Trains of Cars on the Union Pacific Railroad Snow-Bound in a Drift near Odgen—From a Sketch by J. B. Schultz," *Frank Leslie's Illustrated Newspaper* (February 17, 1872). Prints and Photographs, Library of Congress. Snowplows mounted on locomotives were often inadequate in deep snows. In this engraving Chinese and Irish workers provide the power to clear the tracks.

were constructed under the direction of the Central Pacific Railroad's chief engineer, Samuel S. Montague, in 1867–68.

The first galleries were A-frames made of sugar pine. They were soon replaced by flat-roof structures of ponderosa pine because the A-frame tended to shift out of alignment and the sugar pine deteriorated rapidly. Eventually more than 48 kilometers (30 miles) of snowsheds were built in the Sierra. Although the sheds were sometimes swept away by avalanches, or set afire by passing trains, they were the most efficient way to keep the tracks open in the mountains in winter. Their biggest drawback was that they obscured spectacular views of the mountains and canyons. One solution was to build a "summer track" outside the shed that could be used before the snowcover became too deep. It was not until the 1960s that the last of the wooden sheds were removed, replaced by concrete sheds and tunnels on the Southern Pacific Railroad, the successor of the Central Pacific.[1]

Eight types of snowsheds in use on the Northern Pacific Railroad and the Canadian Pacific Railway were discussed in the *Railroad Gazette* in 1890. The article shows that construction engineers had much to learn about the varieties of snow and snow conditions but that they were beginning to look more closely. "Dry snow descends with greater velocity, and its impact upon a structure is severe. Wet snow, on the contrary, though heavier, descends more slowly, and hence is not as destructive in its effects. Snow sheds on level ground are not exposed to slides or large masses of moving snow, and have, therefore, only the vertical snow pressure to resist." The Canadian Pacific solution for mountainsides where the possibility of avalanches was strong involved building a crib on the uphill side to the full height of the shed and filling the space between the crib and the hill with earth. The roof of the snowshed then became an extension of the face of the mountain. Snow fences were placed near the entrances of the sheds to control drifting.[2]

The first snow fences were built in Norway in the 1850s, by farmers trying to capture snow for water, according to Robert Harold Brown, who speculates that forms of snow fences must have been in use even earlier. Citing German sources, Brown suggests that the scientific study of drift control began in that country in the late 1880s and was being conducted in Russia less than ten years later, although he found references to snow fences in France in 1865, and in the United States in 1874. In 1888 the *Railroad Gazette* carried a brief notice titled "Snow Screens on Russian Railroads," which describes the fences as "rough screens of wood, consisting of palisades 2 in. apart, and extending for miles." The article continues, "Still better are hedges formed of fir trees planted closely together and kept well trimmed. These hedges, adopted first by the German railways in East Prussia, have proved very successful." A decade later the Southern Pacific combined snow fences and snowsheds by erecting planks 2.4 meters high and 30.5 centimeters wide (8 feet by 1 foot) on top of the sheds. The screens prevented heavy accumulations and the resulting overhangs on the leeward sides of the sheds.[3]

In 1900 a committee of engineers from five midwestern railroads recommended stationary fences over portable fences; these were solid panels 4.9 meters long and 2.4 meters high (16 by 8 feet). In 1913 the Pennsylvania State Highway Commission began experimenting with railroad-type snow fences, 3-meter boards, 15 centimeters wide (10 feet by 6.0 inches), bolted 6.0 inches apart on a 10-foot frame. The familiar flexible, lath and wire snow fence, with cedar pickets 3.8 centimeters (1.5 inches) wide spaced 6.4 centimeters (2.5 inches) apart and usually painted red, was marketed in the 1920s by the Good Roads Machinery Company of Kennett Square, Pennsylvania (Figure 2.2). Although other types of fences continued to be manufactured, the Good Roads type became standard after E. A. Finney conducted wind tunnel tests at Michigan State University in the 1930s that demonstrated its efficiency. Natural snow fences composed of trees and shrubs continued to be advocated, however. Highway commissioners thought they were less expensive and helped conserve soil moisture. Pine trees were planted in Wisconsin in the late 1920s, supplemented by willows

Figure 2.2. *Snow Fence,* Coos County, New Hampshire (February 1936), Arthur Rothstein. Prints and Photographs, Library of Congress. Rothstein took this photograph while on assignment for the Farm Security Administration. Although the fence does not seem to have held the drifting snow, its shadow inspired Rothstein to make an image in which nature and culture anticipate abstract art.

in the following decade. In Nevada, sagebrush and juniper were hung on wire strung on fence posts along highways, combining some of the features of snow fences and plantings.[4]

Extensive tests of the "Wyoming" snow fence, a structure 1.8 to 3.7 meters (6 to 12 feet) high with horizontal slats 15 centimeters (6 inches) wide and a fifteen-degree slant to the leeward, along a 121-kilometer (75-mile) stretch of Interstate 80 between Laramie and Rawlins in the 1970s, indicated that snow-removal costs could be reduced by 50 percent and accidents cut by as much as 70 percent through the use of the snow barrier. The compacted snow also produced several thousand acre-feet of water each spring, worth hundreds of thousands of dollars on the arid high plains. Although the fences did not reduce the number of days the highway was closed by storms each winter, researchers speculated that 100 percent fencing might produce that result. New technology developed by the Strategic Highway Research Program under the direction of Ronald D. Tabler, who tested the Wyoming fence, has again made the snow fence a popular tool of snow management. An estimate of comparative costs of highway snowplowing and fencing in 1993 claimed that a hundred dollars' worth of plowing can be replaced by a dollar's worth of fencing, but snowdrift control through fencing has never received as much attention as snow fighting with plows and chemicals.[5]

Neither sheds nor fences can prevent snow blockades during major storms, however, and the search for more efficient railroad snowplows continued through the 1870s and 1880s. A writer in the *Railroad Gazette* in 1870 urged the development of snowplows that could lift snow 4.6 to 7.6 meters (15 to 25 feet) before casting it out of the cut in a hillside and could be adjusted to throw snow either right or left depending on the curve. Alluding to a Union general who said that every family ought to have a gunboat, the author urged every railroad to put snowplows on all their locomotives in winter, to fight "the huge, ill-shaped monster" that blockaded trains and caused wrecks. New designs for snowplows appear frequently in the pages of the *Railroad Gazette* and *Scientific American* in the years following the Civil War. In 1891 the *Gazette* noted:

> Not less than 280 patents have been taken out in the United States in the last 30 years, but probably not a dozen of these have ever been actually built and tried in snow. They include devices to plow the snow, to melt it, to explode cartridges in a snow bank, to remove it with knives and spoons and buckets and wheels. Some of these have proved more or less efficient in service, especially those styles now familiar in the northwest, which are invaluable in their field. But as long as a machine plow is too costly for roads which are afflicted with only one snow storm a year, and as it is so expensive to operate that even those roads owning one deem it expedient to use a common plow in dealing with a moderate depth, the old fashioned method must remain much more common, and improvements in the details of push plows will interest a great many operating and mechanical officers.[6]

In the 1870s both the Union Pacific and the Central Pacific used the Bucker snowplow, designed by George Allen Stoddard in 1866. The Bucker took its name from the method of bucking, a combination of manually digging and driving a plow ahead of as many as a dozen locomotives. Sometimes the engines would back up and run at the drifts several times before breaking through. The Bucker plow was an 1800-kilogram (2-ton) iron wedge set at a forty-five-degree angle and mounted on a wooden car. It was ineffective in deep and frozen snow, and often caused the locomotives to jump the track. An improvement was the Russell railroad snowplow, put into service in 1885. The Russell came in three models: a single track, a double track that threw snow to one side only, and a single track wing elevator on which a set of blades or wings could be adjusted to remove snow that piled up along the tracks as the first plow pushed it aside. Russell plows weighed three times more than the Buckers and were less likely to derail.[7]

Most railroads tried to keep their lines plowed by means of flangers, flatcars equipped with removable blades that swept snow aside as the cars were pushed by locomotives. The flanger usually consisted of two pieces, a moldboard plow and a blade. The blade could be raised and lowered a centimeter or so to cut into the snow at different heights depending on conditions and depth. The flangers could be made in local shops and consequently varied slightly from road to road. In addition to equipping flatcars with flangers, most railroads built shelters in the middles of the cars, with either bay windows or a cupola, for crewmen to keep watch for obstacles such as switches along the track. By 1899 the Q & C Company of Chicago had introduced a flanger that kept the blade on the rails by means of compressed air. That year a group of managers met to discuss snowplows and flangers, concluding "that the rotary principle is the right one for plows in heavy snow, but in light snow—say 2 or 3 feet—the wedge-shaped plow gives good results." Flangers, the men agreed, could be used with wedge plows to do the work of hundreds of laborers and save the companies thousands of dollars. Although accidents caused by snowstorms, usually fewer than two dozen each year, were always fewer than those caused by cattle on the tracks and usually fewer than those caused by washouts, they received more attention. Passenger trains blockaded, if only for a few hours, provided dramatic stories for newspapers and have become a prominent part of travel lore.[8]

A correspondent for the *Saint Paul Pioneer Press* in 1876, quoted in the *Railroad Gazette,* described a blockaded train on the Northern Pacific between Bismarck and Fargo, Dakota Territory, in terms of Napoleon's retreat from Moscow. "'Tuesday morning the expedition moves on Fargo and find the cuts all full; but, after a day's hard fight, reach the latter place at 9 p.m., bringing back our material and machinery, expecting engine No. 10, in good condition and ready to be sent back with General Custer in the morning. Not a man injured during the sixteen days of fight on the plains, and no one seriously frosted.'" Four months later Custer may have wished he was still snowbound.[9]

Passengers who were stuck in Reno, Nevada, on a westbound Southern Pacific train for more than two weeks in January 1890 believed that the railroad had become complacent about snowstorms. Not only was it unprepared for the heavy snow that fell in the Sierra that month but it was slow in bringing in the new rotary plows or providing for the stranded travelers. The 700 passengers lived in their Pullman cars, with names such as Samoa and Shenandoah, amusing themselves as best they could. Reno, in its precasino days, was little more than a whistle-stop, with few hotels and restaurants. One enterprising passenger, George T. McCully, published a four-page newspaper, the *Snowbound.* Subtitled "A Souvenir of the Sierra Nevada Blockade 1890," the paper was a mixture of protest and humor. McCully and other passengers were obviously anxious to get to San Francisco, but they knew how to have a good time. At twenty-five cents a copy, McCully discovered a way to turn delay into profit. Although the masthead of the *Snowbound* declared that the paper was published "every week-day afternoon by S. P. Prisoner," there seems to have been only one issue.[10]

A passenger on a train snowbound between Casper and Cheyenne, Wyoming, in the late 1890s described his experience in an unpublished memoir written in the 1950s. Milton F. Westheimer was an educated whiskey salesman traveling in the West to promote his family's product, Red Top Rye. His vivid account is worth quoting at length:

> Once on a return trip from Casper, our train got stuck in a cut between Casper and Glen Rock. The cut was about twenty feet, at the deepest, as I recall. We were stuck there two days and two nights, with almost no food. . . . This trip every seat was taken—all men except for one girl of about twenty-five, a school teacher, who was going East for the Christmas holidays. One passenger was a salesman, and apportioned his candy samples in small pieces every few hours. The expressman had a part loaf of bread and some jelly or jam. He made about twenty-five slices with a little jelly on each. We drew lots to see who was to hand her portion to the touch-me-not school teacher. It fell to me, and I brought it to her on a tin cooking lid. I remember yet her disdainful "No, thank you," but I left it on the seat beside her and withdrew. About midnight the conductor came to me and said, "She's come to her fodder." She had parked herself on the seat next to the stove, and kept the window partly open. The bitter cold kept blowing in, and persuasion could not induce her to close the window; she said she belonged to an "Open Window Club"; wore a large button with the initials "O.W.C." After a few hours, one of the older men went up and slammed the window down with authority. It stayed down.
>
> In relays we passengers and the railroad men would shovel trenches across the track. You see the snow had warmed up around midday and then froze solid toward nightfall. The trenches were two feet wide and crossed the tracks about four to six feet apart. Then the uncoupled locomotive would hit these on the run, and try to dig us out, but with very little success.
>
> Also on the express train was a crate of spoiling apples which had been refused in Casper by the consignee. The expressman strung them, two or three at a time, on wire and roasted them in the stove, and they helped out as food.
>
> Early the second day the brakeman and one sheepherder muffled up heavily and walked twelve miles to the nearest station, which was only a coal and water station with a telegraph operator. From there wires were sent to Cheyenne, from which a snow plow was sent down and dug us out toward morning of the third day, *hardly* any the worse for the experience, but plenty hungry.[11]

Westheimer's adventure was not unique; many trains ran into unexpectedly heavy snowfalls and drifts. A measure of control over snow blockades on railroads was finally achieved by the invention of the rotary snowplow.

Although its principle had been known since the 1860s, the first workable rotary snowplow was not constructed until 1884. Designed by Orange Jull, owner of a flour mill in Ontario, Canada, it consisted of two concentric whirling wheels set perpendicular to the track. The blades of the first wheel cut into the snow and drew it into a fanlike wheel behind. The second wheel rotated in the opposite direction, blowing snow to the side. Both wheels were powered by a steam engine located in a special car, on which the rotary plow was mounted. A locomotive pushed the car into the snow. The prototype was constructed in the machine shop of John and Edward Leslie in Orangeville, Ontario, and tested by the Canadian Pacific Railroad. The Leslie brothers purchased manufacturing rights from Jull, who then invented a new model and established the Rotary Snow Shovel Manufacturing Company in Paterson, New Jersey, making improvements in the blades and engine. Leslie models were sold to several railroads in 1884–85 and tested in Iowa and near Buffalo, New York. By 1887 the Leslie brothers were confidently promoting their product in articles and advertisements (Figure 2.3). Tests on the Oregon Short Line demonstrated that their plow could clear

Figure 2.3. Advertisement for the Rotary snowplow from the *Railroad Gazette* (January 8, 1892). Reproduced from the Collections of the Library of Congress. Snow-removal technology and snow management are linked with nationalism.

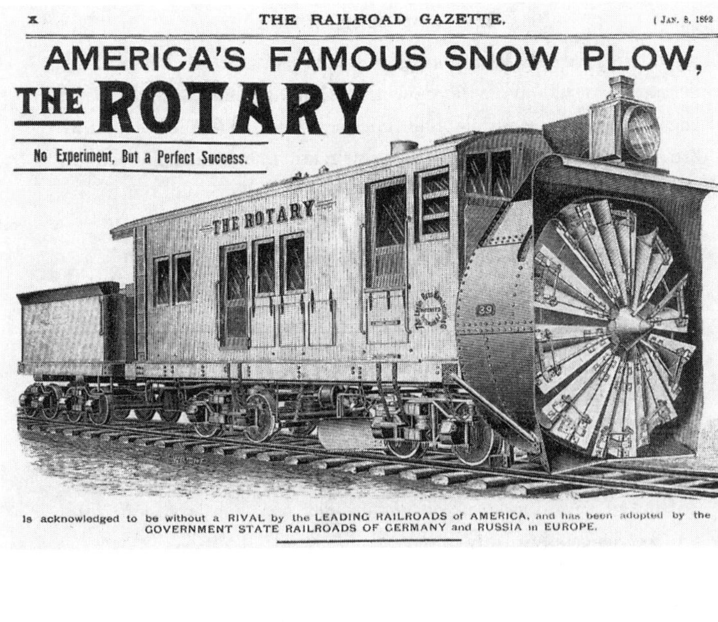
THE RAILROAD GAZETTE.
JAN. 8, 1892
AMERICA'S FAMOUS SNOW PLOW,
THE ROTARY
No Experiment, But a Perfect Success.
THE ROTARY
39
Is acknowledged to be without a RIVAL by the LEADING RAILROADS of AMERICA, and has been adopted by the GOVERNMENT STATE RAILROADS OF GERMANY and RUSSIA in EUROPE.

from 1.0 to 1.5 meters (3 to 5 feet) of snow at 32 kilometers (20 miles) per hour. Moreover, the snow could be thrown as high as 26.0 meters (85 feet) and 122.0 meters (400 feet) from the track, making, as J. S. Leslie wrote, "a grand sight."[12]

In December 1889 managers of the Union Pacific decided to test both Jull and Leslie rotary plows on their narrow-gauge railroad, the Denver, Leadville, & Gunnison. The Leslie plow operated all winter, but the Jull did not arrive in Denver until late March. The 250-kilometer (155-mile) trip from Denver to the test site at the east portal of the Alpine Tunnel was inauspicious. The great weight of the Jull plow broke several rails and damaged two bridges. Tests began April 16, 1890, and were recorded by the photographer William Henry Jackson. The plows took turns attempting to remove several centimeters of ice and a meter or so of snow from the rails. Repeatedly, the Jull machine derailed. The Leslie Rotary also experienced difficulties. Some of its blades had already broken, and others broke in the trial; it also derailed several times. The test proved only that neither plow could work in extremely difficult conditions, but the Leslies gained a public relations advantage, supplying the *Railroad Gazette* with their version of the test on May 23, complete with eight engravings of Jackson's photos. Their article said that the "trial was unfortunate for the Jull and completely successful for the Rotary." Jull responded in the May 30 issue, refuting many of the Leslies' statements about their plow's superiority but admitting that improvements needed to be made on his plow, especially in the construction of its trucks or wheels. Jull describes in detail the changes he planned to make and concludes on a confident note. Jull Centrifugal Snow Excavators were manufactured through 1892, and one continued to be used into the 1940s, but Leslie Rotaries were built into the 1950s, and a few are in use today.[13]

Competition between the Leslie Rotary and the Jull Centrifugal continued for a year in advertising for the Rotary, which contended that "the Truth is bound to come out. Over One Hundred Thousand Dollars in new-fangled Snow Plows condemned to the SCRAP-HEAP. The New Screw and Auger Machine wiped out of existence. Competitive Trials laid bare the hollow claims of all opponents of THE ROTARY." The power of steam engines to stir the public's imagination is well illustrated by this ad. Like the gigantic dinosaurs that paleontologists were reconstructing in the same decade, the snowplows were portrayed as living combatants in a struggle for survival. Cy Warman, the Canadian locomotive engineer turned popular writer, described "The Battle of the Snow-Plows" for readers of *McClure's Magazine* in 1896: "The mad locomotives seemed to enter into the spirit of the fight; at least it was easy to imagine that they did, as they snorted, puffed, and panted in the great drifts." Huffing and puffing, whether by men or machines, remains the sound most associated with snow removal. In cities, where streetcar tracks were usually recessed beneath the pavement, inventors attached rotary rattan or steel brushes to electric-powered cars, adding the swish of snow sweepers to the winter shivaree.[14]

The drama of the snow blockade does not end with the nineteenth century. Trains continue to get stuck. In his autobiography of childhood in eastern Michigan, Edmund G. Love recounts in detail the February day in 1918 when three locomotives and a snowplow were wrecked trying to clear the tracks near his home. The day was especially memorable because it was his sixth birthday and the school burned down, but the snowplow wreck was the biggest event in his young life. Told to do his chores before going to the station to watch the train pass through town, he could only listen as the engines raced into the snow-blocked cut.

> I listened to the roar of the locomotive exhausts for a moment, then walked down off the back porch to the woodpile. Grabbing an armful of firewood, I went angrily back into the house. I was in the kitchen for perhaps two minutes. When I came out onto the back porch again there was no noise at all. In that short space of time, Flushing's only train wreck had occurred. The snowplow was a caboose-type car made of wood. When it hit the drift, instead of knifing through it, the plow crumpled into kindling wood. The first engine behind it ran right through it and slammed into the drift which was packed solid now, like a wall of ice. The second engine climbed right on top of the first engine, hung there a moment, and then toppled over. The third engine went straight up in the air and stayed there, its cowcatcher pointing up to heaven.[15]

Unyielding nature, human recklessness, a cowcatcher thrust skyward like a church steeple—Love's image may be the fantasy of a child playing with toy trains, but a locomotive in snow, whether in a poem by Walt Whitman, a photograph by William Henry Jackson, or a painting by Charles Reiffel, looks big on a winter day, bigger on a snowy night, and biggest pushing a snowplow (Figure 2.4).[16]

IN ADVANCE OF THE BROKEN ARM: SHOVELING SNOW

In 1911 a humorist observed that "the snowshovel will find the boundary line between two lots more accurately than the best surveyor." Americans have long been obsessed by snow shovels and snow shoveling. More than a hundred patents have been granted to inventors of snow shovels since 1871, when the artist Winslow Homer depicted *A Winter Morning—Shoveling Out,* in which two men cut a narrow path through waist-deep snow while a woman feeds birds in the background. People and birds are hardy snow

lovers who refuse to be confined by the snow. Armed with the best shovels the genius of American industry could supply, men and women shared the backbreaking task of taming the snow frontier.[17]

Figure 2.4. *Railway Yards—Winter Evening* (1910), Charles Reiffel. Oil on canvas, 46.0 × 61.4 cm (18⅛ × 24³⁄₁₆ in.), in the Collection of the Corcoran Gallery of Art, Washington, D.C., gallery purchase, Gallery Fund, 1911. The power of snow to turn even the most mundane industrial scenes into beautiful patterns fascinated artists in the early twentieth century.

One of the earliest patentees, William P. Wentworth of Seneca Falls, New York, claimed in 1870 that he had "improved" existing snow shovels by attaching a metal tip to his blade with rivets and by reinforcing his handle where it attached to the blade. Over the next few years, patents were issued for shovels with curved metal tips, detachable metal tips, and serrated metal tips. The attention to the cutting edge implies a need to scrape as well as lift snow and perhaps a need to cut through old, frozen snow. Both of these necessities occur in cities, where clearing and removal of snow are separate operations. By 1889 the two needs were clearly identified in a patent application by a Richmond Hill, New York, inventor with the ironical name Lydia Fairweather, one of two women identified in the snow shovel patent records. Fairweather's claim was "a combined snow shovel and scraper, comprising a scoop, a scraper held on one end thereof, a handle on which the said scoop is pivoted, a key for locking the said scoop on the said handle, and a latch for locking the said key in place." There is no evidence that the Fairweather snow shovel was ever manufactured. All-metal blades had replaced metal tips on most models by 1915, when Marcel Duchamp, the French Dadaist, went into a hardware store on Columbus Avenue near his studio at 33 West Sixty-seventh Street in New York City and bought a snow shovel that he signed and exhibited with the title *In Advance of the Broken Arm.*[18]

Duchamp's whimsy was more than matched by inventors themselves. In 1916 Arthur Huberty of Canton, Ohio, patented a snow shovel on wheels, designed to lift large quantities of snow. The following year, Edgar C. Weaner of Toledo created a snow shovel and loader that combined a wheeled truck with a conveyor belt, both hand powered. The first patent for a plastic snow shovel seems to have been sought by Robert A. Smith of Mahwah, New Jersey, in 1939. Smith claimed that the plastic had low adhesion for moist snow. Snow rakes and scrapers specifically for removing snow from automobiles were patented throughout the 1970s and 1980s, continuing the trend toward more specialized snow-removal instruments.[19]

Perhaps the most interesting variation on the basic snow shovel is the snow scoop, associated chiefly with Maine and Michigan, where, in the Upper Peninsula, it is affectionately known as the yooper scooper. The snow scoop is generally wider than a shovel but has sides, and a handle attached with parallel shafts that extend under the scoop to join with runners. The scoop is pushed along until filled with snow, then tipped forward on the runners to empty it. It works best in conjunction with another innovation of the Upper Peninsula, the raised boardwalks that provide a path from home to street above the snowcover. The scooper is pushed along the wooden walkway, then its load is dumped off the side. Homemade scoops are constructed from 8-liter (30-gallon) oil drums. An 1882 patent by Henry W. Staples of Old Orchard, Maine, suggests early familiarity with the scooper there. Improvements have been patented as recently 1980.[20]

The work of shoveling snow has become something of a cult. Magazine and newspaper articles offer advice each winter on how to shovel without straining the back or incurring a heart attack. It is usually assumed that men will do this work and that they are unprepared to lay down the TV remote control and pick up a shovel. Essayists and poets celebrate the ritual of winter shoveling. David Huddle captured the essence of the phenomenon in a short piece in the "About Men" column of the *New York Times Magazine:* "No man over 40 shovels heavy snow without thinking about a heart attack," he writes. Huddle also muses about the brisk masculine camaraderie among his neighbors after a snowstorm and his resentment when his wife asks if she can help. Shoveling is about proving upper body strength, pride in a job completed alone, and above all, defeating death. Huddle is unhappy when one of his neighbors opens all the driveways on the block with a snow blower, and he wonders how this "act of generosity will affect the delicate web of neighborhood politics."[21]

Outside hardware stores each winter rows of snow shovels stand, beckoning Americans of all sizes and strengths to take up the ancient challenge. Duchamp's Readymade multiplied by Andy Warhol, this is the inevitability and surreality of snow.[22]

If, in the mind of the artist, the snow shovel presaged a broken arm, the view of some physicians is that the snowblower proclaims the severed finger. Motorized snowblowers began to appear in the United States in the 1950s. The earliest models had many features that made them unsafe to operate, and by the 1970s one study suggested that they were a common cause of finger amputations in the northern United States during the winter months. By 1962 there were at least fifteen companies manufacturing twenty-eight models, from a 2.0 horsepower hand-pushed model that cut a path less than the width of the average snow shovel to a 6.5 horsepower, self-propelled monster with two augers and a rotor to scoop and throw snow. Expansion of production in the 1960s and 1970s coincided with continued suburbanization of the population and a series of unusually snowy winters in the Northeast and Midwest. By the end of the 1980s, the machines, renamed snow throwers, had grown predictably larger and more luxurious. One model has heated handle grips. Deadman controls help prevent injuries caused if an operator fails to shut off the engine when the rotors or augers become jammed. As the American population ages, more people are likely to relinquish their shovels for mechanized snow removal. Products of the automobile age, the snow thrower and other weapons for the obliteration of snow, such as calcium-chloride pellets and rock salt, are stockpiled in the arsenals of a cold war that is unlikely to end.[23]

URBAN SNOW: REFUSE OR ART?

As automobiles replaced streetcars and railroads as the primary means of transportation within and between cities, snow became what the geographer John Rooney, Jr., calls the "urban snow hazard." Even before the automobile, snow in the city seemed out of place to some. Downtown crowds, in the immortal words of James W. Watson, trampled the snow with "thousands of feet,/Till it blends with the horrible filth of the street."[24] City streets filled with horse manure and garbage were seldom fit for sleighing or walking. Moreover, what Whittier called the "chill embargo of the snow" was all right for farmers snug in their well-stocked homes, but city folk needed daily deliveries of milk, bread, and newspapers. Men and women had to get to offices and factories, children to school. For many construction workers, snow meant suspension of work and loss of pay. Severe snowstorms disrupted electric light and power, and knocked out telephone communication, all vital to the life of cities by 1900.

Transportation is not, of course, the only aspect of city life affected by snow. It has been argued that snowfall determines the shape and density of cities and that cities can be made livable by adapting to rather than fighting snow. The aesthetic effects of snow on parked cars and parking meters, row house steps and bus stop shelters, kiosks and statues, is beginning to be appreciated (Figure 2.5). Snow in cities has traditionally been perceived as a threat, a disruption of normal life, an instigator of chaos, even when some urbanites welcome the break in their routines. This is not too different from rural attitudes, but the responses of city dwellers to snow are both more frenetic and more personal. Snow is an unwanted reminder of the limits of human mastery of nature. Major snowstorms in cities are reported in terms normally reserved for riots and civil disorders. Cities are "besieged," "vanquished," or "triumphant." Body counts are tallied, emergency services critiqued, and property damage assessed. Reports of looting are covered in detail. Extremes of heroism and selfishness are sought out and photographed for newspapers and television.[25]

If Thomas Gutterbock's calculations are correct, snowbelt cities are denser than sunbelt cities because residents trade off land costs for transportation costs. Suburban living is attractive as long as the commuter can count on relatively few weather-induced traffic delays. While all cities show a tendency toward dispersal, people in a snowbelt city live an average 6.9 kilometers (4.2 miles) from its center, whereas those in a sunbelt city live 13.7 kilometers (8.5 miles) from its core. City size is obviously a factor in assessing the impact of snow, as are regional and cultural differences. After studying seven midwestern and western cities, John Rooney concluded that snow was less of an economic problem in Casper and Cheyenne, Wyoming, than in Green Bay and Milwaukee, Wisconsin, Muskegon, Michigan, or Winona, Minnesota, because the snow that falls in the West is drier and therefore causes fewer driving problems, and because the citizens of Wyoming tend to view snow as a matter more of personal than of municipal responsibility. They complained privately about public services during snowstorms, but they were unwilling to pay for them. Residents of Green Bay, Milwaukee, Muskegon, and Winona, by contrast, were willing to be taxed for municipal snow removal and were relatively satisfied with the results.[26]

Another geographer, Robert Davis, found regional correlations between snowfall and employee absenteeism. In the years of his study, 1978–1982, 0.5 to 1.0 centimeter (1 to 2 inches) of snow in Boston caused a 1 percent increase in absenteeism among midlevel government workers,

Figure 2.5. Snow on Auguste Rodin's sculpture *Monument to Balzac* (1891–1898/cast ca. 1965–1966), Hirshhorn Museum and Sculpture Garden, Smithsonian Institution, gift of Joseph H. Hirshhorn. Snow transforms the urban landscape, inviting us to look at familiar objects in new ways. With the Smithsonian Castle behind it and snow highlighting the dark figure, we notice that Rodin's sculpture is appropriately cloaked and ready to join us in reshaping the fresh snow with footprints.

while the same amount of snow resulted in 15 to 20 percent rates in Atlanta and Nashville. What is unclear from this study is whether southerners are truly daunted by snow or simply less obsessed by work. When a snowstorm struck Jackson, Mississippi, in February 1985, the *New Yorker* called the novelist Eudora Welty, who reported that the airport, schools, streets, and most businesses were closed, but that the mail was being delivered and many of her neighbors were enjoying sledding. Nor do all northerners adapt well to snow. For many, seasonal migration to Arizona and Florida is an alternative to contending with winter. As many as 10 percent of Minnesota's elderly spend five or more weeks away from the state in winter months. These "snowbirds" most often flee rural areas and small towns, where they feel physically isolated and bored.[27] Parts of the country are obviously affected differently, depending on the amount of snowfall, duration of temperature, and wind, as well as on the population's age, wealth, and attitudes toward snow.

Most urban snow hazard is measured in terms of traffic disruptions. Rooney proposes a "hierarchy of disruptions": minimal, nuisance, inconvenience, and crippling. At his third stage, inconvenience, the accident rate reaches 100 percent above average. Crippling disruption is defined by a 200 percent increase in traffic accidents, followed by an absence of vehicular traffic. A study sponsored by the National Science Foundation in 1975 used criteria to define urban snow hazard, focusing on costs, of both actual storm damage and snow removal, and loss of life. Asserting that 41 percent of the total population of the United States lives in the twenty-one high-risk northern states, the authors of the report conclude that reducing urban snow hazards is difficult because of the differences among communities in the crucial variables of local government finances and management of snow removal, mixes of snow-fighting equipment, systems and reliability of weather forecasting, costs of disaster insurance, and the public's willingness to forgo mobility and tolerate disruption. Among the variables, the last is clearly the most complex, since each individual has a different tolerance for disruption, depending on age, health, occupation, and mood.[28]

Social scientists seem bewildered by what they perceive as the lack of public concern about the "urban snow hazard." Why, they ask, is there so little concern for the hundreds of millions of dollars in damage and scores of deaths caused by snow? "Perhaps," muses Rooney, "the esthetic values of snow make us somehow reluctant to wage all-out war against it." An aesthetic appreciation of urban snow is hardly new. Jacob Riis, writing in *Century Magazine* in 1900, concluded that the snow "is beautiful at all times until man puts his befouling hand upon the landscape it paints in street and alley, where poetry is never at home in the summer. The great city lying silent under its soft white blanket at night, with its myriad of lights twinkling and rivaling the stars, is beautiful beyond compare." The engravings accompanying Riis's essay inadequately depict this beauty, however, and it was not until the 1930s, when newspapers and picture magazines began using photographs, that the aesthetics of urban snow become remarkable.[29] Black-and-white photos of snowcapped parking meters and fire hydrants, snowdrifts on park benches and automobiles become part of a ritualized commentary on winter, along with editorials criticizing municipal snow-removal efforts and feature stories comparing storms present and past.

In January 1989 the *New Yorker,* which, through its "Talk of the Town" commentary, cartoons, and covers, has done more than most magazines to celebrate the beauty of snow in the city, published a meditation on an afternoon snowfall in the sculpture garden of the Museum of Modern Art. As the snow covered works by Auguste Rodin, Henry Moore, Tony Smith, Henri Matisse, and Claes Oldenburg, the writer saw a reconciliation of old stylistic wars; all the pieces were "going white together, like schismatic old radicals in the same nursing home." Yet the snow also revealed new insights into each piece, "as if each of the artists, far from being blanketed by a common peace, were using the occasion to hold an introductory seminar on the essential quality of his own work." This snow, the writer concludes, is not what some have thought, something that falls from heaven bringing the peace of a better world, but a kind of specific commentary, individualizing everything it falls on and allowing each object to be different.[30] This interpretation of urban snow easily fits with the environmental philosophy of the late twentieth century that values tolerance of nature rather than dominance of it, but it is also the product of a unique urban context. Few city dwellers experience snow in such cloistered surroundings. Creating a new urban snow aesthetic will require radical changes in attitudes toward both cities and snow.

Redefining snow as refuse in the city when it had been considered beautiful in the country required new rituals of transformation. The first St. Paul Winter Carnival in 1886, sponsored in part by James J. Hill's St. Paul and Pacific Railroad (later renamed the Great Northern Railroad), climaxed with the storming of an ice palace and the surrender of King Winter. Snow, the ritual of the carnival conveyed, was once pure and could still be within the confines of carnival time, but beyond this limited sphere it was filth. A day after the close of the carnival, an editorial writer declared:

> One of the good things which the carnival has brought to St. Paul is a beginning of attention to the condition of some of the principal streets. Along the route traversed by the numerous parades gangs of men were set to work and their labor produced in a short time a very passable thoroughfare. The experience, limited and not wholly satisfactory though it was, has in it a valuable hint for other than carnival times. There is no excuse whatever in a city of this size for permitting the streets to become obstacles to locomotion as they are here in any winter when the snows are numerous and heavy.[31]

The 1886 carnival and subsequent celebrations gave a prominent place in parades and ceremonies to municipal workers, who were becoming the indispensable managers of the technologically deseasonalized city. The process of managing snow on city streets was not without internal conflict, between those who clung to a rural ideal of snow and those who ran the machinery of commerce. On February 12, 1871, the Rochester & Brighton Railway Company began plowing its tracks, and by the morning of the following day mounds of snow prevented sleighing, a traditional recreation from the era of "pure" snow idealized by Currier & Ives and the New England poets. The city superintendent of streets, sympathetic to the sleighing class, sent a gang of men estimated by the newspaper to number between forty and one hundred, to shovel the snow back into the streets. In the ensuing skirmish, the shovelers could not match the pace of the horse-drawn plows, and the streets with horsecars remained closed to sleighing.

That a conscious choice between technologies and attitudes toward snow was necessary is made clear from the story "Trouble in the Streets," in the Rochester *Daily Union and Advertiser* and accompanying letters to the editor. "We understand that the Street Committee disavows all responsibility for the action of the Superintendent, and that he acts entirely upon his own responsibility. . . . No city the size of Rochester can do without street railways. . . . The man who owns lots on a street owns the street. Scarcely one in ten owns a carriage. The street cars are his carriages, and he has a right to use the streets by cars as well as others have to use them by other vehicles." In an adjacent column an angry taxpayer wondered why the shovelers could not be better employed clearing sidewalks and crosswalks for pedestrians.[32]

The struggle to control city streets was far from over, but the clear loser was snow itself. As electric traction replaced horsecars and as automobiles competed with streetcars, snow was evermore a hazard to be removed and dumped into rivers and canals. By 1910 Rochester had so successfully managed snow that it felt good about reintroducing it in the form of snow castles at an ice carnival in Genesee Valley Park. By the 1920s other cities were introducing winter sports in their park and recreation programs, but spontaneous enjoyment of snow in the streets was forbidden. In September 1917 the Rochester Bureau of Municipal Research submitted a lengthy "Report on the Problems of Snow Removal," which included this stern warning: "The days of cutter racing on East Avenue and transportation by runners downtown are now things of the past. . . . It seems clear that a large part of the sleigh traffic from the country has no urgent business on Main Street and other main thoroughfares, and there is ample warrant for believing that trucks and wagons can be used for delivering coal, etc. to buildings on such streets." The annual recurrence of snowstorms meant, of course, that cities had to be vigilant. Mere snow removal had to be replaced by "snow fighting."[33]

A military approach to snow removal preceded the guns of August 1914. Some of the earliest plans to fight snow were literally the result of increased firepower. In 1887 an inventor in Hoosick Falls, New York, proposed utilizing steam plants in cities to heat a system of pipes under streets and sidewalks. A decade later the Snow and Ice Liquefying Company of Paterson, New Jersey, built a snow-melting machine.[34]

Firepower was not the only weapon in the war on snow. Organization and specialization were the twin instruments of the Progressive reform movement and were readily applied to snow management. George E. Waring, Jr., the dynamic commissioner of street cleaning in New York City, put his crews in uniform, dubbed them White Wings, and created a new image for sanitation workers in the 1890s. Shortly before his untimely death he wrote *Street-Cleaning and the Disposal of a City's Wastes: Methods and Results and the Effect Upon Public Health, Public Morals, and Municipal Prosperity,* an organizational manual and a manifesto. The chapter on the removal of snow was written by H. L. Stidham, "Snow-Inspector," who, having inspected past practice and the snows of 1897, pronounced the former inefficient and the latter a threat to public health, especially to the tenement population of the Lower East Side of Manhattan, who needed access to clean streets. "Whether it be winter or summer," Stidham observes, "the people must have this additional room opened up for them, and a delay in the removal of the almost knee-deep snow and befouled slush is at the cost of much sickness, and probably many lives, each winter."[35]

Organization, mechanization, uniformed armies of snow fighters were the hallmarks of snow removal in the first decades of the twentieth century. "By snow fighting," wrote the Rochester Bureau of Municipal Research, "is meant the continuous removal of snow as it falls during the progress of the storm, the work beginning with the first snowfall and continuing without intermission until the end of the storm. This method is quite distinct from that in general use in Rochester, which can best be termed snow removal, since work practically does not begin until after the cessation of a storm." Recognizing that "the problem of snow removal must obviously be considered differently in different cities as its solution is dependent upon such variable elements as climate, population, width of streets, density and character of traffic, location of sewer systems, available disposal places and other local conditions, to say nothing of the financial policy of the municipality," the delegates to what was probably the first American snow removal conference, in April 1914, called for the invention and manufacture of special machinery for scraping, loading, and transporting snow (Figure 2.6). By the early 1920s, Caterpillar, Baker, Cletrac, Mack, and other companies were advertising equipment for snow removal that could also be used for summer road construction and maintenance.[36]

Four distinct themes emerge in the snow management discussions of the 1920s: (1) determining costs and benefits of snow removal; (2) improving plows and other snow-removal equipment; (3) using chemicals to break up ice on streets; (4) establishing legal authority to enforce snow removal from private property. Although these themes are interrelated and often discussed together, it is useful to examine them separately in order to see clearly the ways American cities were deseasonalized.

"Does It Pay to Clean City Streets in Winter?" asked the editors of *American City* in February 1921. To help readers understand that the answer is yes, a photograph of a truck on an unplowed street with chains on its spinning tires accompanies the text: "The chains strike a fearful blow on the snow and ice, and soon reach the pavement and continue the damaging work. Perhaps many engineers will think this is only an incident, but in a number of cities all but the hardest pavements have been damaged to considerable extent in this manner. There is not only the street to be considered—think of the damage to the truck when, as each chain strikes the pavement, the wheel is instantly and momentarily checked in its speed,

SCIENTIFIC AMERICAN
KEEPING A CITY'S STREETS CLEARED OF SNOW. MOTOR PLOW AND SNOW FIGHTERS IN ACTION

Figure 2.6. "Keeping a City's Streets Cleared of Snow: Motor Plow and Snow Fighters in Action," *Scientific American* (December 16, 1916), by Edwin F. Bayha. Snow is being dumped into the sewer through a manhole, a practice later condemned by environmentalists and city managers.

thrusting a great additional torque on the axle." By extending the range of potential damages from tire chains, almost any amount spent on snow removal could be justified. Crude estimates of the value of state highway systems led some engineers to conclude that snow removal cost only half as much as the investment value of roads. In 1926 V. R. Burton, an engineer with the Michigan State Highway Department, developed a sophisticated model for computing the costs of snow removal that considered variables such as topography, wind direction, temperature, and depth and type of snow. Based on his analysis, Burton recommended the use of fast trucks with blades rather than V-plows and heavy tractors, which had to make more frequent runs. Starting early in the storm and keeping traffic moving were the keys to Burton's strategy. He also recommended a gasoline tax earmarked for snow removal.[37]

While cost and equipment were the dominant concerns of the period, the use of sand, salt, and chemical compounds to melt snow and ice was also being discussed. Salt was reportedly in use in Paris in 1887, but its negative effects on the environment were apparently not recognized until much later. In 1921 a brief article in *American City* by Samuel N. Baxter, arboriculturist for the Street Tree Department of Fairmount Park in Philadelphia, called attention to the damage done to the city's trees by the dumping of ice cream salt from delivery wagons. Fifty years would pass before the use of salt on streets again became an environmental issue, and increasing amounts of chemicals were applied to city streets from the late 1920s on. The superintendent of the Motor Coach Department of the Cleveland Railway Company experimented with calcium chloride in 1926, finding it superior to salt and sand in very cold weather when ice on the streets was thick. Howard T. Barnes, professor of physics at McGill University in Montreal, got good results with solutions of 20 percent chloride sprayed on snow with high-pressure hoses, which prevented the grains of calcium chloride from simply making holes in the snowpack, lowering the temperature of the interior and freezing it harder.[38]

The fourth concern of snow fighters was the legality of ordinances requiring citizens to shovel the sidewalks in front of their homes and businesses. Although such laws originated in the eighteenth century and seemed to fall under the police powers of government to provide public safety, their constitutionality was challenged by those who argued that forcing people with property to clean their walks was an unfair and unequal burden analogous to a tax. When the Supreme Court of Maine upheld a Portland statute requiring "the owner, tenant, occupant, or any person having the care of any building or lot of land bordering on any street, lane, court, square, or public place within the city, where there is any footway or sidewalk, shall after the ceasing to fall of any snow, if in the day time, within three hours, and, if in the night-time before 10 o'clock of the forenoon succeeding, cause such snow to be removed from such footway or sidewalk," *American City* reported it as a clear victory in the struggle to rationalize snow removal. Although state courts usually upheld a city's authority in such cases, the fact that such statutes were continually challenged from 1833 through the 1920s strongly suggests that many Americans believed that snow on private property was a private matter. To their minds forced removal of snow was a taking "for public use, without just compensation," a violation of the Fifth Amendment of the Constitution.[39]

The contribution of the automobile to the deseasonalization of the city is undeniable. In 1923 the secretary of the Automobile Club of Berkshire County, Massachusetts, praised the county commissioners for helping organize "shovelling bees" to open the road between Pittsfield and Albany but warned: "The automobile to-day has passed from being a pleasure vehicle to a public necessity, so that county commissioners and city officials must heed the insistent plea that highways and streets be kept open for traffic throughout the year and proper mechanical equipment be provided for snow removal." Four years later a Caterpillar Tractor Company brochure bluntly stated: "Life in Winter soon will go on much as in July." The brochure was illustrated with a cartoon of a malevolent-looking snowman labeled "The Snow Bogy" being pushed aside by a Caterpillar snowplow (Figure 2.7). Soon, the copy continues, the "cruel giant" snowman will "be remembered only as the Ghost of an unenlightened past. The science of Snow Removal has robbed Winter of its terror, and traffic flows as usual—upon *the Opened Road.*" This clever allusion to the familiar American icon of the open road is explicitly intended to replace Whittier's image of "a fenceless drift that was once a road" with one of a plowed highway, deseasonalized, open year-round (Figure 2.8).[40]

SHOVELING OUT OF THE DEPRESSION AND INTO WAR

Lines of unemployed, their backs to the blowing snow, may dominate our mental imaginings of the 1930s, but except for a late March snow in Chicago in 1930, a 43-centimeter (17-inch) storm in New York City in January 1935, and an early February snowfall of 36 centimeters (14 inches) in Washington, D.C., in 1936, the decade of the Depression contained few unusually severe storms. In general, cities and states stretched their snow-removal budgets by investing in more efficient equipment and by systematically studying snow conditions. Applied snow science and technology aided municipal governments in the tilt toward deseasonalization. Although many

Figure 2.7. "The Snow Bogy." Caterpillar Tractor Company promotional booklet, 1927. Snow is demonized and defeated by science in this advertisement, which also appeals to the American ideal of the open road.

Figure 2.8. Snogo rotary plow and 5-ton trucks with reversible blades plowing highway near Berthoud Pass, Colorado, ca. 1935. Clyde E. Learned Papers, Archives Center, National Museum of Natural History, Smithsonian Institution. The original caption reads: "A view to illustrate use of equipment on snow clean-up. The heavy equipment shown is indispensable for major snow removal operation." The photo also shows how trees near highways may be damaged by ice and chemicals sprayed by rotary plows.

men supplemented their meager incomes by shoveling snow, *American City* explicitly rejected a return to hand labor for snow removal in major cities. Chicago's boast that an army of 40,000 men was required to clear the city of the drifts created in the March 26, 1930, storm merely confirmed the contention that the city's dependence on automobile transportation rendered traditional snow-removal methods obsolete.[41]

Superintendents of streets and city engineers had three basic objections to manual snow removal. The first two were that unemployed men were neither physically able nor sufficiently skilled to do the kind of shoveling required on modern city streets. *American City* reprinted a *New Yorker* cartoon to make the second point. In the cartoon one member of a shoveling crew has removed not only the snow but the asphalt and ground around the gas and sewer lines, despite his supervisor's plea, "No, No, McNamara. Just That White Fluffy Stuff on Top." The third objection to hand labor as "welfare work" was less tenuous and more revealing of the attitudes of urban technologists. "Cities should consider carefully the need of purchasing mechanical equipment for snow removal and should realize that skilled labor has been given employment in the production of every machine bought." The article argues that even the overhauling of old equipment is better than a return to hand labor because it will employ skilled mechanics. While conceding that large numbers of extra workers would be needed in some snowstorms, "to save the heavy losses to business which would otherwise develop," the editors rejected the idea that blizzards might be ill winds that blow some good. Snow was an impediment to business and had to be removed at the lowest possible cost.[42]

A new vocabulary of urban snow terms began to appear in the pages of *American City* and other journals devoted to street and highway maintenance in the 1930s. A Minnesota Supreme Court decision holding municipalities responsible for injuries to pedestrians resulting from falls on "artificially accumulated ridges and ruts caused by wheeled traffic, because the accident occurred on a crosswalk forming that part of the pavement which lies between the lines of the sidewalk if extended across the pavement," drew extensive comment from the assistant city attorney of Duluth. Citing H. W. Richardson, U.S. meteorologist for the Duluth District, the lawyer observes:

> If you call Mr. Richardson as your witness or cross-examine him as the opposing party's witness, he will tell you that, although the mercury may be under zero, it is a matter of common occurrence for the sun to cause snow to thaw where it lies favorably exposed to the sun. Mr. Richardson calls this thawing process "solarization." Municipal officers charged with the duty of snow disposal call solarization something else when it combines with wheeled traffic to put ridges and ruts in crosswalks. And injuries sometimes translate this solarization process into something still different—a verdict for cash damages for the plaintiff.

The terms *solarization* and *forensic meteorologist* became part of our vocabulary, and the anticipation of hourly changes in freezing and thawing streets added a better appreciation of the diverse and complicated nature of urban snow.[43]

More and more cities equipped themselves with fleets of light trucks with V-plows, Caterpillar tractors and heavy trucks for drifts, sidewalk plows, front-end loaders, dump trucks, and spreaders for applying sand, cinders, salt, and calcium chloride. In 1931 a consulting civil engineer in Philadelphia, John H. Nuttall, offered readers of *American City* a detailed analysis of the properties of snow and ice and methods for removing them from streets. Nuttall worked out formulas for determining the amount of snow that could be melted in sewers of various dimensions. He noted problems caused by dirty snow and the limitations of mechanical melters and discussed the properties of dry snow, wet snow, and slush, distinguishing further between "snow slush" and "wet slush," the former composed of snow crystals surrounded by water, the latter of crystals in suspension. Fresh snow, Nuttall advised, weighs 2.3 to 5.4 kilograms (5 to 12 pounds) per cubic foot and 5.4 to 23.0 kilograms (12 to 51 pounds) per cubic foot when compacted by rain. Slush weighs between 9.0 and 23.0 kilograms (20 and 50 pounds) per cubic foot. Nuttall's calculations were among the first in the field to recognize the many variables affecting the character of snow—density, temperature, wind, and dirt. By the end of 1932, *American City* was offering advice on the management of seven classes of storms, from heavy rains to blizzards and heavy ice storms. By the end of the decade, experiments indicated that a combination of sand and calcium chloride produced the best surface for controlling skids, but most cities continued to use sodium chloride (rock salt) because it was cheaper.[44]

War news from Europe also raised awareness of the importance of snow, as the Finns fought the Russians in the winter of 1939. When Boston was hit by a blizzard on February 14, 1940, the *Daily Globe* found military analogies appropriate. "Snow Blitzkrieg of 1940" was the caption of one cartoon, while a photo of Governor Saltonstall called attention to the World War I uniform he wore to hike through the snow to the State House. Torn between images of "winter's comic valentine" and the sight of abandoned automobiles that revealed the "desperate vulnerability" of modern civilization, the *Globe* editorialized: "Does it take a blizzard to jolt us out of the drab monotony of our mechanized routines and give us a taste of that powerful intoxicant man's hand-to-hand conflict with nature which has made his life on this earth a thing of vivid interest? Then give us the blizzard." Sixteen thousand men and 1,500 pieces of equipment took five days to clear Boston's streets, a performance that inevitably led to a call for better planning and leadership. "Deep snow pushed the clock back," the *Globe* noted, but news from Finland pushed it forward again.[45]

Another snowstorm on an even more symbolic day marks the transition from Depression to wartime snows. On Armistice Day, November 11, 1940, a blizzard struck the Middle West, killing more than fifty people in Minnesota and nearly one hundred elsewhere. In the following year, on the eve of world war, George R. Stewart's fictionalized account of a snowstorm in the Sierras, *Storm,* became a best-seller, its comforting message that organized intelligence could triumph over irrational forces.[46]

"Snow—Friend of the Enemy" was the title of a bulletin published by the Baker Manufacturing Company of Springfield, Illinois, in 1942, to advertise its snowplows. Baker's simple message was that snow disrupts war production by closing roads and employee parking lots. It is probably a measure of how well cities had learned to manage snow by 1941 that few writers bothered to comment on snow removal during the war. The manpower shortage was offset by mechanization, and wartime rationing of gasoline and tires meant that traffic was less of a problem. Although the metaphor of war on winter was employed occasionally, it was less prominent than during the First World War.

The war years saw an increased use of salt in the name of military emergency. The deputy superintendent of streets and engineering for Springfield, Massachusetts, extolled salt as saving manpower and tires, since trucks had only to spread one layer of salt as opposed to three or four of sand. When automobile owners complained that the use of salt on streets was damaging cars, the city chemist of Rochester, New York, responded by pointing out the savings in safety, but he also developed a polyphosphate compound that he mixed with the salt to minimize corrosion. Detroit and other Michigan cities used rock salt to "do as much as 80 percent of the work of snow removal." Damage to cars may even have been a covert goal of the program, since the state's economy depended on frequent replacement of vehicles.[47]

The campaign to use rock salt to defeat the enemy snow parallels the marketing strategy for DDT and other pesticides as simple panaceas for pest control in the 1950s. Despite the lack of reliable data on the effects of salt on automobiles, asphalt, concrete, and trees, Calgon, Inc., of Pittsburgh, manufacturer of the polyphosphate Banox and the National Aluminate Corporation of Chicago, maker of Nalco 8181-C, convinced hundreds of cities to add their products to the salt spread on streets and highways and to stop worrying about corrosion. Countless tons of salt were spread annually in the name of safety and efficiency before, in the 1970s, serious questions about the long-term effects of salt were raised by environmentalists and engineers.

A close reading of the many articles in *American City* on the virtues of salt reveals other motives and values in its selling. The most powerful force behind the acceptance of salt was the physical expansion of the city and the growth of suburbs. With thousands of new kilometers of streets to maintain and millions of new automobile owners to propitiate, municipal governments needed a quick, but not too dirty, fix. Salt came to replace the mix of sand, cinders, and other abrasives traditionally dumped on city streets to melt snow and increase traction. City dwellers were accustomed to industrial grime, but the suburbs built after World War II were meant to be smokeless utopias of pastel hues. Banox was artificially colored green, to signal that streets were protected "both against traffic mishaps and car damage" and to blend with the neatly trimmed lawns that symbolized the deseasonalized and oasis communities of the expanding "light snowfall cities."[48]

The migration of Americans from snowbelt cities to sunbelt suburbs did not abate the war on snow. There was a demand for bare pavement on all streets and highways, and a combination of technologies made that ideal almost obtainable, for a price. *American City* recommended various kinds of reversible straight-bladed plows equipped with shock absorbers that permitted the blade to ride over obstructions, conveyor and front-end loaders, and salt spreaders. Some cities turned to commercial weather forecasters for specialized forecasts relevant to snow removal. In the early 1950s Detroit, Klamath Falls, Cincinnati, and other cities experimented with radiant heating pipes and coils to keep overpasses, freeway ramps, and sidewalks snow free. While the international Cold War seemed stalemated, the domestic war on cold was going well, even if a few memorable storms, real and fictional, suggested otherwise.[49]

In a fortuitous anticipation of the great New York storm of December 26, 1947, the popular novelist Henry Morton Robinson, best remembered for *The Cardinal* in 1950, published what may be the ultimate Cold War novel, *The Great Snow.* Ostensibly it is the story of a forty-two-year-old patent attorney suffering from myriad family problems who emerges with a renewed sense of life's possibilities from a twenty-day blizzard that leaves 800,000 dead in the New York area. But *The Great Snow* is also an allegory of nuclear holocaust and other fears of the Cold War years. Its use of snow both to isolate the hero and his family and to symbolize order in a disordered world makes it an interesting commentary on its times, and a good example of snow as symbol and plot device in fiction.

The novel opens with the lawyer, Ruston Cobb, preparing to leave Manhattan for his home up the Hudson River for a weekend with his wife, Nolla, a kleptomaniac; his son, Roddy, who is struggling with his homosexuality; his sister-in-law, Berry, with whom he is falling in love; and her companion, the has-been artist Laimbeer. During the course of the storm they are joined by Cobb's daughter, Sicely, and the housekeeper's son, Gunnar. Much of the action of the book centers on Cobb's efforts to rescue Sicely, who is trapped by the storm in Albany, and to save the lives of his family and guests as the storm enters its third week. Roddy dies of pneumonia, and the caretaker, Gunnar's father, kills himself, but Cobb pulls everyone else through, including his rival, Laimbeer, whose infected arm is cured by packing it in snow! "Wonderful stuff, snow," Cobb tells the artist just before the end of the storm. "Variety of uses. Makes houses for Eskimos, protects vegetation, and checks spread of infection in important members."

Earlier in the story, when Laimbeer is pontificating on the loss of unifying symbols in the contemporary world, Cobb suggests, "'Nature. Hills and streams, spring rain, summer clouds, winter snow . . . the blue-white poetry falling in silence over the world.'" To which the artist replies, "'Noble, kindly nature, eh? A bountiful mother stuffing a rosy-nippled tit into Little Boy Blue's mouth while he lies under the haycock, fast asleep?

Sorry, Cobb, but the whole idea has been upset by everything that's happened in the past fifty years . . . the fission-bomb horror hanging over the whole landscape.'" From the opening page—when the first flake of snow stirs Cobb's memories of a milkweed filament, a compass rose, a crystal flacon of perfume, and his mother's diamond brooch, "the basic pattern into which all structures, natural and man made, fell . . . 'a new chip off the old universe'"—through Gunnar's observation that snowflakes are "an attempt on the part of unorganized matter to take on solid form," to the end, when Cobb gazes on the brilliance of New York City in the snow and concludes that men and women will not forget the warning written by the snow, Robinson succeeds in making even the catastrophe of the great snow a hopeful symbol of the future. Unlike George R. Stewart's *Storm,* in which snow is simply a blind force of destruction, or Edith Wharton's *Ethan Frome,* in which snow confuses the characters and erases their memories, *The Great Snow* explores the characteristics of snow in its variety and offers its readers more than the image of snow as negation. One novel cannot, of course, change many attitudes, nor even represent a change in popular thinking, but in the second half of the twentieth century, popular views of snow have become more complex.[50]

New Yorkers who woke up to more than 61 centimeters (2 feet) of snow on the morning of December 27, 1947, probably cursed Robinson, if they thought of him at all. The first editorial response in the *New York Times* was simply awe: "Our city has greatly changed in those sixty years since the Great Blizzard, but man is still at the mercy of the elements as anyone who looked out yesterday on New York's streets could see." Although telephones and radios did not fail, continued the editor, "life [became] disordered then frenzied as distressed workers struggle to get home, and [ended] at last in a great paralysis. . . . It is as if God had suddenly decided to intervene and provide an unheralded day of reckoning for man to take precise measure of himself." By the third day of digging out, the newspaper was less apocalyptic, more inclined to see the gaiety of children sledding and adults building snowmen. Before the new year, there were the usual editorial calls for better planning and snow management.[51]

E. J. Kahn, Jr., a writer for the *New Yorker* magazine, was moved by the human drama enacted in the storm. His account focuses on the closing of the old Albany Post Road where it ran, 6 meters (20 feet) from his house, through Scarborough-on-Hudson, near Ossining. When they discover several cars and trucks snowbound nearby, Kahn and his wife invite the drivers and passengers in for shelter and food. These refugees eventually include three beer truck drivers, two A & P truck drivers, the driver of a tobacconist's truck, a moving van driver and his helper, a mysterious old man who never identifies himself, and two families, one headed by a used-car salesman from Schenectady, the other by an army officer from Virginia—in short, a random sample of automobile-age America temporarily immobilized by snow. As the snowbound travelers talk, Kahn learns that one of the salesmen's sons is named Gilroy, close enough to the legendary Kilroy, who wrote his name wherever American troops went during the war, to lend a mythological note to the experience. The road is reopened, and the drivers depart, leaving the Kahn home somehow better integrated into the nexus of humans and machines that characterizes modern society. Kahn's grandfather's memories of the snowdrifts of 1888 no longer seem so powerful.[52]

In 1947 snow liberated Americans from the grip of the past and the fears of the future, restoring normalcy and bringing generations together. The midwestern blizzard of 1947 also inspired a story by Mari Sandoz about a school bus with eight children and their teacher lost in the storm for eight days. The plucky children, ages six to sixteen, and the teacher survive by using their knowledge of the terrain, gathering willows for firewood and cooking a frozen calf they find along a fence line. The story breaks the tradition of frozen women, since the teacher neither sacrifices herself nor exhibits any special heroism. The excesses of nature again become the enemy, and the military can be called upon to use its power to save lives and property threatened by snow, as it was in January 1949, when much of the country, from the Sierras to the Great Plains, experienced heavy snowfall and blizzard winds. President Truman ordered the Nebraska National Guard and parts of the Fifth Army to get food and medical supplies to the Middle West in Operation Snowbound. In Operation Haylift, the Air Force flew C-82 flying boxcars filled with feed to drop to cattle in remote areas in Nevada. The winters of 1950–51 and 1951–52 also brought record snows to parts of the West and Middle West. Old-fashioned winters seemed to be returning, and the popular press touted the idea of a new ice age. Photo-stories in *Life* magazine and local newspapers helped raise snow consciousness and make Americans ever more demanding of accuracy in weather forecasting.[53] One result was a minicrisis in November 1953.

THE FLURRY FLURRY AND POLITICAL SNOWS

On November 6, 1953, an early snowstorm moved up from western North Carolina and dropped more than 60 centimeters (2 feet) northwest of Harrisburg, Pennsylvania. Unfortunately for the Weather Bureau, New York City also received about 8 centimeters (3 inches), most of it unexpected. That a minor storm, even misforecast, should cause the bureau great internal agony and soul-searching seems inexplicable, but a possible answer lies in the context of the times. The Weather Bureau had grown in size and scientific importance under its dynamic chief, Francis W. Reichelderfer, a former navy pilot and meteorologist who was one of the first Americans to recognize the importance of Vilhelm Bjerknes's theories of the effects of polar cold fronts and air masses on global weather. Reichelderfer also appreciated the value of good public relations. In 1939, the year he became chief, the bureau initiated automatic telephone service in New York City, Washington, Newark, Baltimore, Detroit, and Chicago. Under his direction the bureau prepared four public forecasts a day rather than two. Radio, and later television, turned weather reports into a form of daily entertainment and helped to raise public expectations about the accuracy of forecasts.[54]

The rapid pace of scientific and technological improvement in meteorology during World War II may have deceived government meteorologists into believing they were infallible. C. G. Rossby and his colleagues at MIT developed mathematical models for forecasting, and Reichelderfer obtained John von Neumann's support in using the ENIAC electronic computer to perform his calculations. The development of radar made it possible to track the movement of storms minute by minute. Improvement in aircraft and the development of rockets equipped with cameras that could be parachuted back to earth provided panoramic aerial views of clouds. Nuclear weapons testing led to fallout forecasting, and in the early 1950s Irving Langmuir and others were confidently predicting weather modification. Possibilities for climate control appeared limitless. At last meteorology seemed a science of prediction and control. Its language was being redefined. Then a small cloud appeared on the horizon.[55]

The cloud became a mass of clouds over North Carolina, moved north, and brought early snow to Pennsylvania and parts of the East Coast. There was considerable damage from winds and high tides. In a published review ordered by Reichelderfer, a Weather Bureau official pronounced the storm unique, while conceding that the combination of a moist low-pressure system coming up from the south and a cold front moving down from the Gulf of St. Lawrence at 6 kilometers (20,000 feet) bore some similarity to storms in the autumns of 1920, 1923, and 1950. A series of internal memos admitted that the prognostic charts had placed the storm too far east, had overemphasized a cold air barrier between the two fronts, and had lagged twelve hours behind actual developments. In requesting a "case study of the November 6 cyclone," Reichelderfer added, "Consideration may also be given to development of a new terminology for handling 'incipient' and 'borderline' elements in the forecast."[56]

A large part of the problem was semantic. The meteorologists in New York and Washington, D.C., defended themselves by blaming the media and by insisting that the 9.2 centimeters (3.6 inches) measured in the snow-box was much more than actually fell in the streets. The crucial issue was the use of the term *snow flurry.* The press claimed that the Weather Bureau had predicted flurries and heaped several column inches of ridicule on the hapless weathermen. I. R. Tannehill, chief of the Division of Synoptic Reports and Forecasts, attempted an explanation. "I am convinced," he wrote to Reichelderfer on November 19, "that a great deal of unnecessary trouble comes from the Weather Bureau's efforts to be too specific about how much precipitation will fall. Some 10 years ago we became infatuated with the idea of making quantitative precipitation forecasts, either by listening too long to the high-pressure boys or by paying too much attention to the uninitiated. Most of the time we make ourselves look rather stupid." Part of the problem was relations with the media, Tannehill conceded, but the bureau had to learn how "to bury them with . . . so much stuff to read that they haven't time to look in the Farmer's Almanac." New York City, as the center of news publishing and broadcasting, was crucial, but the core of the problem was changing definitions:

> According to what you read in the papers, they frequently use the term "snow flurries." Just what we mean by a snow flurry and what the public understands it to mean is a question. We recently put out a terminology and after a long hassle we defined snow flurries as follows "fall of snow of short duration with clearing between occurrences. Total accumulation of snow expected to be small." Maybe this is okay, but I wonder. The Weather Bureau Glossary put out in 1945 says that a flurry is "a shower of snow, brief, and accompanied by a gust of wind." In the public mind the word flurry does carry a connotation of wind. Webster's dictionary says that it is "a light snowfall accompanied with wind." Thus the public gets the definite impression that a snow flurry is something like a gust of wind with a few flurries of snow in it. I always thought that too until my mind became confused in reading Weather Bureau instructions. So, when the New York forecaster forecasts a snow flurry, the New York public expects to see a gust of wind come around the corner of the Flat Iron Building with 6 or 8 flakes of snow in it. We at one time said that terms used in the forecasts should be taken from the dictionary. Later we changed it and we didn't tell Webster.
>
> As time goes on I have come to detest this term "snow flurries." We have come to the point where "snow flurries" get more laughs than Jimmie [*sic*] Durante or Milton Berle.[57]

Tannehill's tirade makes some good points but misses the essential fact that professional meteorologists and television weatherpersons, many of whom were, in 1953, weather "girls," were increasingly serving different functions. Moreover, meteorological research was outstripping the ability of local forecasters to interpret the complex data.

The taxonomy of snow evolves in response to changing concerns. In 1940 the Weather Bureau was transferred from the Department of Agriculture to the Department of Commerce and began to focus more on snow as an impediment to air and automobile travel. There was a new emphasis on visibility. A "light" snow became one in which objects could still be seen more than 1 kilometer (0.625 mile) away, regardless of the final accumulation. Duration and wind were also factored separately, to the point where *flurry* was redefined as "a shower of snow" without wind. This was the definition the New York meteorologists had in mind when they issued their forecasts based on a reading of the synoptic data they were receiving from radar. Tannehill was right to complain that "we need someone in that office who will look out the window and coordinate with the map . . . to keep up with the public," but his explanation was not totally satisfactory to his chief. Responding to another memo, Reichelderfer archly commented: "Your last paragraph suggests that we be on the lookout for this first snowstorm every year. . . . Perhaps we need something more. Perhaps on November 1st each year we should send livid cards or posters to be placed conspicuously on top of the current weather map each morning so that the forecaster would have to remove it before he could see the map—the poster to read somewhat as follows: ARE YOU SURE THAT THIS IS NOT THE DAY TO EXPECT THE ANNUAL FALL 'BOOMER'?"[58]

Perhaps the Weather Bureau redeemed itself in the eyes of the public when it began to publish the magnificent satellite photographs of the earth in the 1960s, perhaps the public simply learned to live with probabilities and percentages of confidence, or perhaps they found in their elected public officials a scapegoat riper for sacrifice. For whatever reason, the snowstorms of the past thirty years have often been identified with mayors and governors whose cities and states have suffered, rather than with the meteorologists whose predictions failed to provide sufficient warning. The record 58.5-centimeter (23-inch) snowfall in Chicago, January 26–27, 1967, illustrates this point. As reported by the *Chicago Tribune,* the initial responses to the storm included the familiar reports on hardworking street cleaners, the holiday mood of discommoded travelers and commuters, and accounts of weathermen unable to get to their stations.

"Many Are Stalled, Few Are Frozen," blared one headline, while the editors quoted Edwin Markham, "The pity of the snow, that hides all scars," suggesting that the last word be changed to "cars." But there were scars, too. After a month of continuous snow removal, a rising death toll, huge business losses, and sporadic looting, many citizens of the Windy City began to question the efficiency of Mayor Richard Daley's administration. The report of an internal investigation into the handling of the snow emergency, which concluded that the 1967 storm was a once-in-two-centuries occurrence, was seen as an inadequate response. Corruption had been exposed. The riots at the 1968 Democratic party convention completed the destruction of the mayor's image. As the anthropologist Mary Douglas has argued, when nature is dragged into the moral code, meteorological disasters acquire political dimensions beyond issues of bureaucratic inefficiency.[59]

Two years later less snow but even more unfavorable publicity fell on Mayor John Lindsay of New York City. The *New York Times* reporter Israel Shenker, alluding to the song popularized in *The Graduate,* described the city as filled with "Sounds of Silence." His story quotes a psychoanalyst who had been snowbound on the turnpike: "'As we sat in our car in the snow . . . my first thought was that there's something about snow which means death. I felt like a spectator at my own funeral.'" On the other hand, Shenker reflected, "New York [was] washed clean by snow and overcome at times by a spirit of camaraderie that would melt with the snow." Melting began between pages 18 and 30, where the editors criticized Lindsay for failing to mobilize the snowplows early in the storm, for waiting until the middle of the day to call a meeting of his Emergency Control Board, and for returning to routine business before the streets were cleared. As uncollected garbage piled up, so did attacks on the mayor. "The whole dismal story of the mishandling of this emergency," the *Times* editorialized on February 13, the fourth day after the storm, "was the product of inadequate foresight, bureaucratic complacency fed by misleading weather reports, lack of needed areawide leadership and incredible business-as-usual attitudes in many quarters, private as well as governmental. Now it is not too early to begin rectifying the appalling weaknesses that have been uncovered, weaknesses that in a future emergency could threaten the safety and lives of even more New Yorkers than suffered after Sunday's snowfall."[60]

Criticism, like snow, fell on the just and the unjust, but most of it stuck to Mayor Lindsay. His response was *Planning for Snow Emergencies,* released ten months later in an attempt to bury his critics in thirty pages of "white paper." Typically, the white paper reviewed past storms, analyzed street cleaning practices, discussed mobilization procedures, and compared New York with other cities in the United States and Canada. The conclusions were also predictable: Large snowfalls (more than 38 centimeters or 15 inches) are rare, occurring on an average of every twelve years, and do not justify additional investment in equipment or manpower; weather forecasts are subject to "a margin of error"; the classification of snow emergency routes is not logical; and decisions to mobilize rest on individual judgments that are frequently wrong. The solutions? Reclassify the streets, improve the organization, and educate the public.[61] New York got more than 38 centimeters of snow in February 1978 and again in 1994. If these storms are any indication, improved forecasts and a more tolerant public have lessened the criticism and the impact of heavy snows.

Governor Michael Dukakis joined Mayor Lindsay as a court jester in what the *Boston Globe* called "The Age of Shovelry," when the snow of February 7, 1978, accumulated to more than 60 centimeters (2 feet) in a twenty-four-hour period in many places in Massachusetts and neighboring states. Although the *Globe* praised the governor for quickly mobilizing the National Guard and imposing a driving ban that continued for several days, many voters saw the photos of Dukakis operating his antiquated snowblower as contrived and chafed at the restrictions on private vehicles and his sanctimonious suggestion to those "suffering from cabin fever or getting a little stircrazy sitting around the house" to "remember that we have some people in the state that barely escaped with their lives; that are sitting homeless and in very, very desperate straits and in churches and in schools and just think about them as you wonder about your own problems because you haven't been able to move around." Not satisfied with this advice, Dukakis gratuitously told Bostonians to "go out and take a walk or shovel some more snow." One woman who took a walk recalled an incident that was probably typical:

> I was living on Commonwealth Avenue in Allston, where the entire street was lined with 3–5 floor brick apartment buildings. When the snow finally stopped, I decided to take a two-block walk to the supermarket. As I passed the building next to mine on the way to the store, I saw an elderly woman (probably in her early 80s) trying to maneuver the steps out of her building. The snow was piled very high and had not been shoveled; the only path was from the foot steps of those who had already walked on it. The woman was clearly having great difficulty getting out of her building, so I walked up to her and asked if she needed any help. Her voice sounded fragile and frightened. She said she was walking to the supermarket because she had run out of food the day before, and welcomed my assistance.
>
> She slipped her arm through mine and held as tightly as she could. We were walking very slowly; each step was an effort not only for her, but for me, because of the depth of the snow. We were having a simple conversation about the storm when she began to slip; instinctively, she said, "oh, this fucking snow!"

While many cursed the snow, the governor, and the mayor, who like mayors before and since was vacationing in Florida when the blizzard arrived, the *Globe* drew on regional pride to note that "the impression that Tremont Street conveys is precisely how Childe Hassam painted Boston: gentle, stately, and a little obscure."[62]

Dubbed "Awesome!" by Governor Dukakis, this storm was nearly matched by the Presidents' Day snowstorm in Washington, D.C., a year later. Letters to the editor of the *Washington Post* complaining about the city's inability to deal with a 12.5-centimeter (5-inch) snow were already filling columns when another 46.0 centimeters (18 inches) fell. The editors' response was philosophical. No use complaining, they wrote, "The message was fairly clear . . . a knock in the head to the human pretensions of controlling one's *own* destiny, that whole I-am-the-master-of-my-fate business, which in the world of Washington of course takes on the added dimension of an ambition to be master of everybody else's fate in the country as well." The paper remained resolutely cheerful in the wake of a 43.0-centimeter (17-inch) snowfall four years later but lost patience when a third quadrennial snowstorm deposited 35.5 centimeters (14 inches) on January 22, 1987. Under the headline "Breakdown—Total Breakdown," the *Post*'s editors asked sarcastically, and with more than a hint of irony given rumors of Mayor Marion Barry's drug habits, "What did the city government use to get rid of the snow—spoons and matchbooks?" A reporter was assigned to do a story on the psychological problems caused when families are confined to their homes. After interviewing several psychoanalysts, he concluded that "in some ways, being stuck in the Great Blizzard is like forced psychotherapy on a massive scale." Mayor Barry, in the great tradition of East Coast mayors, was away during the snow but returned to develop a well-publicized "Comprehensive Emergency Snow Plan," complete with new signs for banning parking during snow removal that read:

SNOW EMERGENCY ROUTE
NO PARKING DURING EMERGENCY

Whereas many cities used graphics—snowflakes in Bozeman, Montana, a snowplow in St. Paul, Minnesota, a snowman in St. Louis, Missouri—Washington, a city of big memos and little snow, used words.[63]

"EVEN AS WE SPEAK": THE LIVABLE WINTER CITY

Cities and snow will never be a happy combination, but judging from the editorial pages of newspapers in Boston, New York, Chicago, St. Paul, and Washington, D.C., an accommodation has been made. Snowstorms now provide a welcome respite from the everyday routines of work. They are seen as useful reminders of the limitations of technology. They are even occasions for local pride. The ultimate expression of this new attitude is the livable winter city movement, which has two contending points of view, one that emphasizes accommodation to winter, the other seeking ways to control it. The first was pioneered by the urban philosopher Frederick Gutheim, who, in an essay in *Architectural Record* in 1979, suggested zoning northern cities to allow snow to remain untouched in some neighborhoods for aesthetic effect and recreation. He deplored the "dark and dirty urban scene" produced by the current snow-removal techniques and even proposed making snow in cities to brighten their appearance and encourage nonmotorized transportation. The benefits of urban snow are its encouragement of community spirit and cooperation, improved building design, and better development of public transportation. Building design is the main focus of the other faction of livable winter city advocates, who emphasize indoor activities. In Canada "windscreen" buildings and indoor malls shut out snow, although some cities continue to hold ritualized outdoor winter carnivals.[64]

"The survival of winter cities," according to Norman Pressman, depends on reconciling the two positions. His advice: (1) do not overprotect people from nature, and (2) offer as much protection as possible. Again, solutions are sought in landscape and architectural design. Rocks and trees arranged to create snow breaks, houses and shelters painted in bright colors, microclimate areas, and infrared heaters in outdoor recreation areas are among the designs offered to alleviate physical and psychological problems associated with living in the Far North. Construction of the Alaska oil pipeline led to further studies that considered the need for privacy when it is impossible to go outside, as well as the differing needs of men and women, adults and children. Yet many of the commentators on livable winter cities seem to reflect a bias against snow, assuming that those who dwell in snowscapes many months of the year suffer from a poverty of visual stimuli. Those who have experienced long periods of snow, and who love it, recall vividly the reflection of lights on snow, the warm glow of candles inside paper bags in snow, and the reassuring sounds of snowplows passing on nearby highways. For careful observers such as the writer Annie Dillard, snow is almost overstimulating. The illusion that snow is lighter than the sky makes her feel as though she is walking upside-down. Perhaps it is this power of snow to reverse the familiar order that makes it so appealing to some and so threatening to others.[65]

Red Grooms is one of the few artists to visualize the livable winter city. In several of his cartoonlike paintings and lithographs, Grooms depicts snow falling in New York City. *Purple Umbrella* (1964) focuses on the legs, coats, and, of course, umbrellas of pedestrians caught in a flurry of large snow crystals. Grooms's patterns and colors make an urban forest; boots are dark trunks springing from shoveled pavement, while piles of snow form small Manhattan Alps. Some of his works locate snow in specific neighborhoods. In *Eighth Avenue Snow Scene* (1965), several pedestrians jump from the unplowed street into sidewalk drifts. Taxis and other cars seem unimpeded. In *No Gas: Slushing* (1971), snow falls on Lafayette Street in Greenwich Village, crowded with traffic (Figure 2.9). The bright colors of buildings, signs, and coats are carnivalesque. The snow occasions a festival of light in which the largely African American crowd is not exotic but intrinsic. They have made their city livable by incorporating snow into their lives. The same is true in *Snowy Day on Walker Street* (1976), in which the proprietor of a glass store throws salt on the sidewalk while people hurry by in a light snow. The dark green building is like a forest through which Oriental and Occidental immigrants pass in the process of Americanization.

Figure 2.9. *No Gas: Slushing* (1971), Red Grooms. Color lithograph on Arches paper, 56 × 71 cm (22 × 28 in.). © 1997 Red Grooms/Artists Rights Society. Grooms's whimsical view suggests that snow contributes to the carnival atmosphere of urban life.

The livable city reenacts a historical tableau. Snow frontier and western frontier are mirrored in Grooms's image.[66]

Willard Scott, who began weather reporting for WRC-TV in Washington, D.C., in 1967 before moving to the NBC *Today* show in 1980, signals the transition to local station forecasts with the words "Here's what's happening in your world, even as we speak." Scott's tag line provides the listener with a playful ritual for restoring weather, including snow, as a resource. The cultural critic Andrew Ross suggests how this happens in his essay "The Work of Nature in the Age of Electronic Emission." Ross asks how people use weather programming to make sense of their everyday lives; how the "National" Weather Service, with its national maps, has helped to create an American identity based not on social and political life but on nature. Although he provides no clear answers, Ross forces us to consider the familiar expression "everybody talks about the weather" in new ways. "Even as we speak" resonates, I think, with echoes of "We the People" and "The People, Yes." Scott and his equally jovial backup, Al Roker, remind us that snow will fall, even in the deseasonalized city. They and the more serious, scientific weathercasters place us in the contexts of nature and culture.[67]

A character in Don DeLillo's novel *White Noise* puts it succinctly: "I realized weather was something I'd been looking for all my life. It brought me a sense of peace and security I'd never experienced. Dew, frost, and fog. Snow flurries." Weather words are a litany chanted to provide a sense of continuity in the midst of change. Since 1934, when Jimmie Fiddler began making weather forecasts over radio station WLBC in Muncie, Indiana, and 1941, when the whimsical puppet Wooly Lamb made his first weather reports on WNBT-TV in New York City, the electronic media have developed formats and formulas for translating the complex data of meteorology into entertainment. They have, perhaps, succeeded in form better than in content. Robert Henson, in his readable book, *Television Weathercasting: A History*, cites a 1978 study that "found that viewers retained very little of the information in evening weathercasts. Only half remembered any part of 'tomorrow's forecast.'" Yet most were satisfied with the weather segment. As late as 1983, only 60 percent of American cities subscribed to weather services; the remainder relied on TV and radio reports. It has been suggested by some communications experts that commuters, isolated in their cars and already agitated by traffic, tend to exaggerate the urgency in the voices of radio weather reporters. The distinctions among "winter storm watch," "winter storm warning," "winter weather advisory," and "traveler's advisory" blur at 7:00 A.M. and 60 miles an hour. The writer Cullen Murphy speculates that the trend toward hiring professional meteorologists as weatherpersons on television and the introduction of concepts such as windchill factor leave viewers with a feeling of gloom.[68]

Even as they speak, however, we are reassured about our national identity, our local environment, and our personal stake in the weather. "When Push Comes to Shovel," the theme of the 1993 Annual North American Snow Conference of the American Public Works Association, sums up in a phrase the long contest between snow and humans in the age of rail, automobile, and air transportation. The urbanization of North America has further exacerbated the conflict. But the deseasonalization of the city remains incomplete; and we are beginning to appreciate snow where it touches and is touched by human beings. We now need a new, richer vocabulary of urban snow to describe the honeycombed piles pushed up around parked cars, or the snow that freezes to traffic signs, obscuring parking regulations. Tentative beginnings have been made. Snow compacted by a snow-processing machine used to make snow roads and runways in the Arctic and Antarctic is called *peter snow*, after the Peter Junior Snow Miller, and residents of the Dakotas call the mixture of snow and dirt that blows off the prairies *snirt*.

People enjoy coining new terms for snow conditions. Entries in a contest to name the phenomenon experienced by parents who discover that schools are closed because of dire forecasts when no snow has fallen included *dis-snowlusionment* and *snowphistry*. A New Hampshire driver told me of her game of snow turds—looking for football-sized, exhaust-blackened chunks of snow that have fallen from the fenders of moving cars and trucks and swerving to crush them. The game's challenge lies in guessing which "turds" will gush pleasingly under the tires and which, ice hardened, will puncture an oil pan or dislodge the muffler. We need roads strewn with snow turds and tailpipes for snowphisticated fun.[69]

Snowmen and Snowmanship

Recreation in the Snow

3

Ancient nomadic snowman has rolled round.
His spoor: a wide swathe on the white ground
signs of a wintry struggle where he stands.
P. K. PAGE, "THE SNOWMAN"

SNOWBALLS AND SNOWMEN

Long before the bare pavement and scorched snow policies of city street superintendents, humans asserted their dominance over snow by shaping it in their own image. Perhaps children were the first to learn the plasticity of snow when they were left to amuse themselves while their mothers were working and their fathers hunting. Certainly the association of children with snowballs, snowmen, sledding, and other play in the snow is as old as it is strong. Yet adults also play in the snow, as the popularity of skiing in the twentieth century shows. More and more cities are sponsoring winter carnivals with elaborate displays of snow and ice sculpture. Folklorists study snowmen as ephemeral art. Andy Goldsworthy uses snow to make serious art, and Bill Watterson's comic strip "Calvin and Hobbes" frequently uses snow to comment on art and aesthetics.[1]

One strip has the little boy, Calvin, making a snowman despite the warning from Hobbes, the stuffed toy tiger with whom he converses, that it will melt. "This time I'm taking advantage of my medium's impermanence," Calvin replies. "This sculpture is about transience. As this figure melts, it invites the viewer to contemplate the evanescence of life. This piece speaks to the horror of our own mortality." In another sequence Hobbes asks why Calvin has built a snowman holding a snowball in his stick hand. "He's contemplating snowman evolution," says Calvin. "Obviously, if he evolved from a snowball, it raises tough theological questions for him." "Like the morality of throwing one's precursors at someone?" asks Hobbes. "Sure," answers Calvin, "and what about shoveling one's genetic material off the walk?"[2]

When snow is used in play, two of life's most unpredictable elements are joined. Both are complex and evanescent. Each disrupts everyday routines. What is the distance from a child's snow fort to an ice palace? How has snow been turned from refuse to a resource by the ski industry and with what cultural and environmental consequences? These are the questions this chapter seeks to answer.

"On March 5, 1770, there gathered in the square before the Boston custom house a crowd which John Adams later described as 'a motley rabble of saucy boys, negroes and mulattoes, Irish teagues and outlandish Jack tars.' Before them stood the main guard of the Twenty-ninth Regiment. . . . As in all such affrays, it was difficult later for eyewitnesses to agree on how the shooting began. It is clear enough that the soldiers were receiving a heavy bombardment of snowballs and rubbish when they opened fire." With a snowball fight enshrined in history as the opening salvo of the Revolution, we are again reminded of the importance of snow in the formation of American national identity, as well as the close connections between snowball warfare and real war.[3]

New Englanders Thomas Bailey Aldrich and Henry Adams clearly saw the links between snowballs and minié balls. Aldrich, recalling a winter-long battle between the North End and South End boys of Portsmouth, New Hampshire, in the 1850s, mentions using snowballs filled with bird shot and marbles in a battle that eventually led the town constabulary to destroy their snow fort. Adams describes a half-day snowball fight between the tough kids of Boston's South End and the boys of the Boston Latin School that included his older brother Charles and two young men who died fighting for the Union a dozen years later. "As long as snowballs were the only weapon," Adams writes,

> No one was much hurt, but a stone may be put in a snowball, and in the dark a stick or a slungshot in the hands of a boy is as effective as a knife. One afternoon the fight had been long and exhausting. The boy Henry . . . had felt his courage much depressed by seeing one of his trustiest leaders, Henry Higginson—"Bully Hig," his school name—struck by a stone over the eye, and led off the field bleeding in rather a ghastly manner. As night came on, the Latin School was steadily forced back to the Beacon Street Mall where they could retreat no further without disbanding, and by that time only a small band was left, headed by two heroes, Savage and Marvin. A dark mass of figures could

be seen below, making ready for the last rush, and rumor said that a swarm of blackguards from the slums, led by a grisly terror called Conky Daniels, with a club and a hideous reputation was going to put an end to the Beacon Street cowards forever. Henry wanted to run away with the others, but his brother was too big to run away, so they stood still and waited immolation. The dark mass set up a shout, and rushed forward. The Beacon Street boys turned and fled up the steps, except Savage and Marvin and the few champions who would not run. The terrible Conky Daniels swaggered up, stopped a moment with his bodyguard to swear a few oaths at Marvin, and then swept on and chased the flyers, leaving the few boys untouched who stood their ground. The obvious moral taught that blackguards were not so black as they were painted . . . and ten or twelve years afterwards when these same boys were fighting and falling on all the battle-fields of Virginia and Maryland, he wondered whether their education on Boston Common had taught Savage and Marvin how to die.[4]

Daniel Carter Beard, author and illustrator of children's books and one of the organizers of the Boy Scouts of America, was alarmed by the level of violence in snowball fights and developed an elaborate set of rules for "Snow-Ball Warfare" that he published in a children's magazine in 1880 (Figure 3.1). Beard provided detailed instructions for the construction of a snow fort, ammunition sleds, and wooden shields, and for the electing of captains and the choosing of sides. Defenders and attackers carried flags, the object of the "war" being to capture the opponents' "colors." Prisoners could not be made to fight but could be put to work making snowballs and repairing the fort. Deserters could have their faces "washed with snow" and be put to work with the prisoners, but, Beard wrote: "No water-soaked or icy snowballs are allowed. No honorable boy uses them, and any one caught in the ungentlemanly act of throwing such 'soakers,' should be forever ruled out of the game. No blows are allowed to be struck by the hand, or by anything but the regulation snow-ball, and, of course, no kicking is permitted."[5]

Beard's attempts to regulate traditional games are part of a larger movement occurring in the 1880s to preserve a waning rural and Anglo-American past and to teach immigrant children American ways. Folklorists, psychologists, and others collected and codified activities they considered appropriate for schoolyards and the newly organized playgrounds. Unsupervised play was defined as "doing nothing" and "just fooling." As children continued to have snowball fights on their own, there is some evidence that they shifted their targets from rival gangs to objects of adult culture. The novelist and essayist Annie Dillard recalls from her Pittsburgh childhood in the 1940s that "in winter, in the snow, there was neither baseball nor football, so the boys and I threw snowballs at passing cars." Dillard describes a morning when she and five boys pelted a car and were chased by its driver,

Figure 3.1. "Storming the Fort," *St. Nicholas Magazine* (January 1880), by Daniel Carter Beard. Beard offers instructions for building snow forts and conducting snowball warfare in an attempt to impose order on spontaneous winter play.

who, when he finally caught them, could only sputter, "'You stupid kids.'"[6]

Beard did not stop with instructions for building snow forts. In a chapter titled "Snow-Houses and Statuary," adapted from an article by Samuel Van Brunt, Beard tells boys and girls how to make snowmen and other figures. Snowmen, as Avon Neal and Ann Parker observe, are usually jolly figures, constructed by or for children, and are familiar characters in children's fiction. Beard's snowman is a comic Frenchman with a "waxed" mustache made of two icicles. In 1890 John Champlin and Arthur Bostwick observed, "An image often made by boys is that of a man fitted with an old hat, with cinders for eyes and a clay pipe in its mouth, and then used as a target for snowballs" (Figure 3.2). Michael Gold, growing up in the tenements of the Lower East Side of Manhattan, built a snowman on a vacant lot on Delancey Street: "His eyes were two coals; his nose a potato. He wore a derby hat and smoked a corncob pipe. His arms were flung wide; in one of them he held a broom, in the other a newspaper. This Golem with his amazed eyes and idiotic grin amused us all for an afternoon."[7] John Sloan visualized a similar scene in his 1914 painting *Backyards, Greenwich Village,* in which a girl, pale as a snowman, watches two cats, one of which is watching two boys build a snowman (Figure 3.3).

Sloan's arrangement of the elements—girl, cats, boys, snowman, window frame, fenced yard, fresh laundry, drifting snow—and their possible permutations: male-female, active-passive, human-animal, sentient-inanimate, nature-culture, spectator-performer, and so on—is a virtual anthology of snow symbols and a commentary on their transformation in the city.

Although Sloan's snowman is unfinished, Neal and Parker correctly note that the snowman's face usually gets the most attention from its creator. Their book amply illustrates the wide variety of materials used for facial features and decorations on snowmen: chunks of coal, shriveled apples, stones, bolts, electrical fuses, bottle caps, champagne corks, and flashlight batteries for eyes; twigs, pebbles, rusty horseshoes, and discarded false teeth for a mouth; carrots, corncobs, sticks, and clothespins for a nose; nutshells, Christmas wreaths, cake pans, Jell-O molds, discarded clothing, and, of course, hats for added decoration. Gender, class, and ethnicity of snow images are significant, too. Although Neal and Parker illustrate two female snow figures, most snowmen look androgynous but are presumed to be male. Woody Allen cleared up this ambiguity in his movie *Radio Days,* when he had two boys decorate their snowman with a carrot penis. Reports of anatomically correct nude snowmen regularly appear in the press.[8]

Hats, traditionally tall stovepipe models or sometimes derbys, were associated with wealth and social pretensions. Pipes,

Figure 3.2. "The Snow Man—Happy Days," from a stereograph by George Baker, Niagara Falls, N.Y. (ca. 1888). The photographer has staged this scene to communicate something of the innocent joys of childhood, but children often made snow effigies of authority figures and then destroyed them. These children seem more interested in the cat on the sled.

Figure 3.3. *Backyards, Greenwich Village* (1914), John Sloan. Oil on canvas 66.0 × 81.3 cm (26 × 32 in.), Collection of Whitney Museum of American Art, New York. Sloan's painting echoes the photograph in juxtaposing snowman and children, children and cats, snow and buildings. Snow's power simultaneously to conceal and to reveal remains a constant theme in art and literature.

too, denoted class, depending on their style. In Richard Smith's 1934 song "Winter Wonderland," popularized by the Andrews Sisters in the 1950s, the snow-parson is built by the lovers to perform a mock wedding. "Frosty the Snowman," also a hit song of the 1950s, leads children on a chase to the town square, where Frosty ignores a policeman's command to stop, then disappears. His manic behavior confirms that snowmen share the characteristics of their medium—paradoxically hard and soft, substantial yet ephemeral, jolly but with a hint of what Watterson's Calvin sees as "Deranged Mutant Killer Monster Snow Goons." As targets for snowballing, snowmen can obviously represent anything the thrower imagines, but on the whole the traditional snowman is a symbol of authority to be attacked, and his destruction by sun and wind is welcomed.[9]

The Canadian poet P. K. Page, in the poem quoted at the beginning of this chapter, sees snowmen as Everyman, doomed to transience, who under the influence of the sun seemed to inch forward and back "in a landscape without love." As they melted they had "greyed a little too, grown sinister/and disreputable in their sooty fur." Howard Nemerov writes of the melting process as the "Journey of the Snowmen," who, "Becoming featureless/As powerful Pharaohs" turn to water "To glitter in the gutters/And snakedance down all hills/And hollows, on the long fall/That makes the sewers sing." In many ways the simple snowman with his lumps of coal and carrot carries more cultural meanings than the artistically crafted snow sculptures and ice palaces of winter carnivals.[10]

ICE PALACES, WINTER CARNIVALS, AND SNOW SCULPTURES

The first modern winter carnival in North America was held in Montreal in 1883. The celebration included sporting events, parades, Indian dances, costume balls, and an ice palace 27 meters (90 feet) long with towers 15 meters (50 feet) high. Leadership in organizing the carnival came in part from the Montreal Snow Shoe Club, which had been organized in 1843 and had been sponsoring races and social events since the 1850s. Snowshoeing, emblematic of Canadian identity, experienced phenomenal popularity after the Confederation in 1867. "By 1885," notes one historian of the club, "there were 1000 members in the M.S.S.C. alone and there were some twenty-five thriving snowshoe clubs in the city." The Montreal Winter Carnival was an expression of civic pride with an eye to attracting tourists and investment. The culmination of the celebration, the storming of the ice palace by participants armed with fireworks, contained elements of even earlier Saturnalias.[11]

The winter carnivals held in St. Paul, Minnesota, over the past century show how such ceremonies function, how the paradoxes of snow serve the problematics of urban culture. According to a history of St. Paul published in 1890, the idea for the first carnival, in 1886, originated with George Thompson of the *Dispatch,* one of the city's newspapers. The idea appealed to businessmen anxious to promote St. Paul as the retail center of the entire northwestern United States. The Great Northern Railroad had recently made the capital of Minnesota the third largest rail center in the country, but the city seemed remote and the climate inhospitable to most Americans. When a smallpox epidemic caused the cancellation of the Montreal Carnival, A. C. Hutchinson, architect of the Montreal ice palaces, was invited to design a larger structure for St. Paul, and a civic celebration to improve the city's image was planned (Figure 3.4).[12]

Following the Montreal example, the St. Paul Carnival Committee

Figure 3.4. Ice Palace, St. Paul Winter Carnival, 1887. Minnesota Historical Society. Although this photograph appears to have been retouched, the crowd is dwarfed by the towering Ice Palace, symbol of winter's power to stir the imagination.

publicized the ice palace as the main attraction and held a cornerstone-laying ceremony thick with symbolism. The president of the Carnival Association fused blocks of ice from Stillwater, Minnesota, and Fargo, Dakota Territory, to indicate the tributary relationship of these communities to St. Paul. Although most of the ice blocks used in building the palace were cut from the Mississippi River and nearby Lake Como, all the communities in the region, especially those linked by the railroad, were urged to contribute something to the carnival. The parade that concluded the initial ice palace ceremony and the events that followed in the first twelve days of February may be seen as a complex text containing many messages. A comparison of the 1886 event with carnivals in 1916 and 1982 reveals changes in emphasis and the symbolic functions of snow. Finally, the St. Paul carnival must be placed in the contexts of college snow sculpture festivals.

The first parade, January 15, 1886, preceded the carnival by two weeks. It was led by the police department in winter uniforms. The parade marshal and his aides followed. Next was the Great Western Railroad band, a sleigh with the carnival queen (the daughter of the Carnival Association president) dressed in the red wool jacket of the Nushka Toboggan Club, and a carriage containing the governor of the state and several mayors. One hundred twenty-five uniformed members of the St. George Snowshoe Club with snowshoes strapped to their backs were jeered by spectators who shouted: "Take off those Mother Hubbards. Take those bladders off your feet." An equal number of Wacouta Toboggan Club members escorted twenty members of the Minneapolis Toboggan Club. Six members of the club carried a toboggan with a little boy in club uniform. Next came the Nushka Toboggan Club, thirty-five men carrying Chinese lanterns and chanting the club's name in a kind of cheer: "N-U-S-H-K-A, Nushka." According to a reporter for the *St. Paul Pioneer Press,* the crowd responded: "R-A-T-S, Rats." Other clubs followed, including the Ice Bears, dressed in white duck and firing shotguns every few minutes.

The Scandinavian Ski Club, carrying 3.5-meter (12-foot) "snowshoes," were followed by a sleigh from Boston in the form of a ship that held forty passengers and required a team of eight horses, a smaller sleigh with local dignitaries, marching units from the fire department and the post office, a band, infantry, cavalry, and artillery from the militia, a Sons of Veterans drum corps, and 150 members of the Grand Army of the Republic (GAR). The final units in the parade were thirty-five horseshoers, described as "the only contingent from the working element," and thirty members of the Auld Scotia Curling Club.[13]

The next two weeks were filled with activity as the ice palace was completed and the thirteen-day festival organized. The carnival began on Monday, February 1, with a reception for visiting clubs and the formal opening of the toboggan runs and skating rinks, promenades and concerts. The Ice King, played by Gen. R. W. Johnson, arrived on the third day in a parade much larger than the first. The National Guard Toboggan Club dressed as devils, and several thousand more GAR members joined the procession. The fourth day began with curling matches and concluded with an assault on the ice palace by the Fire King and his forces. The evening ended with the defeat of the Fire King. The fifth and sixth days were filled with skating tournaments, masquerades, parades of dog teams and Indians, horse races, and entertainments by the St. George Snowshoe Club. Sunday the carnival was suspended. Monday, the eighth day of the festival, saw toboggan races, ice polo, snowshoe races, and entertainment by the Winnipeg branch of the St. George Club. The ninth and tenth days were devoted to Indian sports, a demonstration by the Scandinavian Ski Club, and the illumination of the ice palace by electric lights. A parade on the tenth day was composed largely of floats sponsored by local merchants. On the eleventh day the GAR held its own parade and in the evening stormed the ice palace, forcing the Ice King to surrender. In the final two days there were snowshoe contests between whites and Indians, more costume dances, a children's day, and performances of the operetta *H.M.S. Pinafore* by the St. Paul German Society. The *Pioneer Press* proudly reported that sixty-two clubs with 4,740 participants took part in the festivities, which were viewed by 50,000 spectators, of whom 30,000 were visitors to St. Paul.

A float-load of semioticians might spend a winter analyzing the carnival, but its cultural meanings are not hard to see. The 1886 St. Paul Winter Carnival was about more than markets, climate, and regional identity, although these were important. The inclusion in one exhibit of relics of the ill-fated Greely Arctic expedition and editorial comment on the conflict with Britain and Canada over sealing in the Bering Sea make the issue of identity more explicitly political. St. Paul's celebration not only mocked the eastern skeptics who wrote that the city's winter weather was too harsh to permit industrial and commercial development but put them on notice that the future of America lay to the north and the key to that expansion was the northern rail route. Frequent references to ethnicity in stories about the parade point to an awareness of the need to resolve conflicts, to balance pluralism and the melting pot. Jean Spraker points out the similarity between the carnival and the Centennial Exposition in Philadelphia and Wild West shows in their treatment of Indians as historical curiosities. The citizens of St. Paul seem to have followed the example of Montreal in organizing private clubs for tobogganing and snowshoeing, excluding Native Americans from membership. Holding separate races between whites and Indians also followed Canadian practice. Snowshoeing and tobogganing appear to have been adopted by the city's elites as American, while skiing and curling remained identified with more recent immigrant groups. Germans seem to have identified themselves with popular theater rather than sport. The parade of the farriers suggests that they still held a high status among working men in a horse-dependent society. Blacksmiths were important to the railroads as well. Some class antagonisms are hinted at in the jeers that met the St. George and Nushka marchers.[14]

The carnival was both a playful expression of local customs and a commentary on their changing meanings. The emphasis on play, recreation, and leisure is significant. The United States was entering a new era of leisure. The carnival sanctioned the "gospel of relaxation" that the sociologist Herbert Spencer had called for in 1882. On the Sunday observed during the festival, J. L. Scudder of the First Congregational Church entitled his sermon "The Relation of Amusements to the Church of Christ" and told his congregation that "the man who has no play in him is only part a man." The citizens of St. Paul legitimized their first winter carnival with appeals to church and state. Moreover, they fused the two in the symbol of the GAR. The marching veterans of the holy war to save the Union, the flag-bearing uniformed social clubs, the surrender of the ice palace to the GAR forces rather than to the Fire King, all attest to the war's influence on the carnival.

Members of the generation that had risen to prominence in that war were reaching their pinnacle of political and economic power. The winter carnival, with its palace-fortress, was a perfect symbol to unite the men who had fought for the nation and for their children, who could only imagine war in their snowball assaults on snow forts.[15]

A key figure in the 1886 festival was W. A. Van Slyke, who served as general manager. Van Slyke had come to St. Paul at the age of nineteen in 1854, the year the capital opened. He began his own dry goods business in 1856, organized a company of volunteers in the Civil War who served in the Mississippi and Tennessee campaigns, and returned to prosper in both the dry goods and grain commission businesses. He had been a member of the Chamber of Commerce since its organization in 1867 and was serving as commissioner of parks at the time of the carnival. It was said that "every celebration is unfinished without his hand." None of the subsequent carnivals was as elaborate as the one he staged. In one grand month, St. Paul made a metaphorical and metonymical transformation from frontier to urban culture. The changes in identity involved in growing from a settlement of a few thousand inhabitants to a city of 120,000 in the space of thirty years, and the crises of leadership faced by the men who had fostered this growth, were partially resolved by building and destroying an imaginary city of ice and snow. The ice palace, illuminated by new technology, clearly functioned both as a symbol of change and as an expression of real technological and political power. In later carnivals, the ice palace was captured by the forces of warmth; finally it disappeared altogether. The whole ceremony became ritualistic and formulaic. But in 1886 the festival was closer to spontaneous play. The Civil War veterans returned for one last snowball fight.[16]

The driving force behind the 1916 carnival was not a local businessman but Louis W. Hill, son of James J. Hill, who had brought the Great Northern Railroad to St. Paul. Anxious to promote tourism to the city and the northwestern United States served by the line, Hill convinced local businessmen to join in his version of civic celebration and to elect him president of the Winter Sports Carnival. The change of title is significant, since organized sports figured much more prominently in this carnival than in its predecessors. It is significant, too, that the Fire King played a more important role than the Ice King, reversing the meaning of the first carnival, which had honored winter.

Hill's concept of carnival is revealed in a speech he gave to two hundred northwest retailers at a banquet during the celebration. "Men who cannot forget their business and get out and take part in the winter carnival are not the kind of men we want in St. Paul," he told them. "Wherever I go there is a carnival. Why not? It is all in a day's work. Work fast, think fast, don't loaf, don't stand around. Take part in sports and do it with all your might. Do something to boost your city and country." Carnival as part of a day's work is as straightforward a statement of capitalism as Thorstein Veblen's *Theory of the Leisure Class.* Production was only half the purpose of business; the other half was consumption. As advocates of skiing were suggesting, unused snow was wasted snow. With slightly more than twice its population of 1886, St. Paul now boasted 164 winter sports clubs with 9,690 members, also about twice the earlier numbers. The city was ripe for Hill's message. It had a new armory and auditorium in which to hold sporting events; the weather was no longer the key to a successful carnival. The Fire King had triumphed over King Borealis in the furnaces and electrical heating units of the Twin Cities.[17]

The 1916 carnival began promptly at 8:00 P.M. on Thursday, January 27, and ended at Hill's bidding on the stroke of midnight ten days later. The main parade, on Friday the twenty-eighth, consisted of marching clubs and floats. Newspaper accounts complain that the crowds prevented the marchers from performing fancy drills but praise a boy who put up a ladder on one street corner and sold more seats on the rungs than he could accommodate. Clearly, he demonstrated the spirit of James J. Hill better than the old robber baron's own son. The floats, many of which were motorized, were chiefly advertisements for department stores, bakeries, the telephone company, and local manufacturers. An automobile parade of more than 200 cars was held separately. Each day's events consisted of trapshooting, horse racing, curling, and pushball and tug-of-war contests between school and club teams. Races between homemade motorcycle-powered sleds, forerunners of snowmobiles, were held with some success. On Saturday night the Fire King dethroned the Ice King in an assault on the ice palace with its 3.7-meter (12-foot) walls, greatly diminished from the 30.5-meter (100-foot) high Gothic castle of 1886, which had measured 32.9 meters (180 feet) in length and 48.8 meters (160 feet) in width. King Boreas was restored to his throne on the last night in a ceremony that was supposed to climax in the naming of a queen. For reasons not clear in the newspapers, the king refused to choose and named all 108 candidates queens.[18]

There is, I think, a clue to the deeper meanings of the 1916 carnival in the king's failure to name a queen. On one level, the carnival was again about economic power and technological change—the ice palace was even attacked by balloons and dirigibles. But on another level, this carnival was a comment on the place of women in society. Photographs of the numerous queen candidates filled the pages of the *Pioneer Press* during the days of the festival. Several articles commented on women's participation in sporting events, and the society pages were filled with plans for balls and other events organized by women. Yet the police announced that women would not be allowed to attend boxing matches, and much amusement was expressed over the newsboys' "queen" candidate, Reuben Fredkovsky. Moreover, a front-page story the day before the carnival began told of a young woman who had found work in the White Enamel Refrigerator Company by posing as a boy. The article described in detail how she had deceived her landlady, boss, and co-workers by cutting her hair, wearing men's clothes, and going out with the boys to pool halls. The tone of the story wavered between shock and amusement.

Certainly this story did not cause the Ice King's cold feet, but gender-role confusion in St. Paul's new industries, and the rapidly changing image of women in the movies, may have. It is also significant that once during the carnival King Boreas tried, without success, to get the movie crews who were filming it to pick the queen. Part of the play of this carnival was sexual, and F. Scott Fitzgerald's short story "The Ice Palace" is an insightful interpretation.[19]

In Fitzgerald's story a young woman from Georgia visits a man in St. Paul during the winter carnival. She is introduced to all the snow sports—tobogganing, snowshoeing, skiing—watches the parade, and visits the ice palace, where she becomes lost in the labyrinth and feels the cold breath of death. The opaque message seems to be that the northern climate is too strenuous and the outlook of northern people too tragic for this southern belle. In 1916, with the Great War raging in Europe, men were eager to demonstrate their physical dominance. The emphasis on trapshooting, boxing, pushball, tug-of-war, and motor-sled racing seems to have been calculated to preserve masculinity in the face of the threat from women all too ready and willing to take over men's jobs.[20]

Nostalgia, declining public participation, and increased spectatorship have marked the carnival since its revival in 1946. The most significant change has been the loss of the ice palace. After diminishing in size annually until 1967, when it was made of packed snow, it was eliminated entirely in 1984, partly because insurance companies refused to write liability insurance at affordable rates. The 1982 carnival coincided with a record snowstorm that left as much as 51 centimeters (20 inches) on the ground but hardly disrupted events, which were mostly held indoors. People were advised to watch the parade from the "skyways" that had been built across the main streets from store to store to make St. Paul a "livable winter city." The theme was "Family Frolic," its logo incorporating a Norman Rockwell ice-skating scene. Events held during the two-week carnival were eclectic, fewer than half related to winter in any way. Backgammon, tennis, chess, squash, softball, basketball, Ping-Pong, bowling, Scrabble, and swimming tournaments dominated. Skating, sledding, skiing, and hockey competed for audiences with a hairdressers' carnival of fashions, an Afro-Hispanic variety show, a cat show, and an orchid show. Winter carnival was finally deseasonalized, the snow forts of childhood forgotten.[21]

Where the spirit of play prevails, on college campuses and at winter resorts, snow sculpting survives, an annual reminder of the uninhibited freedom offered by snow. The Dartmouth Winter Carnival, begun in 1910, claims to be the oldest of the college snow festivals, and Dartmouth's students have produced imaginative snow statues, as have the students of Michigan Technological University, who began their carnival in 1964. Sculpture contests are held annually at Lake Tahoe, Breckenridge, Colorado, and other snowy places. In the winter of 1993, students in an art class at St. Michael's College in Vermont created almost life-sized Easter Island heads.[22]

The tradition of snow sculpture endures in both the folk art of the snowman and the work of professional snow and ice modelers. The mythic origins of American snow sculpturing may be found in the story of young Larkin G. Mead, Jr., a clerk in a Brattleboro, Vermont, hardware store, who, on the night of January 2, 1857, built a 2.4 meters (8 feet) tall "Snow Angel" holding a tablet and stylus. Later, when Mead had become a renowned sculptor, he carved a smaller copy in marble for All Souls Church in his hometown. The story implies that snow is an American artistic medium, allowing the poorest child to demonstrate creative talents and rise from obscurity, if he or she is lucky enough to live above the snow line.[23]

"SEAMING A VIRGIN FACE": SKIING AND THE REDEFINITION OF SNOW

Readers of *Cosmopolitan* magazine in 1896 were introduced to Mt. Shasta and Mt. Hood by an article that told them, "No one has ever viewed the grand scenic panorama to be seen from their tops . . . or finally taken a wild, swift, exhilarating and almost uncontrollable ride down the steep declivities of snow, without acquiring for them an intense admiration that remains undimmed until death." The almost uncontrollable ride was not in this case on skis but on the seat of the pants with an alpenstock as a break![24] The combination of vertiginous play and reverence for nature that characterizes this cascade in the Cascades highlights the paradox of winter sports in general and skiing in particular. Can mountain snow inspire awe as it is being packed on trails? When snow is used as a resource for sport, will recreation ultimately wreck creation?

The historian E. John B. Allen places the origins of American skiing in the context of the back-to-nature movement of the late nineteenth century. Inspired by the Nordic ideal of outdoor physical exercise, young men and women took up climbing, "coasting," and jumping on skis. The snow-covered countryside provides "freedom and . . . intimate intercourse with Nature," wrote W. R. Rickmers in 1904, and he reeled off a list of twenty kinds of snow:

For the ski-runner this great variety of ground is increased a hundredfold by the different states of the snow, which he learns to distinguish in the course of his outings. The changes snow is capable of are wonderful to behold, and the observant tourist never ceases to discover some kind or condition which is new to him. There is soft, flaky snow, fresh fallen from above; there is downy, fluffy, powdery, sandy, dusty, floury, crystalline, brittle, gelatinous, salt-like, slithery, watery snow; there is snow as hard and white as marble, and snow with a thick crust which breaks into big slabs; there can be a layer of soft or powdery stuff on a hard sheet, or a thin, glassy film over loose snow. I have seen it in thin shells, the size of half-crowns, rustling under the ski like the leaves of an autumn forest, or, again, in the form of long, streaky crystals, like asbestos. Often it lies pat and smooth over the rounded hill, at other times it will be a frozen turmoil of waves, ridges, and grooves.[25]

Rickmers's nomenclature of snow on the ground was copied the following year by Theodore A. Johnsen in the first American ski book, *The Winter Sport of Skeeing,* but these categories were superseded in the 1920s and 1930s by less poetic terms more descriptive of the density, water content, and age of the snow.[26] The change accompanied a shift from Nordic- to Alpine-style skiing, which emphasizes downhill speed and slalom, or "gate," racing, requiring mechanical lifts to return skiers to the top of the runs as quickly as possible. Greater use of the best trails or ski runs, in turn, necessitated snow grooming. Skiers were increasingly removed from snow as it had fallen. Acceleration on the ski slopes was also dependent on rapid movement to them. Railroads opened the snow frontier, putting distant mountains within reasonable travel time from cities. Automobile and air travel continue the transformation.

From the beginning, the tactile experience of rapid movement over snow inspired wordplay. In 1916 George Wharton James, a frequent promoter of the Southern Pacific Railroad in California, humorously compared skiing to being shot from a gun: "I glided, *glid,* or *glode* for quite a number of yards; then, as my speed increased, I swayed, *swid,* or *swode,* and finally fell, *fill,* or *fode,* and at once chaos reigned." James cautioned his readers that "snow is not always the same, neither are all surfaces as level as they ought to be for a new beginner. After snow has been partially melted and then frozen it is very different material from what it is when soft and flaky . . . [and] when one comes to an irregularity in his pathway, as, for instance, where the snow has partially melted around the trunk of a tree, all roads seem to lead to that tree."[27]

The experience of walking up a mountain in order to slide down, frequently falling during both ascent and descent, was sensuous and, for some, sensual. In 1919 Walter Prichard Eaton, a popular novelist and playwright, introduced an anthology, *Winter Sports Verse,* by comparing skiing to sex:

"No coquette, to be sure, not the fascinating Beatrice herself, can provide more excitement and variety, and in a shorter space of time, than a pair of skis on a mountain side." In his poem "Skis," Eaton is practically orgasmic:

> [I] drew in
> One last deep breath of stinging air, and slipped
> My skis across the rim: then farewell breath,
> And almost vision, too, as tears rolled down
> My cheeks, while past my face the riven air
> Tore by, and all the hillside flew to meet
> My flying figure with a low-hissed song—
> The song of rapid runners cleaving snow!

Another poet followed in a similar vein:

> Straight run our ski down the untracked steep,
> Seaming a virgin face.[28]

As the poets strained for metaphors to convey their enjoyment of speed and freedom, their verse became comic, but their sentiments remained true. Skiing was, for them, sacred play. Giddiness and ecstasy were the poles between which these skier-poets slalomed. Others were planting the gates of profit.

The opening of W. Averell Harriman's ski resort in Sun Valley, Idaho, in 1936 is frequently used to symbolize the beginning of the modern era of skiing in the United States, a period of Alpine-style skiing with Austrian-led ski schools, luxurious hotel accommodations, and mechanized lifts. The origins are earlier, however, and involve continuing debates over the use of public lands for skiing and the uses of snow. The creation of a ski area in the Mt. Hood National Forest serves as an illustration. The Forest Service, a bureau of the Department of Agriculture since 1905, has several areas to administer in the national forests: timber, watershed, wildlife, range management, fire control, and recreation. Beginning early in 1921, the owner of Cloud Cap Inn, a small hotel on Forest Service land on Mt. Hood, Oregon, began to correspond with the secretary of agriculture about expanding his facilities and encouraging tourism. Two years later the Forest Service decided to improve the inn and told the owner, who had the support of some local politicians, that he could bid on the development. After another two years of inaction, W. B. Greeley, chief of the Forest Service, rejected the hosteler's application as inadequate. At about the same time, L. L. Tyler, a Portland developer, applied for a permit to construct an aerial cableway to the summit of Mt. Hood. On March 5, 1927, Greeley denied the request, citing the need to preserve western wilderness.[29]

On November 7, 1927, Secretary of Agriculture W. M. Jardine appointed a ten-member committee of Oregon businessmen and academics to make recommendations on the development of Mt. Hood. Julius L. Meier, director of the Oregon State Chamber of Commerce and later governor, was chairman. C. M. Granger, the district forester, was secretary. On August 29, 1928, the committee unanimously reported in favor of the hotel and tramway project submitted by Tyler and recommended the creation of permanent recreation councils "in states having National Forest areas suited to the purposes of recreation." Less than two months later, Secretary Jardine rejected the committee's report in a long and detailed letter to Meier. Citing his recent visit to Mt. Hood, which, he said, "deepened my sense of the magnitude of the scenic and recreational resources which the region offers," Jardine argued that national forests had to be developed in the national interest and that advances in air transportation might obviate the construction of tramways by providing aerial views of the mountains. Nevertheless, the secretary accepted the committee's suggestion that the north slope of Mt. Hood and the Mt. Jefferson region be designated "Wilderness Areas."[30]

Jardine concluded his letter with a stinging rebuke to the committee for submitting its opinions without "supporting evidence and data on which rested your conclusions, or the reasoning by which you reached it." While urging the Oregon Chamber of Commerce to undertake further studies of the development of Mt. Hood, Jardine appointed his own committee, composed of Frederick Law Olmsted, Jr., landscape architect and member of the National Capital Park and Planning Commission; John C. Merriam, president of the Carnegie Institute; and Professor Frank A. Waugh of Massachusetts Agricultural College and author of a 1920 Forest Service report on Mt. Hood that recommended strict federal control of development.[31]

There are important implications for the future of Forest Service ski area development in Jardine's letter. First, it shows that the spirit of the Progressive conservation movement in which the Forest Service was begun continued into the 1920s. The appeals to expert opinion and to national rather than local planning are clear indications of the values of this federal bureaucracy. Second, it confirms that the concept of wilderness area, where no economic activity is allowed, was part of the Forest Service outlook long before it was made law. Third, in cases of conflict between a district forester and a secretary of agriculture, the preferred strategy was delay through further study. Despite considerable pressure from Oregon congressmen and Portland newspapers who were near hysteria over the prospect of Washington's Mt. Rainier getting a tramway before Mt. Hood, Jardine refused to be hurried. A clue to his thinking may be found in a letter he wrote to Merriam while the new committee was conducting its study:

> I am counting heavily on your personal collaboration in this matter. It presents in current and acute form a problem which in time will de-

> mand attention in many other National Forest areas. Our action in this case doubtless will constitute a precedent, will mark a new epoch in the popular conception of the relation of land areas to social advancement. For that reason I particularly need your assistance especially in considering the educational and scientific aspects which appear to be involved.[32]

While Merriam, Olmsted, and Waugh investigated and wrote their report, Herbert Hoover was elected president, and a new secretary of agriculture, Arthur M. Hyde, was appointed. Hyde received the hundred-page report, "Public Values of the Mount Hood Area," in May 1930. The most interesting part of the report is the attempt to assign economic values to the "recreational, inspirational and educational" uses of the forest. Recognizing that it is necessary to choose between two mutually exclusive uses of a given piece of land, the committee explained that one difficulty lies in assigning a precise value to long-term uses, such as inspiration, in contrast to the known value of timber. "From the point of view of social economics," they report, "the basis for such a relative evaluation of that alternative use of a Forest area which has the vaguer and more uncertain even though possibly greater value is essentially an estimate of whether a sufficient number of people would obtain such good from it that IF they could be confronted with the necessity of making economic sacrifices rather than forego that good the aggregate economic price they would willingly pay for it would exceed the economic values obtainable from the alternative use of the land." Awkwardly phrased, certainly, but clearly indicative of the problems of managing a commons. The report goes on to consider how a hierarchy of values can be determined and how the public can be induced to pay "at least the full economic cost of what they get."[33]

After a lengthy discussion of the regulation of concessionaires in the forest, the final twenty-five pages of the report considers the advantages and disadvantages of the cableway to the summit of Mt. Hood, a tramway below the timberline, a new hotel on the site of the Cloud Cap Inn, and a new road to the hotel. The chief advantage of these installations was, of course, access for a greater number of visitors, many of whom were not physically able to climb to the summit. The principal drawbacks were equally obvious; the construction would alter the character of the mountain, intrude on the view itself, and bring "the sophisticated and man-dominated region of every-day life into the borders of an ultimate and essentially untamed alpine wilderness of rock and snow and ice." The committee agreed that more people could derive pleasure through access to the timberline area than could be taken to the summit by cableway and that the value of their experience would be reduced by the visibility of the cableway. Moreover, those visiting the summit by cableway would stay a shorter time. Therefore, two members of the committee opposed its construction, despite local support, arguing that "if instead of aiming consistently at the very best results that can possibly be got out of the area as a whole, a beginning is now made by sacrificing some of those best values for the sake of an immediately popular detail a precedent will have been set up for the gradual frittering away of the extraordinary potentialities of the area." The third member believed that "mechanical transportation to one or more of the several high snow-clad peaks of the northwest" was inevitable and that putting it on Mt. Hood might spare other mountains.[34]

The secretary of agriculture ignored the majority report and authorized both the cableway and hotel construction proposed by Tyler and his Cascade Development Company, but he referred details to Regional Forester C. J. Buck, who, with the support of Assistant Forester L. F. Kneipp, blocked plans for a "conspicuous" hotel. Their rejection brought a letter of protest from Tyler, who included plans for the new Cloud Cap Inn from architect Carl L. Linde of Portland. His drawings show a six-story cruciform building with a tower, resembling the Emerald City of Oz. Kneipp wrote that he was aware that "rejection of architectural designs submitted by local agencies and believed by them to be adequate, may expose us to charges of bureaucratic stubbornness, nevertheless the importance of the subject is such that we must insist upon types of architectural development which will conform best to the spirit and purpose of the area."[35]

The election of Franklin Roosevelt and the growing popularity of skiing created the conditions for development of publicly funded winter recreation. A new generation seized the opportunity to lobby for federal projects. On February 6, 1935, for example, James A. Mount of the Portland Winter Sports Association wrote to Sen. Charles McNary urging the appointment of an administrator for winter recreation and outlining five duties:

1. Plan to make winter recreational areas available. Select best and most practical spots which may be reached most easily by the public.
2. Assist in the planning of proper types of winter chalets.
3. Develop areas for winter sports. Clear slopes of snags. Construct ski trails. Provide for public protection.
4. Encourage the development of winter sports areas. Supervise construction of jumps, slalom courses, toboggan slides, ice ponds, etc. Supply an accurate government bulletin of information about the winter sports areas and how to use them. Supply local foresters with authentic information to help in developing their areas. Make authentic government cinemas showing winter sports in the various forests and parks.
5. Gather data on the use of the forest areas and parks. Use in determining how much government effort should be spent for winter sports activities.

What stands out about this wish list is its comprehensiveness. If skiing was to expand beyond a handful of well-traveled sports enthusiasts and shake the image of Scandinavian folk festival competition, if it was to be truly democratized and Americanized, it needed access, facilities, accommodations, safety regulations, promotion, and, above all, "data." Whereas the earlier generation of Mt. Hood developers had chafed under the Forest Service's emerging planning process, winter sports advocates were, like everyone in competitive sports, avid record keepers and fact gatherers. Information, they knew, was the first step in expansion.[36]

Beginning with the use of Civilian Conservation Corps (CCC) workers to clear ski trails in New Hampshire and Vermont in the mid-1930s, the federal government subsidized American skiing in untold ways. Access roads, ski trails, warming huts were just the beginning; in the West the Forest Service followed the Park Service in leasing land for private development, then provided roads, maintenance, and other services at public expense. The great snow rush was on.

Timberline Lodge may be called the Sutter's Mill of the snow bonanza. With development of an inn on the north side of Mt. Hood effectively blocked, attention turned to the south face. A Mt. Hood Development Association was organized January 15, 1934, by Jack Meier and included among its members James Mount of the Portland Winter Sports Association and Floyd V. Horton, assistant regional forester. On December 17, 1935, the WPA approved financing and construction of Timberline Lodge under Forest Service supervision. The lodge was to be managed by the development association under a ten-year contract. The architect Gilbert Stanley Underwood designed what the historian Alan Gowans calls the "last great architectural monument to the Old Progressive impulse," which valued the American landscape, local building materials, and democratic populist attitudes (Figure 3.5). As Linn Forrest, a young Massachusetts Institute of Technology graduate employed by the CCC, put it: "The shape of the central lounge was inspired by the character and outline of the mountain peak. The steepness of the roof was determined by the heavier snow loads at that elevation. . . . It was our hope not to detract from the great natural beauty of the area. The entire exterior was made to blend as nearly as possible with the mountain side." Construction began in June 1936, and the lodge opened February 4, 1938, although it was dedicated by President Roosevelt in September 1937.[37]

Timberline Lodge, now on the National Register of Historic Places, is also a monument to interior design. Under the direction of Margery Hoffman Smith, assistant Federal Arts Project (FAP) director for the state of Oregon, who had been trained in art and design at Bryn Mawr College, Parsons School of Design, and the Art Students League in New York, the wrought-iron gates, hand-carved decorations, and handwoven rugs reflected

Figure 3.5. Timberline Lodge, Mt. Hood, Oregon, ca. 1940, photograph by Ray Atkeson. Copyright © Ray Atkeson, courtesy Rick Schafer Photo Library, Portland, Oreg. Snow blankets the lodge and trees, joining culture and nature. The first skiers break a trail between them.

the ideals of the Arts and Crafts movement and FAP Director Holger Cahill. The design motifs of Indians, wildlife, and pioneers reflect the ideology of the New Deal mural projects, the memorializing of American history as an epic struggle of people with the land resulting in a new national identity in which nature continues to play an equal role. Outside the lodge, nature cooperated by offering "not acres but square miles of skiing terrain. Should these areas become crusted or too fast, the open but smaller meadows below the Lodge, being protected from wind by belts of trees, offer good skiing."[38] In the light of subsequent Forest Service actions relating to resource management, the promotion of Timberline seems naive or, worse, self-serving, but a deeper reading of the record leads to a different conclusion.

Contrasting philosophies and cultural values were in conflict from the beginning of federal involvement in the development of skiing. The conflict can be illustrated in the correspondence of Robert Marshall, chief of the Forest Service Division of Recreation and Lands from 1937 until his untimely death in 1939. Marshall was a precocious writer and wilderness advocate, with a forestry degree from Harvard and a Ph.D. in plant physiology from Johns Hopkins University, who believed in strict federal control of public lands. Thus, when a regional forester advocated construction of a chairlift at Timberline, the response from Marshall's associate was strongly worded:

> The Division of Recreation and Lands feels that a chair lift is incompatible with the development of Mt. Hood. . . . Mr. Marshall's note attached thereto express[es] this view.
>
> One of the principles laid down in the recreation report and now being incorporated in the recreation section of the manual is: "The Forest Service will develop or permit the development of such facilities as will aid in the enjoyment of those types of recreation appropriate to the forest environment. It will especially discourage developments which tend to introduce urbanization into the forest." The Division of Recreation and Lands does not consider a chair lift as a necessary convenience at a resort maintained for skiing. In fact, it considers a chair lift as a decidedly sophisticated improvement which is inappropriate in the forest environment.

The construction of a rope tow was permitted, and a chairlift had already been constructed on nearby Multorpor Hill, but Marshall's office was engaged in the larger task of trying to define "forest environment" and to prevent the "dangerous precedent" of further development.[39]

Marshall was in conflict with traditions within the government as well. When the forest supervisor of the North Pacific Region announced that he would "accept applications by properly chaperoned, acceptable organizations for overnight use of the [Stevens Pass] ski hut dormitory, which will accommodate, roughly, up to 20 sleeping bags or cots," and that "in case of conflicting applications for certain week ends, the Supervisor will determine the user, and will give preference to winter sport groups, character building, educational, religious, and other organizations," Marshall was quick to respond. "I do not believe in mid-twentieth century America it is any longer necessary for most groups using our ski huts to be chaperoned," he wrote.

> Similarly, it seems to me that labor unions, junior unions or farmer groups should have as much preference as the groups you mentioned. Incidentally, in view of the fact that less than half of the American citizens belong to any religious organizations, it does not seem to me that we have any conceivable authority to give preference to this minority as

opposed to the majority who are associated with no organized religion. It would be doing violence to the spirit if not the letter of the First Amendment.

But the paragraph in the supervisor's press release that most offended Marshall was one indicating that the Forest Service was only maintaining ski huts until they would be taken over by private business. Marshall was deeply committed to public ownership and management of ski facilities and was outraged when Jack Meier, as president of Timberline Lodge, Inc., published pamphlets promoting the National Ski Championships at the facility without mentioning Forest Service ownership. Marshall may have realized in his final months that he was waging a losing battle against commercial development and the mechanization of skiing on public lands, but the contrast between Timberline and the other celebrated ski resort development of 1936 suggests that the attempt to democratize skiing by offering the experience without the costly trappings was worthwhile.[40]

Averell Harriman's ski resort near Ketchum, Idaho, opened December 21, 1936. It was called Sun Valley—a name that rejects Bob Marshall's reverence for wild places filled with snow—by the Florida real estate promoter Steve Hannagan, to offset what he thought was the negative image of winter in Idaho. Snow had its place, but, like the St. Paul organizers who decided to have King Boreas defeated by the Sun King during the winter carnival, the promoters of mechanized skiing came to praise snow, not to be buried in it. The official histories of Sun Valley emphasize its European elegance and Hollywood glamour, but the more interesting story concerns how W. Averell Harriman, chairman of the board of the Union Pacific Railroad, financier, and later politician and diplomat, turned his extraordinary energies to building a ski resort; how he attended to the smallest details of the resort's design, including snow management; and how the success of Sun Valley affected the development of skiing in America for half a century. The recent opening of the Harriman papers at the Library of Congress makes it possible to write this history.[41]

Harriman, who had skied in the Alps, was well aware of the growing interest in skiing in the United States. On October 2, 1935, he informed the president of the Union Pacific that he was dispatching Count Felix Schaffgotsch, an Austrian skier, "to make a trip over our territory and advise us whether any of the existing places are susceptible for development." Schaffgotsch toured ski developments near Mt. Rainier, Mt. Hood, Yosemite, Lake Tahoe, Salt Lake City, Yellowstone, and Denver. On January 18, 1936, he telegraphed Harriman from Ketchum: "Perfect place, any number of excursions, ideal snow and weather conditions." In mid-February he reported that Ketchum was the "most attractive spot I have seen in the U.S.A., which can hardly be beaten by any place I know in Switzerland or Austria for a winter sport resort." Among other assets, Schaffgotsch listed these: "far away from any big town, where the chances for the Sunday crowd are practically limited" [*sic*]; and "end of the railroad, and very few people can drive a car in, as the desert between Shoshone and Hailey is usually blocked with snow drifts."[42]

The requirement for isolation and control is paramount in the correspondence with Harriman during the planning of Sun Valley. Steve Hannagan's letter to Harriman, March 28, 1936, begins by noting that Ketchum is not easily accessible, but "it may be just as well, from the central classification of visitor you expect, to have the resort isolated. Certainly it makes it controllable." Hannagan then develops his vision: "This is one city in which roughing it must be a luxury. . . . It must have every modern convenience." A heated swimming pool, ice-skating rink, motion picture theater, "mechanical devices . . . to take people to the top of the slides," all these were imperative. Hannagan outlined his public relations campaign to make Sun Valley widely known, to attract "those who have an ardent, active, participating interest in outdoor sports," and "another group who like snow and the outdoors." Unlike that of Mt. Hood, the goal of Sun Valley was not the democratization of skiing. "Although the very nature of the community being established makes it selective," Hannagan continues, "it will be necessary to further restrict clientele through careful perusal of reservation requests. A good hotel manager will accomplish this. The price and distance will set a standard. But not all people with money are desirable. Therefore it is necessary to cull all reservations with judgement."[43]

In the following weeks, Harriman bought land, let construction contracts, and compiled statistics on ski areas served by rival railroads. His reasons for selecting Gilbert Stanley Underwood, architect of the Forest Service's Timberline Lodge, to design the Sun Valley Lodge are not revealed, but the contrast between the buildings heightens the irony. Rejecting the massive stone walls and high, steep roof of Timberline, Underwood created a three-story structure of concrete with a relatively flat roof and large picture windows, through which guests could view the skating rink as well as the mountains. The effect was streamlined, like one of the Union Pacific's new passenger cars. The lodge had 144 double rooms, some suites, a beauty parlor, a doctor's office, and a Saks Fifth Avenue boutique with ski shop. Marjorie Oelrichs Duchin, wife of the bandleader Eddy Duchin and friend of Harriman's second wife, Marie, was placed in charge of interior decoration. Using colors to contrast with the snow, she gave the lodge an Art Deco look that did not please the movie producer David O. Selznick, who also complained about the lighting and the exterior color. In the following two years, Sun Valley opened the less expensive Challenger Inn and the first of several private chalets, bringing maximum guest capacity to more than 600, twice that of Timberline. The design of the new buildings was Tyrolean, with small windows, high-pitched overhanging roofs, and wood and stone construction. When offered a long list of indigenous plants, animals, and Indian tribes for naming the chalets, Harriman selected "Pine," "Cedar," "Spruce," and "Balsam," with the admonition "that the signs be kept on the simple side and not too 'faked rustic' or 'arty' such as one might find in a dude ranch or auto camp."[44]

Harriman was equally attentive to the management of snow. After hiring W. P. "Pat" Rogers in 1938 to manage the resort, he wrote, "Should outside doors open in instead of out as planned? Wherever we have roof protection shown over entrance doors, should these not be gabled so as to avoid snowslides off the roof or formation of icicles over the entrance?" Rogers explained the necessity for outward-opening doors in case of fire and thought that they "would not have difficulty with snow slides due to the lower pitch of roofs and icicles will form only on the eaves . . . lengthwise of the building." In 1938 Harriman ignored an offer from E. Bucher of the Swiss Snow and Avalanche Institute in Davos to study snow problems at Sun Valley, but by January 1940 he had one of the ski school instructors sending him weekly snow reports. He instructed Rogers to issue snow reports that were "of an optimistic character and yet not inaccurate from the angle of the expert skier who will probably only want to come to Sun Valley when conditions are to his liking." Don Fraser used a reporting system with sixteen types of snow, but in May, Christopher LaFarge wrote Harriman urging "honest and disinterested information about snow conditions," and the resort's doctor attributed twelve fractures to "bad snow . . . a condition in which the snow was of such quality that when the ski toe or heel entered it, the ski became fixed." In November the staff was informed that "Mr. Harriman suggests that future early season reports should make no reference to Dollar, Ruud and Proctor [mountains] until there is skiing on them. He feels we have got to get away from the idea that still prevails in so many people's minds that there is no skiing at Sun Valley until Dollar Mountain is buried in snow . . . when Baldy has good skiing to offer there is no reason why we should apologize for Dollar not having skiable snow."[45]

Although he admitted to Alfred Biddle that "most subjects about skiing are Greek to me," Harriman made two crucial decisions: first to mechanize the slopes, and second to hire Austrian ski instructors. The Harriman Papers are disappointingly incomplete on the construction of the world's first chairlift on Dollar Mountain (Figure 3.6). All published accounts attribute the invention of the chairlift to Jim Curran, a Union Pacific engineer, who is said to have based his design on machinery used in loading banana boats. Photographs show pylons supporting a cable from which chairs are suspended by metal poles. The cable, powered by a gasoline engine, continuously pulls the chairs uphill. The chairlift was an improvement on the rope tow, introduced in North America in 1933, which required

Figure 3.6. The chairlift up Dollar Mountain, Sun Valley Resort, Idaho, 1936. Photo courtesy of the Sun Valley Resort. Averell Harriman wanted his ski resort to rival European facilities, and Jim Curran, a mechanical engineer for the Union Pacific shops in Omaha, came up with a practical design.

skiers to grip a rope as they were pulled uphill. Chairlifts allowed the skiers' feet to leave the ground and were thus faster and safer over steep, uneven terrain. They were smaller and less expensive than aerial tramways and funiculars, which had first been constructed in the Alps. Beating his competition in the mechanization of skiing was important to Harriman, who instructed the Union Pacific shops in Omaha to develop a lift that would move 100 skiers an hour. The response was swift. On May 15, 1936, William Jeffers, a UP vice president, received a coded telegram from one of the chief engineers announcing the testing of two lifts capable of carrying 240 persons per hour.[46]

The Dollar and Proctor lifts were bigger, longer, and faster than any others in the United States. The use of code names conforms with Harriman's desire to keep the Sun Valley development quiet until all the elements were in place. Once successfully installed at Sun Valley, lifts and other forms of mechanized transport for skiers proliferated. In September 1937, Harriman requested information on the "sno-motor," an oversnow machine developed by the Forest Service on Mt. Hood for pulling heavy loads. The

Iron Fireman, as the Forest Service called the machine, seemed a promising way to take as many as twenty-five skiers up mountains without lifts, but the lack of snow at lower elevations would preclude its use. Another problem was securing permission to use the sno-motor on Forest Service and private land not owned by Sun Valley. Thus, the Union Pacific's chief engineer recommended building a double reversible passenger tram to the top of Bald Mountain. The appeal of an aerial tram for UP executives was that it could be operated year-round, providing a sightseeing attraction in the summer. Towers for a tramway required less clearing of timber, thereby reducing the avalanche hazards according to the local forest ranger. Once Secretary of the Interior Harold Ickes was convinced that it was legal to allow ski lifts to be built on Bureau of Land Management property, further mechanization of the hills around Sun Valley was inevitable.[47]

Harriman's other decision, to employ primarily Austrian ski instructors, must be placed in the context of what John Allen calls "the Battle of the Ski Schools" in order to understand the conflicts and controversies that enveloped the Sun Valley ski staff during the Harriman years, 1936 through 1964. By choosing Schaffgotsch to select the location for his resort, Harriman placed himself in the vanguard of the change from Scandinavian- to Alpine-style skiing. As early as 1930, Peckett's Inn in New Hampshire advertised the Austrian or Arlberg technique, developed and promoted by Hannes Schneider, which dictated a crouching, knees-together stance and turns in which the skis are kept parallel, a form appropriate to plunging rapidly downhill.

As the desire for speed replaced the enjoyment of a day spent climbing as well as descending, the attitude toward snow changed as well. Skiers became accustomed to the rutted, hard-packed runs down from lifts. Schaffgotsch recruited nine instructors from Salzburg in the summer of 1937, and Hans Hauser was placed in charge of the Sun Valley School. The Austrians became known for their charming accents and success teaching wealthy clients in private lessons. Others complained that they were abusive and overbearing when teaching large groups. After two years' experience with the Schneider method, Harriman remained convinced that no American would be accepted as head of a ski school but that "American boys should be added to the staff in increasing numbers from time to time."[48] The popularity of American instructors at Aspen, Winter Park, and other Colorado ski resorts after World War II signaled an end to European dominance, even at Sun Valley, by 1950.

The commercial triumph of Sun Valley over Timberline, of mechanized skiing over mountaineering and cross-country skiing, left a legacy of environmental problems yet also a deeper understanding of snow and ways of managing it. *Skiing* magazine, in its 1989 annual guide to resorts, contrasted Sun Valley and Timberline in terms that each had emphasized fifty years before:

> Oh sure, the food and facilities are lavish and its history the stuff of Hollywood, but the real Sun Valley serenade is the sweet music of sustained pitch. From crowd-pleasing greens [easy trails] . . . to the black diamond bowls [most difficult runs] in between, runs tumble off Baldy like baseballs rolling off the world's biggest pitcher's mound. Nearby, Dollar/Elkhorn is a beginner's playing field *par excellence,* and Sun Valley's grounds crew keeps it all groomed as smooth as a little leaguer's chin. . . .
>
> On the field, the area [Timberline] is more playground than power hitter. Blue/green trails wind through the woods and lead to a few short drops, while above timber the Magic Mile is just made for ballroom skiing.[49]

The success of both lavish and modest ski resorts resulted in problems of overcrowding, overexpansion, and innovations such as helicopter transportation to remote snowfields (heli-skiing). The success of Timberline and the pressure to develop skiing in the West brought the Forest Service and other federal and state land-managing agencies into the business of ski area development, providing both hidden subsidies to the ski industry and national guidelines in areas such as landscaping and safety. As Sun Valley was joined by Alta, Aspen, Squaw Valley, and Taos, and as older eastern resorts like Stowe, Cranmore, and Jay's Peak cut new trails and built lifts, a pattern of skiing was established that inevitably restricted skiers to supervised and crowded slopes. Concomitantly, these heavily used areas had to be repaired, groomed, and, when natural snowfall was inadequate, covered with synthetic snow.

DESIGNER SNOW

Experiments in snowmaking at ski areas began in the winter of 1949–50, when engineers from Tey Manufacturing Company, Milford, Connecticut, and Larchmont Farms, an irrigation equipment supplier in Lexington, Massachusetts, combined their spray nozzles, pumps, air compressors, and pipes to make snow at Mohawk Mountain, Cornwall, Connecticut, and Split Rock Lodge, White Haven, Pennsylvania. Two years later a permanent snowmaking apparatus was installed at Grossinger's Hotel in the Catskills. A patent for making and distributing snow was issued to Wayne M. Pierce, Jr., in 1954 and assigned to Tey but later sold to Larchmont (Figure 3.7). In Pierce's method, the basis for most snowmaking at present, a mixture of air and water under pressure is sprayed into the ambient air when the temperature is a few degrees below freezing. The cooling effect of the airstream converts a portion of the water particles into ice crystals, which form a cloud in which some of the water spray sublimates onto the ice particles and

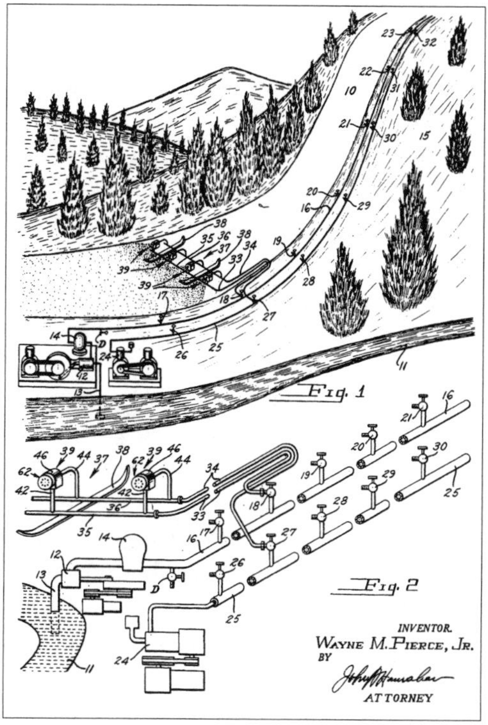

Figure 3.7. Patent drawing for the snowmaking apparatus invented by Wayne M. Pierce, Jr., 1950. U.S. Patent Office. Synthetic snow at ski resorts has lengthened the skiing season and extended skiing into southern states but raised problems where water is in short supply and snow grooming may damage soil and vegetation.

precipitates as snow. Pierce claimed that he could manufacture dry, powdery snow; wet, heavy snow; and an ice glaze by adjusting the ratio of water to air.[50]

Forty years later there were more than 1,200 snowmaking systems in operation in more than twenty countries. In the United States, virtually every ski resort makes some of its snow. In the East the median percentage of a ski area covered by synthetic snow in 1991 was 73 percent, in the West, 26 percent. Ski-trail location and design at many resorts are dictated by snowmaking requirements. In the 1980s, even Sun Valley installed snowmaking equipment on 60 percent of Baldy's groomable terrain. Snow can be made when the temperature is as high as 8°C (46°F) if the humidity is low. All it takes is lots of water—570,000 liters (150,000 gallons) for 0.4 hectare (1 acre) of snow 0.3 meter (1 foot) deep—and money—$2,700 for that amount of snow in 1992, excluding the $62,500 per hectare ($25,000 per acre) cost of installation.[51]

Snowmaking is eagerly pursued at eastern ski areas hoping to compete with those in the West by lengthening their seasons, but the discovery that water resources are limited has resulted in a new awareness of the complex ecology of winter recreation. As early as 1963, Joe Tropeano, one of the inventors of the Larchmont snowmaking equipment, warned readers of *Ski Area Management* that without careful planning water scarcity would limit future development. Synthetic snow is between twice and five times wetter and denser than natural snow, but, depending on weather conditions and grooming, the artificial surface can provide satisfactory skiing. In the 1990s another ingredient in the manufacture of faux snow was introduced. Snowmax, invented at the Eastman Kodak laboratories, uses a protein from a naturally occurring bacterium, *Pseudomonas syringae,* as a nucleus for ice crystals. By mixing the protein with the air and water in a snow gun, Snowmax claims to "produce more snowflakes, at higher temperatures and with less damage to the environment."[52]

Pseudosnow, mechanized snow grooming, computerized monitoring of air and water temperatures to automatically trigger snowmaking and build databases, each of these developments narrows the difference between a ski area and an amusement park, and helps to create the same pollution, traffic congestion, and snow management problems experienced by cities. As the director of Camelback Ski Corporation in the Poconos says, "A ski area is nothing but a great big machine." Even on public land, what began as ski areas in the 1950s had, a decade later, become winter resorts and, by the 1970s, year-round recreation complexes. The law professor Joseph L. Sax, writing about the successful effort to prevent Walt Disney Productions from building an Alpine ski village on National Forest land in Mineral King Valley, northeast of Bakersfield, California, concludes that "the Mineral King dispute nicely illuminates a difference between an entrepreneurial and a public policy perspective on dealing with questions of recreational demand." Policies on the use of public land are created by laws passed by elected representatives, presumably in response to popular demand, ascertained by petitions, polls, and observed behavior, and administered and enforced by public employees. All costs are paid by tax revenues. Profit is not a motive. The entrepreneur, by contrast, does not merely meet a demand but stimulates it to maximize profit. By the 1960s, the costs of providing the kind of experience that Sun Valley had made synonymous with skiing were so great that they could only be covered by real estate development, restaurants, and souvenir and other retail shops. The experience of skiing became just one of a number of demands met by private resort development. When these private resorts have permits to use public land (22,500 hectares or 55,500 acres of national forests at 167 sites in 1982), entrepreneurial and public values inevitably conflict.[53]

Until the passage of the National Environmental Policy Act of 1969 (NEPA), conflicts were usually resolved locally, as the histories of Timberline and Sun Valley show. Some regional foresters gave priority to "forest values," others to economic development. In 1960 the Multiple Use–Sustained Yield Act authorized the Forest Service to manage the forests "to produce sustained yield of products and services." Timber was harvested and tourists, cultivated. Yet as early as 1966, resort developers were feeling the pressure of environmental groups. When the Disney company proposed an "Alpine Village" in Mineral King, with twenty ski lifts capable of serving 20,000 skiers daily, the Sierra Club led the opposition that ultimately prevented the project.[54]

Another legacy of ski area development is water pollution. After an outbreak of diarrhea at Aspen was traced to a defective sewer leaking into the water supply, *Ski Area Management* informed its readers of various waste-treatment systems, including using treated wastewater for snowmaking. The pollution of streams and rivers by ski resorts was one of the strongest motivations for the passage of Vermont's 1970 environmental control bill. The law effectively stopped ski area expansion for a few years by prohibiting even treated sewage from entering a "pristine stream," defined by the Vermont Water Resources Board as any stream above 457 meters (1,500 feet) elevation with a flow less than 0.04 cubic meters (1.5 cubic feet) of water per second. Sewage is not the only pollutant, however. Since ski trails from lifts require cutting trees, bulldozing hillsides, and planting new shrubs to create "edges" to the run, ski areas require extensive "turf management" to prevent total erosion of the slopes. This management includes extensive use of pesticides and fertilizers. Moreover, many ski areas use chemical compounds to prevent softening and melting of snow. The cost of ski area development in terms of pollution, soil erosion, and the loss of wildlife habitat is just beginning to be widely discussed and understood.[55]

Perhaps the strangest ecological discovery of recent times is that of Ron Hoham and his associates in the Department of Biology at Colgate University. Studying algae in the snow at several sites in New England, they discovered two new species of yellow-green and green algae, one of which, designated *Chloromonas* sp.-B, was found only in the manufactured snow on ski slopes. Hoham's hypothesis is that the species may be distributed from ski area to ski area on the bottoms of skis, surviving at room temperature through the summer and autumn, when the skis are stored. He jokingly proposes to call the new species *Chloromonas skiensis.* Snow algae are near the bottom of the complex food chain that stretches upward through insects such as collembola, which are eaten by spiders, which are consumed by subnivean mammals such as shrews and voles. The evolution of a hardy microbe capable of living in dense synthetic snow may be a tribute to the resilience of nature, but its appearance also suggests the extent to which ski areas have altered mountain ecology.[56]

Because of western states' dependence on water from the snowpack, the most profound alterations have occurred there. The anthropologist Sylvia Rodriguez has been part of one of the few long-term studies of the social and environmental impacts of ski resorts on watersheds. Regarding the

Rio Hondo, which originates in the Taos, New Mexico, ski basin, she concludes that there has been a 50 percent reduction in the number of organisms in the river and that removal of water from the river has damaged downstream agriculture. This damage extends beyond the loss of crops, since the farmers are Hispanic villagers whose community structure is built upon irrigation and an ancient system of land and water rights. Federal and state environmental laws are more effective in protecting endangered species than human cultures, but all life is threatened. Twenty-four years after the National Environmental Policy Act and Vermont's environmental protection law, environmental groups could only delay, not prevent, Sugarbush ski area in Vermont from diverting water from Mad River to its snowmaking facilities. The trout may survive, but traditional life in the valleys of the Rio Hondo and the Mad River is disappearing because of economic pressures.[57] In some areas of the West, recreation is now a bigger business than agriculture, but both were built on the fickle foundation of snow.

The scarcity of snow in both eastern and western ski areas in dry winters, the possibility of global warming, and conflicts among agencies involved in snow management have resulted in numerous technological and administrative innovations beyond snowmaking. Portable snow barriers have been designed to control wind erosion of ski runs. Snow fence drift lines are used to catch snow on some small slopes and cross-country trails. Such practices have been labeled snow farming and reflect a conservation ethic toward snow rarely seen in the past. Snowmaking on the ground at ski areas and snowmaking as part of cloud-seeding weather modification are both demiurgic attempts to improve nature. When snow enhancement proved unreliable, however, weather modification shifted to storm dispersal. It is somewhat easier to break up a snowstorm than to create one. Snow dispersal, it was argued, could save money in street and highway snow cleanup and prevent airport closings. When federally funded experiments in snow dispersal began in western New York State in the late 1960s, ski area managers complained to Congress about the theft of their snow. A different and ultimately more costly kind of snow theft takes place on the ground. It has been estimated that an average skier displaces 907 kilograms (1 ton) of snow each day. As skier after skier crosses the surface of a ski run, the snow is packed harder and denser and pushed downhill. Density and texture are the fundamental parameters of skiing. A variety of techniques for altering them, for snow "grooming," have been introduced in the past forty years.[58]

Grooming is subdivided into three basic operations—compaction, smoothing, and chopping—all intended to increase snow durability and skier safety and enjoyment. Early studies indicated that snow "should be compacted to densities of about 500 kg/m^3 and to a hardness which resists penetration by a steel shovel," a condition better achieved by trampling with boots and by oversnow vehicles. *Ski Area Management* attributes the first snow-grooming machine to Steve Bradley, one of the developers of Winter Park, Colorado, who in 1952 built a roller of electrical tubing, screen-door springs, aircraft cables, and bicycle wheel rims that could be pulled by a skier or a small tractor to chop and smooth the moguls caused by heavy ski traffic. By the 1960s ski area managers had a choice of rolling and sliding packers designed by engineers at the University of Massachusetts and were using the rammsonde, or ram penetrometer, an instrument used in soil evaluation, to measure the hardness of snow on the ski slope. Catamount, Tucker, Polaris, and other manufacturers of oversnow vehicles offered a wide variety of snow-grooming apparatus. In 1993 Beartrac, the snow vehicle division of LMC, formerly a division of Thiokol and once owned by the automaker John DeLorean, was producing seven snow vehicle models and snow tillers, compactors, and harrows in several sizes. A ski industry analyst candidly compared ski trails and superhighways, with merge lanes, signage, median strips, landscaping, and unobstructed views. With trails 30 to 122 meters (100 to 400 feet) wide, islands of trees and glades of shrubs are removed, because "skiers aren't as interested in them any more, they are difficult to groom in winter, they can't handle the traffic and they are difficult to maintain in the summer."[59]

The analogy between ski trails and superhighways is also apparent in the emergence of new specialties in medicine and law dealing with skiing trauma, negligence suits, skier responsibility statutes, and safety legislation. The International Society for Skiing Safety has, since the 1970s, addressed the biomechanics of skiing; boot, binding, and ski design; and the morbidity and mortality of skiing. To the ordinary hazards of sliding downhill on thin strips of wood, metal, or fiberglass have been added injuries stemming from collisions, lifts, and skier-caused avalanches. *Ski Area Management* began a legal advice column in 1970. The law is still evolving. One of the earliest suits resulting from a skiing injury arose when a woman skiing on Mt. Mansfield in Stowe, Vermont, hit a snow-covered tree stump and broke her leg. The court rejected her claim that the lift company and the hotel were responsible, citing Justice Benjamin Cardozo that "one who takes part in such a sport accepts the dangers that inhere to it so far as they are obvious and necessary." Obstacles buried in the snow were, in 1947, an inherent part of the sport, but with the growth of skiing and the emphasis on grooming, the assumption-of-risk standard changed and the courts applied traditional negligence standards. Judges made a distinction between the dangers of skiing and those of riding lifts, and redefined *skier* as a person utilizing a ski area, not one participating in the sport of skiing. The business of mechanized skiing required new occupational health and safety standards.[60]

State safety codes regulating lifts and skier responsibility statutes were enacted in the 1970s and '80s, and ski areas used disclaimers of liability issued with lift tickets and rental equipment to protect themselves from suits.

Nevertheless, in *Sunday v. Stratton* (1974), a Vermont court awarded $1.5 million to a skier who suffered permanent quadriplegia after he struck a snow-covered bush 1 meter (3 feet) off the main trail; in 1987 a Colorado jury awarded $5.0 million to an expert skier paralyzed by injuries suffered while skiing at Aspen. Awards such as these, even when overturned on appeal, forced ski area owners to use safety patrols to police the slopes. Skiers also sue one another for injuries caused by recklessness. Juries find for the plaintiffs in only 23 percent of skiing accident suits, but when the accident is caused by a ski lift, plaintiffs win in 74 percent of the cases. A combination of improved skier education and enforcement of regulations reduced the number of skier injuries from the 1950s to the 1980s, but serious injuries and fatalities have risen to record numbers in the 1990s.[61]

In the long run, the impact of snowmaking and grooming may be greater on the cultural than on the natural environment. Snow is being redefined and reclassified, and skiers have altered their expectations and their techniques to adjust to pseudosnow. A 1972 industry-sponsored test to determine the "skiability" of the products of different snowmaking machines gave most of them good ratings, calling the snow either "sugar" or "mashed potato." *Sugar snow* usually means a kind of crumbly ice crystal, while *mashed potato* is, according to the magazine *Snow Country,* "wet, heavy snow so thick a shovel will stand up in it." Neither sounds very appealing, nor does *white asphalt,* a term the director of planning at Sugarbush in Vermont uses to describe the layer made by snowmaking machines on which to build natural snow or later coatings of artificial snow. Today newspapers use a glossary of fourteen terms to describe skiing conditions in the East. Current usage defines *powder* as "dry, cold, powdery snow; the perfect snow for . . . all winter sports" but also creates a category known as *packed powder,* a "powder snow, either natural or machine-made, that has been packed down by skier traffic or grooming machines." There are also conditions labeled "machine groomed," "death cookies" ("nasty ice chips frozen to the snow surface, often left behind by grooming machines"), and "kitty litter" ("rough debris caused by shoddy Sno-Cat work").[62]

AVALANCHE HUNTING

Injuries and deaths from another snow hazard—avalanches—have increased as skiers have attempted to avoid the crowds on groomed trails in favor of backcountry skiing. Avalanches are the most dramatic events involving snow. According to the Colorado Avalanche Information Center, 367 persons died in avalanches in the United States in the years 1950 through 1991; the majority were skiers and climbers. Miners and railroad builders were the first to encounter avalanches in the mountains of the west-

Figure 3.8. An avalanche hunter fires a World War I French 75-mm howitzer to knock down a potentially dangerous snow slide. Photography by Ray Atkeson (ca. 1950). Copyright © Ray Atkeson, courtesy of Rick Schafer Photo Library, Portland, Oreg. Monty Atwater, with the cooperation of the U.S. Forest Service, the Utah National Guard, and others, began shooting down avalanche hazards in the Wasatch Mountains in the 1940s. The use of explosives to control avalanches continues today.

ern United States and Canada. In 1891 the *Railroad Gazette* reported on the research of Vincenz Pollack of the Austrian State Railroads, who classified avalanches into "mass avalanches" and "dust avalanches," the former composed of old snow and ice that breaks loose when disturbed by melting underneath or the weight of new snow on top, the latter made up entirely of new snow, which dissipates as it descends. According to Pollack, the Swiss began constructing walls to protect villages from avalanches as early as the sixteenth century, and the alpine railroads were using rows of stakes driven into the mountainside in an attempt to hold the snow. North American railroads, as described in Chapter 2, built snowsheds, while highway maintenance crews have adopted the ski resort practice of controlling avalanches by using explosives to bring down potentially dangerous accumulations of snow (Figure 3.8). Control of avalanches in the United States and some of the research behind the control were, from 1937 to 1985, chiefly in the hands of the Forest Service, still another hidden subsidy to private ski businesses.[63]

The U.S. Forest Service and the Weather Bureau initiated avalanche studies at Alta, Utah, in the Wasatch National Forest, where a ski area also began development in 1937. Since there was little recreational skiing during the war, the Alta studies were limited to observing weather conditions and avalanches, but Swiss research had continued, and it was the Swiss who introduced the use of explosives to control avalanche hazards. In 1945 Montgomery "Monty" Atwater, a forty-one-year-old veteran of the U.S. Army's ski troops, Tenth Mountain Division, arrived in Alta to write a book and was hired by the Forest Service to renew avalanche studies. Energetic and ambitious, Atwater was more successful at mythologizing himself than at creating a permanent avalanche study center. Nevertheless, he supported the research of Wasatch Forest Supervisor Felix C. Koziol, Snow Ranger Edward LaChapelle, and Ron Perla before the Forest Service closed the Alta center and placed all avalanche research at the Rocky Mountain Forest and Range Experiment Station in Fort Collins, Colorado, in 1971.[64]

Atwater began his education in avalanche physics and mechanics by skiing the ridges around Little Cottonwood Canyon and the Alta area. A standard procedure for releasing avalanches before they threaten life and property was for a snow ranger to ski over the suspect slopes, beginning at the top. If an avalanche was not triggered by the ranger, who hoped to be above it and who had a partner to rescue him, the run was declared safe. Atwater survived a number of avalanches, vividly described in his memoirs, and eventually learned that they are the results of an extremely complex interrelationship among types of snow crystals, temperature, wind, topography, vegetation, and snowpack metamorphism. Each of the variables, except topography, may change from moment to moment. This is what makes avalanche prediction so difficult even with constant monitoring. In general, an avalanche needs large amounts of snow and a starting zone slope steeper than 30 degrees. Most avalanches begin above the tree line and follow a track down a gully or fan out over a slope until they culminate in a "runout" zone. Large tracks in North America are about 3000 meters (1.8 miles) long. Avalanches with relatively dry snow may travel as fast as 160 kilometers (100 miles) per hour and flow over gullies and around trees. Wet snow moves more slowly, tends to flow in channels, and pushes trees and buildings out of its path. Both are deadly, largely unpredictable, but exhilarating for survivors, as writings about avalanches attest.[65]

John Muir called his ride in a "snowslide," "the most spiritual of all my travels," and Atwater said his "principal sensation was one of wild excitement." As early as 1895, the novelist Mary Hallock Foote combined a traditional melodramatic end to her story of doomed lovers, "The Cup of Trembling," with environmental realities when she had a character comment on "snow-slide weather." "Hain't you been hearing how things is lettin' go? The snow slumpin' off the trees—you must have heard that. It's lettin' go up above us too. There's a million ton of snow up there a-settlin' and a-crawlin' in this chinook, just a-gettin' ready to start to slide. We fellers in the mountains know how 't is. This cabin has stood all right so far, but the woods above was cut last summer."[66] Deforestation is another reason avalanches are increasing.

In Tom Lea's 1960 novel *The Primal Yoke,* an avalanche claims the life of Hank Spurling, a marine veteran of World War II who returns to his home in the Tetons, where his father is a guide. His older brother was killed in combat. He tries to rebuild his life by helping his father but falls in love with the daughter of one of his father's clients, a woman with a past more troubled than his own. She leaves him to return to her husband, and Hank tries to forget her by becoming a ski instructor at Sun Valley. After her husband's suicide, she comes back and they resume their passionate affair. Things finally seem to be going right when her father's plane crashes in the mountains and Hank leads a team to recover the body. Here is the moment he reaches the plane:

> He stepped into the sun's clean and living dazzle, looked up, and saw the cornice fall. He had one snowshoe off, he had the binding loosed on the other, when the thunder shook.
>
> He was running out upon the tilt, thrashing knee deep, when the monster slipping caught him sideward. The white blind monster grabbed him, threw him, swallowed him.
>
> He had time flailing he had time ice needling terror sharp in his lungs O he had time when you don't care is when *they don't get you* but I care now I care *I care.*
>
> Then in tearing shatter whitelighted as the sun he did not care. . . .
>
> [The mountains] claimed him before he got away.

Lea's bleak depiction of the power of the snow avalanche, like Henry Robinson's depiction of a blizzard (see Chapter 2), symbolizes postwar America, powerful but unstable, metamorphosing under the pressures of the nuclear age, the loss of innocence, and the threats posed by liberated women. No wonder Atwater would exclaim: "The mountains had personalities too. . . . Rustler . . . superior, looming over the highway, was feminine, a blond bitch, beautiful and vicious."[67]

Skiers began to fall victim to avalanches in the late 1930s. Sun Valley's first fatalities occurred in 1952, when a ski school instructor and one of his students were lost along with two unattached skiers. Pat Rogers assured Harriman that it "could not be analyzed other than 'an Act of God.'" The act of God defense was made by lawyers for Alpine Meadows ski area in California in a suit brought after seven persons were killed by an avalanche in a parking lot in 1982. A jury decided that the resort had exercised ordinary care in avalanche forecasting and control and awarded no damages. Nevertheless, the definition of reasonable care is changing, and ski areas are expected to do more to control avalanches than they have in the past. Avalanches are controlled by forecasting their occurrence and warning those in danger, by modifying the terrain with structures or vegetation above and in the track, by compacting the snow with grooming machines, and by deliberate release of snow by explosives delivered by hand, cable systems, or artillery such as the avalauncher, a compressed-air mortar that can lob a 1-kilogram (2-pound) charge almost 2 kilometers. The dangers of avalanches can also be reduced by mapping avalanche hazards for land-use decision making and by educating skiers about snow conditions.[68]

As knowledge of avalanche conditions improved, the old classification of "loose snow" and "slab" avalanches—"dust" and "mass"—was expanded by the Swiss scientist Marcel de Quervain in 1955 to ten types based on five criteria: type of rupture—slab or loose snow; position of sliding surface—in snowcover or on the ground (full-depth avalanche); state of humidity—dry or wet snow; form of track—unconfined or channeled; and form of movement—airborne powder or flowing. In 1965 an additional criterion, the triggering factor, distinguished between spontaneous internal release and external trigger, either natural, such as a rainstorm, or artificial, such as the weight of a skier or an explosive charge. By 1973 de Quervain, Edward LaChapelle, and other members of the working group on avalanche classification of the International Commission on Snow and Ice proposed a morphological classification that recognized eight criteria—(1) manner of starting; (2) position of sliding surface; (3) liquid water in snow; (4) form of path; (5) form of movement; (6) surface roughness of deposit; (7) liquid water in snow debris at time of deposition; (8) contamination of deposit (soil, rock, branches, trees)—with twenty-four subcharacteristics. The triggering factor was dropped, and several characteristics of the deposit zone were added, partly because wet snow and debris increase the difficulty of rescue work and avalanche clearing. Avalanche snow also has hydrological importance in places where it may retard snowmelt and delay meltwater flow into a watershed.[69]

Avalanche forecasting based on a knowledge of snow metamorphism, climatology, and terrain is not an exact science, even with monitoring systems that provide continuous data on changes in temperature, snow density, wind, and other factors. Understanding the factors contributing to an avalanche and deciding what to do about the hazard involve two very different epistemologies. Physical and mechanical factors can be quantified and analyzed, whereas warnings based on probable risks are subject to cultural and individual interpretation.

The wide differences among the snow-stability rating systems in the United States, Canada, Switzerland, and France are revealing. A four-class system for warning the public is used in the United States: low hazard, moderate hazard, high hazard, and extreme hazard. Canadians describe the same snow conditions using good stability, fair stability, poor stability, and very poor stability. The differences in psychology are striking. The use of *hazard* rather than *instability* might discourage all but the most daring skier, whereas the Canadian oxymoron *very poor stability* might seem more stable than unstable to the inexperienced skier. The Swiss distinguish between local and widespread hazards, while the French have an eight-part classification that distinguishes between natural and accidental avalanche release: minimum hazard, low hazard, local accidental hazard, general accidental hazard, moderate natural hazard, high natural hazard, avalanche situation, and extreme avalanche situation. David McClung and Peter Schaerer, who helped create the Canadian scale, object to the word *hazard* when rating snow stability, preferring to limit its use to the potential effects on people and property once avalanches start. But the risks to skiers, resorts, and transportation systems from unstable snow conditions are both obvious and difficult to quantify.[70]

It may be impossible to reduce avalanche hazards because of attitudes of winter recreationists. Without some risk and challenge, all play becomes boring. Avalanche hazards are to the backcountry skier what speed is to the downhill skier. Charles Bradley, a veteran ski mountaineer, summed up the attitude of many ski tourers when he wrote: "The main risks, of course, come from the unknowns. They are large, numerous, and ubiquitous. But who would want to remove them all? They are part of the salt of life." A 1988 survey of callers to the Utah Avalanche Forecast Center found that while most experienced backcountry skiers carried beacons, shovels, and other rescue equipment, few knew how to test snow stability accurately and many misinterpreted the warning system. Paradoxically, more accidents involving avalanches occurred when the rating was moderate than when it was high or extreme, because skiers were less vigilant. Another study estimated that over 70 percent of avalanche victims would have described themselves as "experienced" backcountry skiers. Offering skier education, closing slopes when hazards are observed, and zoning to prohibit building in avalanche paths will prevent some deaths and injuries, but the growing popularity of ski touring and the continued use and misuse of snowmobiles will cause others.[71]

Snow, natural and synthetic, is beautiful, dangerous, surprising, mocking—challenging humans and eluding their mastery. This is why there will always be tensions between those who seek in snow play fragmentary experience and paradox and those who demand the total control of a snowtopia. Research shows that both wilderness users and downhill skiers value the feelings of creative accomplishment, exhilaration, freedom, harmony of mind and body, and enjoyment of nature. Make-believe is also part of the experience, "seaming" what seems a virgin face. Whimsical ski trail names help create a fantasyland geography with runs such as Betty's Buzz, Utter Abandon, Pipe Dream, Cloud 9, Peek-a-Boo, and Ecstasy.[72]

Snow arouses playfulness. Who can resist catching falling snowflakes on the tongue or scooping up snow to mold into balls? Snow, like play, is extraordinary, a break from everyday routines; it stimulates imagination and fantasy. Those who play in snow—who sled, sleigh, snowshoe, and ski—know instinctively that snow is more than either resource or refuse, metaphor or scientific fact. It is as impossible to define *snow* as it is to define *play.* The word *snow* is a riddle for phenomena with unnamed and unnumbered dimensions. There are as many ways to know snow as there are snows, from the powder that blows away as a skier descends to the avalanche that buries everything in its path. No two experiences with snow are exactly alike.

Opening the Snow Frontier

4

"Water, water," it will utter—
the famous last words of snow.
Into the earth the water will go.
TOM HANSEN, "POEM OF FALLING SNOW"

SNOW INTO WATER

"Whether you boil snow or pound it, you can have but water of it," observed the seventeenth-century British poet George Herbert, and even those who spend their lives praising the beauty of snow crystals or explaining their metamorphosis have to agree that at some split second snow disappears into a vapor or a liquid that lacks the quintessential mystery of snow.[1] The snow sciences that developed in the twentieth century—specialized branches of hydrology, glaciology, meteorology, chemistry, physics, and biology—employ instruments and theories that allow them to see the invisible qualities of snow and to predict the future of a snow crystal and a snowpack. They vastly expanded our knowledge of snow while reducing its importance to that of a frozen reservoir. Ironically, the application of science and technology to measure snow and to make it, to boil and pound it, helps us appreciate snow in new ways. In a sense, Herbert was wrong; boiling and pounding can be utilitarian or artful, snow can be cooking or cuisine.

In their professional societies, snow scientists preserve the original sense of wonder experienced by the pioneer surveyors of the snow frontier, who combined their science with aesthetic appreciation. This aesthetic dimension of their work is rarely expressed in poetry or art, although they enjoy a well-crafted snowscape as well as anyone. Rather, they express it in the playfulness of their meetings and in the comic awards they bestow on one another for misforecasts, misinterpretations, and general foolishness. This chapter continues the theme that it is impossible to work with snow without playing with snow.

Contrary to the belief expressed by Charles Dana Wilbur in 1881, rain did not follow the plow onto the Great Plains. Water in arid regions comes either from underground aquifers, where it is trapped in layers of porous rock, or from streams and rivers. On the plains, in the Great Basin, and in California, the ultimate source of most freshwater is snow. It has been estimated, for example, that 51 percent of California's water comes from the snowpack covering only 12 percent of the state's area.[2]

The importance of water to all life is understood but often taken for granted. Water is still a bargain in most cities. In Washington, D.C., for example, I spend an average of $2.00 a day for electricity, $1.50 for natural gas, but only $0.25 for water. My water-sewage bill tells me that I consume about 227 liters (60 gallons) of water daily, about two beer barrels full, a little less than the national average. Estimates of the total amount of water available in the United States and the amounts consumed by individuals, industry, and agriculture vary considerably; part of the problem stems from the translation of inches of precipitation into gallons of water and gallons into other units of measurement, such as acre-feet. All estimates agree that per capita consumption has increased dramatically over the past 200 years, that a huge amount of water is wasted annually, and that water shortages will occur more frequently in many parts of the country in the future. Snow management for water conservation is one of the necessary strategies for survival.[3]

The hydrologist Luna Leopold estimated in 1970 that 46 percent of all water consumption was agricultural, 46 percent industrial, and 8 percent public. Municipal use in street cleaning, firefighting, public buildings, and parks amounted to 38 liters (10 gallons) per capita daily, while the average person used 11 liters (3 gallons) to flush a toilet, 38 liters (10 gallons) to wash dishes, 75 to 115 liters (20 to 30 gallons) to wash clothes, the same amount to shower, 115 to 150 liters (30 to 40 gallons) to take a bath, and 500 liters (130 gallons) to water a lawn in the summer. A dripping faucet wastes 15 liters (4 gallons) a day, a leaking toilet can waste more than 115 liters (30 gallons), and municipal systems waste at least 20 percent of their water supply in leaks and faulty measurement. Far more water is wasted in irrigation, where use is measured in acre-feet (1.2 million liters, 325,850 gallons), the amount needed to cover an acre of land with water to the depth of 1 foot. An acre-foot would supply an average family of four with enough water for more than three years, and in Washington, D.C., they would pay about $1,400 for it. A farmer in the Central Valley of California pays less than $15, however. If farmers paid what water is worth, they would waste less and grow fewer water-consuming crops, but the price of groceries would rise. If I had to carry 227 kilograms (500 pounds) of water (the weight of my daily consumption of 227 liters (60 gallons), I too would use less.[4]

The apparent perfection of the hydrologic cycle—the return of precipitation to the atmosphere through transpiration and evaporation—and the relative abundance of water in the United States continue to limit our appreciation of the interrelation of precipitation, land use, and population. One difficulty, as the geographer Yi-Fu Tuan points out, is the time scale involved. Some features of the cycle take a few seconds, as when rain or snow falls into the ocean, whereas others take millennia, as when aquifers are created by percolation of rain and melting snow. What has taken aeons to create may be destroyed in a few years.[5]

In 1851 the great Swiss scientist Arnold Guyot attributed the hydrologic cycle to God's beneficent wisdom. At the same time, he was drawing up instructions for the Smithsonian's network of voluntary weather observers. These include the first description of a snow gauge and methods for accurately measuring snowfall. Ironically, measurement, with its initial evidence of abundance, encouraged waste. Guyot's essay on snow gauges is important not only because it is the first scientific recognition in the United States of the contribution of snow to the water supply but also because his selection of instruments and his instructions for using them set the pattern for the subsequent Weather Bureau snow measurement system. Although measurements of the depth of snow had been made throughout North America by various observers since the seventeenth century, there was little attention to such matters as drifting, total annual accumulation rather than single storms, and the water content of snow. Although Guyot recognized differences between wet and dry snow, he was confident that "as a general average, it will be found that about ten inches of snow will make one of water."[6]

This 10:1 ratio became fixed in the writings of meteorologists for the next fifty years and was modified only when evidence from California and Canada demonstrated that the ratio could vary from more than 50:1 to less than 2:1. In the Canadian Rockies, for example, a water equivalent of 12:1 is preferred for predicting spring runoff, while in the Northern Sierra Nevada a ratio of 6.5:1 was proposed as early as 1901.[7]

The transfer of the U.S. Weather Service from the army to the Department of Agriculture in 1891 facilitated the study of snow in several ways. First, the number of weather stations expanded from about 450 to almost 3,000. Second, placement of meteorology in Agriculture underscored the interrelations of weather, water resources, soil, forests, and land use. For fifty years the Department of Agriculture functioned as a bureau of environmental science, integrating the work of botanists, zoologists, entomologists, meteorologists, and others and producing, in cooperation with land-grant universities, some of the remarkable advances in farming that made the United States the breadbasket of the world. Third, it opened a career for Charles Frederick Marvin, who joined the Weather Service in 1884, shortly after graduating from Ohio State University as a mechanical engineer. From 1888 to 1913, he served as head of the Instrument Division and from 1913 to 1934 as chief of the Weather Bureau.[8]

A year after the transfer of the bureau to the Department of Agriculture, Marvin published the first of many instruction booklets for using rain and snow gauges. The rain gauge described in 1892 had a funnel 20.0 centimeters (8.0 inches) in diameter and a receiving tube with a diameter of 6.4 centimeters (2.5 inches) and a length of 51.0 centimeters (20.0 inches). It was not self-registering. Marvin recommended two methods of measuring snowfall. The first was simply to remove the funnel and tube and allow the snow to fall into the overflow cylinder that surrounded the measuring tube. The contents could be melted and either poured into the tube directly or mixed with a tube full of warm water and measured by subtracting the amount in the full tube. The latter method was preferred because less was lost by evaporation. When it was impracticable to melt the snow, Marvin followed Guyot's 10:1 snow to water ratio.[9]

Instructions for Using Marvin's Weighing Rain and Snow Gauge, issued in 1893, shows several new developments. This pamphlet describes a self-registering instrument with a container 21.6 centimeters (8.5 inches) in diameter and 28.0 centimeters (11.0 inches) deep—the large size, Marvin writes, "being necessary for the collection of snow." The gauge also had a number of new elements to offset the effects of wind. The recording cylinder was activated by a battery-powered electric current and was capable of measuring changes by the minute. As innovative as Marvin's instrument was, it still required daily attention and was limited to relatively small amounts of precipitation.[10]

Observations made by the operators of two platform scales in 1894 anticipated later developments in the measurement of snow, but they went virtually unnoticed at the time. In early April of that year, a 2.4-by-3.7 meter (8-by-12 foot) scale in Nittany, Pennsylvania, was covered with an 84-centimeter (33-inch) snowfall weighing 744 kilograms (1,640 pounds), or 83 kilograms per square meter (17 pounds per square foot). At the same time, Edward Perry of Bel Air, Maryland, reported that his 2.4 meter (7 foot, 10 inch) by 4.6 meter (15 foot) scale had 690 kilograms (1,520 pounds) of snow, or 12.93 pounds per square foot. "Although the results may be rather crude," observed Cleveland Abbe in the *Monthly Weather Review,* "it suggests an excellent method of getting at the average quantity of precipitation in case the snow falls without melting and without much drifting. It also gives some idea of the weight that must be supported by roofs in cases of similar heavy snowfalls."[11]

Too much water rather than too little prompted the first discussion of the use of a snow survey to predict spring runoff. In May 1897, L. G. Carpenter of the State Agricultural College at Fort Collins, Colorado, and F. H. Brandenberg of the Weather Bureau reported on the contribution of heavy snows in the Rockies to floods along the Missouri and Mississippi Rivers, and Brandenberg later issued a special snowfall report on the Arkansas River basin. By 1900 snowfall reports were being issued by the Weather Bureau section directors in Idaho, Montana, New Mexico, Utah, and Wyoming, as well as Colorado. Brandenberg's Colorado report mentions increased irrigation in the state and calls attention to the diminished water supply from snowpacks that were melting earlier because of deforestation and overgrazing. Moreover, he links the availability of water from melting snow to both soil and general weather conditions. He expresses concern about accurate reporting of snow conditions and the wise use of this information. Brandenberg's observations, like the 1897 report on reservoir sites in Wyoming and Colorado by Capt. Hiram Martin Chittenden, contributed to the growing demand for federal reclamation projects to make the deserts bloom.[12]

The snow surveys, which began in the first decade of the twentieth century, are both central and peripheral to the larger conservation movement of the Progressive period. Central, because water is an essential natural resource, and more than half of all the water in the western states originates as snow. Peripheral, because until recently snow seemed the least threatened, most abundant, and an easily managed resource. The standard histories of the early conservation movement, *Conservation and the Gospel of Efficiency: The Progressive Conservation Movement, 1890–1920,* by Samuel Hays and *Wilderness and the American Mind* by Roderick Nash, quite properly focus on individual and institutional concerns about the decline of natural resources such as forests. The distinctions these and other historians have drawn between preservationists, who seek to protect parts of nature from any human development, and conservationists, who believe in managing the environment in order to maximize its benefits, hardly apply to water and never to apply snow. Water was so scarce in the West that efficient management was the only acceptable strategy. As Marc Reisner has pointed out, to not use water in the West is to waste it. The passage of the Newlands Reclamation Act in 1902, which began an era of dam building and irrigation projects, was, as several scholars have noted, virtually inevitable. The act promised and delivered water for settlement of the western states. The demand for water for irrigation and electric power led directly to the snow surveys.[13]

Donald Worster's *Rivers of Empire: Water, Aridity, and the Growth of the American West* directs our attention to the crucial role of water in sustaining all the resources of the West and tells the history of the manipulation and mismanagement of laws governing the use of water. In an earlier work Worster distinguishes between what he calls arcadian and imperialist tendencies in scientific studies of nature. Arcadians depict nature as a symbiotic community in which humanity is merely one element; imperialists praise human dominance of nature and emphasize stewardship. Both positions are attempts to describe the natural world and to manage it in rational ways. It would be easy to see the development of snow surveys and snow management as part of the imperialist approach. But such a view would be misleading because it is too simple. In the ninety years between the first national reclamation act and the Water Law of 1992, Americans have changed their priorities concerning the use of water and their strategies for managing it. The remainder of this chapter looks at those changes through the prism of snow.[14]

THE MOUNT ROSE SNOW SAMPLER AND THE ORIGINS OF SNOW SCIENCE

The basic idea of a snow survey to determine the volume of snow and its water equivalent over a drainage area occurred independently to four men in different parts of the country at roughly the same time. Each put forth his claim to being first, but only one persisted in perfecting his system and in promoting himself as the principal inventor of the snow sampler. It is impossible to separate the history of snow surveying from the career and personality of James Edward Church. Church's importance to the history of snow goes beyond the invention of a snow sampler and the creation of surveys. It is his total involvement—the word *devotion* is not too strong—with all aspects of snow that makes him a pivotal figure. What began as a recreational and aesthetic interest for Church the winter hiker, camper, and photographer developed into a scientific career and climaxed in a geopolitical philosophy in which he saw the study of snow playing a part in bringing about world peace. Church had an advantage over his rivals in that he was a skillful writer who used both popular magazines and scientific journals to express his ideas. He is interesting not just because he is accessible but because he synthesizes Worster's arcadian and imperialist perspectives on the environment. Church's success in combining various attitudes is all the more interesting because it was so self-conscious. "I had gone to the hills for pictures and pleasure," he recalled in 1937, "but to the public I was merely a great fool. So the humanist decided to become a scientist and a 'hero,' yet still take his pictures on the side."[15]

Church was born in Holly, Michigan, about 50 kilometers (30 miles) northwest of Detroit, on February 15, 1869, and graduated from the University of Michigan in 1892. In that year he accepted an appointment as instructor in classics at the University of Nevada in Reno, where he taught for more than forty years. Mildly obsessed by the dream of becoming a polar explorer, he turned to winter hiking and camping along the eastern slope of the Sierra from Mt. Whitney to Mt. Lassen. He was an early member of the Sierra Club; one of his first publications was a description of a 110-kilometer (70-mile) hike from Susanville, California, to Fall River Mills in the winter of 1897.[16]

He spent the years 1899 through 1901 studying for a Ph.D. in classics at Ludwig-Maxmillians University in Munich. In addition to the opportunity to do some climbing in the Alps, the experience provided two valuable rewards that served Church well in his later career as a hydrologist—a knowledge of statistics, which he had used in his dissertation on Latin epitaphs, and the title Doctor, which distinguished him from all but a few of the scientists of his time. Dr. Church returned to Nevada with new ideas about winter mountaineering (Figure 4.1).

In 1904 Church's interest in weather forecasting was stimulated by an article by Alexander McAdie, of the U.S. Weather Bureau in San Francisco, that called for the establishment of meteorological observatories on the highest peaks in the Sierras. The following March, Church and a companion attempted to climb Mt. Whitney in the snow to test the possibility of checking instruments weekly during the winter. Although they failed to reach the summit of the 4,556-meter (14,948-foot) peak, Church was con-

Figure 4.1. James E. Church with snowshoes and camera, ca. 1908. Church Collections, Special Collections, University of Nevada–Reno Library. Church's passion for winter hiking led him to develop the snow sampler and to calculate spring runoff on the basis of the water content of the snow.

Figure 4.2. Dr. Church (looking up at instruments) and friends at the Mt. Rose weather station in the 1920s. Nevada Historical Society. Church inspired great loyalty and enthusiasm in his students and friends, many of whom helped him establish the science of snow surveying.

vinced that a weather station could be maintained through the winter at slightly lower elevations, including the 3,285-meter (10,778-foot) summit of Mt. Rose, about 32 kilometers (20 miles) from his home in Reno. In June 1905, Church and five volunteers began construction of the first weather observatory on Mt. Rose (Figure 4.2). The Weather Bureau furnished maximum and minimum thermometers and a rain gauge, and Church and his crew built a wooden shelter. "On October 14th," Church wrote, "a thermograph and a barograph were installed, capable of recording every fluctuation of temperature and air pressure during a period of eight days, and of indicating by means of a perpendicular stroke the highest and lowest subsequent temperatures and pressure." Enlisting the aid of a botanist from the Nevada Agricultural Experiment Station to make a survey of plants near the station, Church the amateur meteorologist was on his way to becoming Dr. Church the snow scientist. When some colleagues in the university's School of Agriculture asked him to investigate the relationship between forests and depth of snow, Church the snow surveyor was born.[17]

The immediate benefit of this observatory was a frost forecast that could be made twenty-four to forty-eight hours in advance; with it orchards could be developed on the eastern slope. Caught up in the excitement of providing useful information from his avocational interests, Church began putting more of his time and energy into meteorology. Continually improving his equipment, he soon had a barograph capable of recording for five weeks and a wind gauge with a similar self-recording apparatus. The passage of the Adams Act, which provided annual appropriations for original research by agricultural experiment stations, resulted in a $500 grant, and on June 30, 1906, the observatory officially became the Department of (Mountain) Meteorology and Climatology of the Nevada Agricultural Experiment Station. Church wrote with justifiable pride that the Mt. Rose observatory was the highest meteorological station in the United States and unique in that its instruments could function for long periods without attention. Summing up his feelings in the *Sierra Club Bulletin* in 1907, he was still a wilderness romantic. After describing the hard work of building the station, he concluded: "But that is physical. To the spirit, as it revealed itself at midnight and at noon, at twilight and at dawn, in storm and in calm, in frost-plume and in verdure, the mountain became a wonderland so remote from the ordinary experience of life that the traveler unconsciously deemed that he was entering another world."[18]

The world he was about to enter was that of science and politics. As Church monitored the weather, he became aware of the economic importance of snow. A very wet winter in 1906–7 had caused the power companies to lower Lake Tahoe prematurely, thus reducing the amount of water available for the area around Reno in the summer. The only exit of water from the lake is the Truckee River, which flows north and east 113 kilometers (70 miles) to Pyramid Lake. A dam had been built at the headwaters of the Truckee as early as 1865, and by 1906 there were several small hydroelectric power plants along its banks. The Newlands Reclamation Act funded construction of more dams for an irrigation project east of Reno. A U.S. government treaty promised the Pyramid Lake Paiutes enough water to maintain their fishing at the mouth of the river. Conflict developed among the many users of Lake Tahoe's water, and all of them—property owners at the lake, the power company and its customers, farmers, and Indians—had crucial interests in predicting the rise or fall of the lake. The level of the lake, in turn, depended on the amount and water content of the winter snows.[19]

In the winter of 1908–9, Church invented a sampler that both measured the depth of snow and determined its water content by weight. The basic element of this snow sampler was a stainless steel tube, composed of shorter tubes joined by screw couplings, which could sample a snowpack almost 3 meters (10 feet) deep. The tubes were 4.4 centimeters (1.75 inches) in outside diameter. Soldered to the end of one tube was a cutter 3.8 centimeters (1.50 inches) in diameter that increased to 4.0 centimeters (1.60 inches) so that the lip inside the head could help hold the snow in the tube after it was withdrawn from the snowpack. The tubes were slotted and engraved with a scale in inches for easy measurement. When filled with snow, they were weighed on a spring scale whose dial was designed to indicate the depth of water rather than the weight of the snow. Church's prototype was built of galvanized sheet iron in the machine shop of the Engineering Department of the University of Nevada, where he received invaluable help from H. P. Boardman and S. P. Fergusson, inventor of the weighing rain-and-snow gauge used by the Weather Bureau since the 1880s. When sheet iron proved too flimsy, Church ordered stainless steel tubing from Edgar T. Ward and Sons Seamless Tube Company of America in Boston. He announced his invention in February 1909. The response was mixed, suggesting that rivalries among the inventors of similar instruments were strong.[20]

Church's first rival had already announced his invention in the April 1903 *Monthly Weather Review.* Charles A. Mixer, an engineer at the Rumford Falls Power Company in Maine, described his method of collecting samples of snow by "forcing a cylinder down to the ground, then shoveling down around it, inserting a sheet metal bottom and lifting it out"; he claimed that he had taken his first sample in March 1900. Although he failed to describe how he determined the water content of his sample, it does not appear that he weighed it. He also neglected to give the dimensions of his cylinder. It seems likely that he melted his samples, following the method recommended by Guyot fifty years earlier. Mixer did, however, take several samples to determine average water content over several months and covering a wide area of the Androscoggin River basin.[21] A more serious claim for the invention of the snow sampler and the snow survey system was made by Robert E. Horton in 1905.

Like Church, Horton had been born in Michigan, in Parma, 120 kilometers (75 miles) southwest of Holly. Six years younger than Church, he received his B.S. degree from Albion College in 1897, worked on the U.S. Deep Water Ways Survey, then became an engineer and hydrographer for the U.S. Geological Survey in Utica, New York. It was in this position that he became interested in controlling streamflow and thus in the effects of cold and snow on waterways. He soon discovered that there was a considerable difference in water content between loose, freshly fallen snow and compact, accumulated snow. In 1903 he began experimenting with a tin tube about 7.6 centimeters (3 inches) in diameter that he thrust into the snow to obtain a sample, then weighed to determine the water content. Horton was a well-trained hydrologist, and, after working out comparisons of winter precipitation and runoff for several drainage areas, he drew five conclusions, the most important being that the Marvin rain-and-snow gauge, a commonly used device, seriously underestimated the total amount of precipitation and water supply. Horton also emphasized the importance of temperature changes during the winter and the role played by topography and vegetation in determining the density of a snowpack.[22]

On April 21, 1909, Charles Marvin, who would soon become chief of the U.S. Weather Bureau, wrote Church that as far as he knew, the first measurement of snow by sampling tubes was by Mixer and Horton, and their samples were not weighed. "The use of the weighing sampler for measuring snow," Marvin continued,

> was first mentioned to me at an informal committee meeting held at the office of the Weather Bureau in Washington early in 1908. The device was subsequently quite fully described in a paper giving the general principles of snowfall measurements, submitted by me to the Chief of Bureau in August of 1908. Owing to statements made to me by those familiar with snow conditions in the west that it was necessary to be prepared to measure beds of snow ranging from 20 to 30 feet in thickness, I considered it impracticable to attempt to take out a whole section of snow in one sample. The plan which seems best to me under this view was to construct an apparatus so that small samples of snow could be taken and weighed, some at the top, others part way down, and still others at the bottom. We have actually made up one or two pieces of apparatus of this character. The spring balance for weighing is provided with a special dial, graduated in either 50 or 100 parts which show the percentage in the snow sample taken. In the case of the dial graduated to 50 parts, the scale is so constructed that if the sampler is filled with pure water, the index will make two rotations of the dial corresponding to 100 percent of water equivalent of the snow.[23]

Church must have read this letter with mixed emotions. Clearly, Marvin, a more formidable rival than Mixer or even Horton, had invented some kind of snow sampler. On the other hand, it was equally clear that the Weather Bureau's chief of instruments had no firsthand knowledge of snow conditions in the West and that his administrative duties left him little time to perfect his apparatus.

Church realized that his snow sampler was original in at least three ways: (1) the small diameter of its tube; (2) the joining of sections to obtain samples in snow up to 6 meters (20 feet) deep; and (3) the slotting of the tube to allow direct observation of the sample and to facilitate removal—all of which were the results of specific snow conditions in the Sierra Nevada. Deep, wet snow lying in steep canyons and ravines over dozens of watersheds covering hundreds of square kilometers required special instruments and techniques for measuring and for estimating the potential spring runoff. By 1917, when he published his fullest account of snow surveying, Church made clear that his sampler grew out of the conditions of the Lake Tahoe basin and that other instruments might be better suited to other areas. Nevertheless, he was obviously proud to announce that the Glacier

Commission in Switzerland was using his invention, the Mount Rose Snow Sampler.[24]

Naming his instrument the Mount Rose Snow Sampler was shrewd. While other gauges bore the names of their inventors or were simply called snow density apparatus, Church evoked some of the romance of winter climbs and called attention to the uniqueness of the place where snow surveying was born. While never denying that Mixer, Horton, and Marvin had invented snow samplers independently, he created a story about the origins of the Mount Rose sampler that placed it centrally in a scientific context transcending the immediate concerns of the Weather Bureau. At the opening of the first meeting of the Western Interstate Snow Survey Conference on February 18, 1933, Church offered a condensed version of his origin myth:

> Snow surveying, like most inevitable things, had many births, though few survivals. It sprang from the need of the people. It was born in Europe [he had earlier identified Alfred Angot, director of the Bureau Central Météorologique de France, and an anonymous Russian as inventors of portable snow-measuring devices], in the study of the density of snow. It was born again in the East in the study of stream flow. It was born a third time in the West, on Mount Rose in the Sierra Nevada, in a dispute regarding the effects of forests on the conservation of snow, but quickly turned to the flaming problem of the flooding of Lake Tahoe.
>
> . . . Nature furnished the facts. The Sierra became a great laboratory for the study of weather, and snow cover, and snow density, and evaporation, and melting. The human need for food and power gave purpose to the undertaking, for the rapid increase in population was forcing one drop of water to do the duty for two.[25]

Legitimizing his contribution to snow surveying by placing it in the western scientific tradition as well as the more immediate political milieu of the Progressive conservation and reclamation movement gave Church a personal victory over his rivals even though the sampler was redesigned and renamed the Federal Snow Sampler in the 1930s.[26]

The instrument itself was only part of Church's innovation. More important was the method of using it in a survey. In a paper presented to the Pan American Scientific Conference in Washington in 1915, he outlined two general methods of snow surveying: by seasonal percentage, and by areas. Church developed the former to determine the variations in spring runoff in the Tahoe basin, a watershed of 880 square kilometers (340 square miles), far too large to be sampled in detail. His method involved establishing fixed courses, or sampling points, in characteristic parts of the watershed, taking a few dozen samples along each course during late March and early April, when the snow was wettest, then comparing the measurements of snow depth and density to the "average" as estimated from earlier records. Since it was relatively easy to measure the level of Lake Tahoe, and even to estimate the amount of water lost from surface evaporation, Church was able to refine his correlations between snow measurements and runoff in just a few seasons. By trial and error, by careful review of the existing literature, and by consulting with his colleagues in engineering at the University of Nevada, Church perfected his system of surveying between 1909 and 1915. Scientific study of evaporation, soil absorption, snow metamorphism, and ablation continues to add to our understanding of the relationships between snowfall and runoff, but the basic methods of the snow survey in use today were worked out eighty years ago by James E. Church, Horace P. Boardman, professor of civil engineering, S. B. Doten, director of the Agricultural Experiment Station, and others.[27]

INTERCEPTION: THE FOREST VERSUS SNOW CONTROVERSY

At the same time the Weather Bureau was struggling to achieve some degree of standardization in snowcover measurement, new data were pouring in from the Great Basin. Church on the western edge and J. Cecil Alter on the eastern rim were doing annual snow surveys. In 1910 Alter, a Weather Bureau observer in Salt Lake City, suggested to his section director, Alfred H. Thiessen, that they conduct a survey of the snow in a small watershed south of Provo. Either Alter was unaware of Church's sampler and ongoing survey at Lake Tahoe or he chose to ignore them, prudently adopting instead the Marvin sampling tube and working out his own survey techniques.

Topographical and environmental differences between the western and eastern edges of the Great Basin contributed to the differences between the Utah and Church survey methods. In the Wasatch and Uinta mountains, small watersheds feed streams into the Sevier River, providing water to irrigate most of central Utah. In contrast, the Lake Tahoe basin is a single watershed strategically located between the fertile Sacramento Valley and the irrigable deserts of Nevada. Faced with surveying an area almost as large as the city of Los Angeles, Church could, with his seasonal percentage method, predict the maximum water available in Lake Tahoe up to four months in advance.

Alter and Thiessen, by contrast, began their surveys with intensive studies of small watersheds. Maple Creek is 27.8 square kilometers (10.75 square miles); Big Cottonwood Canyon near Salt Lake City, where surveys were begun in 1912, is 125.6 square kilometers (48.50 square miles). In both surveys, Thiessen and Alter took several hundred measurements using the Marvin sampler and estimated the total acre-feet of water rather than a percentage of the norm. Each of the early surveyors was aware of the novelty of this experiment in water-supply forecasting. Like Church, the Utah surveyors had to learn about snowshoes and skis, snow blindness and avalanches, waterproof clothing and emergency shelters. In March 1912, Alter reported to Thiessen that "my waterproof boots kept my feet dry, when coated daily with hot tallow and beeswax . . . my cravenetted trousers and coat kept me quite dry and comfortable." Also like Church, both Alter and Thiessen recognized the importance of publicity and used local newspapers and national magazines to convince the public and the Weather Bureau of the importance of their work.[28]

When the reports from the surveyors were unusually detailed, Thiessen forwarded them to Washington. Assistant Observer A. A. Justice submitted a six-page description of his survey of City Creek, made in late March 1914. After describing the canyon and explaining how he tried to take measurements in places containing an average snow layer, he remarked:

> Near the crests of all ridges where the wind has a clean sweep the snow drifts greatly. During the time the work of the survey was in progress a storm occurred during which the wind blew strongly from the southwest for two days. Where previously the crests had been smoothly rounded with a layer of snow now great drifts and perpendicular walls appeared. These walls of snow were in places 30 or 40 feet high and faced the leeward side of the mountain, being capped with an overhanging ragged edge something in appearance like the swell on the ocean as it breaks near the shore. On the windward side for a distance of one or two hundred feet the ground was swept nearly bare. No attempt was made to measure the snow in such places where drifts were evident. On the northern slopes pine trees are numerous and owing to the fact that pines catch a great deal of snow on their boughs which later evaporates or is blown to the ground forming rings of deep snow around the trees and at some distance from them, thus disturbing the uniform depth of the snow layer, it was not considered desirable to have measurements over areas where pines grew. . . .
>
> With few exceptions the best places for measurements were found among the quaking aspens. These trees seem to grow in protected places where the snow lies on the ground late enough in spring to furnish moisture for their summer growth. While forming a wind break to an extent the aspens do not collect the snow on their branches causing drifts as the pines do. In taking measurements through these trees it was found that as long as the trees were of uniform size the depths were about the same, the greater the trees the greater the depth of the snow.[29]

Aside from the vivid description of the conditions in which the surveyors worked, this report addresses the question of the influence of forests on snowpack formation and retention.

Conventional wisdom, from Noah Webster in 1799 through George Perkins Marsh in 1864 to Bernhard Fernow in 1893, held that forests help to retain snow on the ground, control flooding, and act as natural reservoirs. Within the Weather Bureau, however, there were dissenters from this opinion. The differences again seem related to Donald Worster's distinction between imperialist and arcadian approaches to the management of nature. Alter, Marvin, and Benjamin C. Kadel, who succeeded Marvin as chief of the Instrument Division, were at odds with the foresters, who, in 1905, had been moved from the Department of the Interior to the Department of Agriculture. Intradepartmental rivalry between the Weather Bureau and the Forest Service continued.[30]

As early as 1911, Alter dismissed the idea that trees help to preserve the snowpack. "Contrary to popular supposition," he wrote, "the snow that clings to the mountain sides throughout the summer and keeps the irrigation streams running steadily in August and September, is not hidden in the cool shadows of forests, but lies out in the open, where the summer sun and the mountain winds may attack it freely." While acknowledging that "these icy stores of snow" are limited to the north and west slopes above 3050 meters (10,000 feet), after the end of May, Alter goes on to argue that snow falling on the branches of trees usually evaporates before it reaches the ground, thus actually reducing the amount of water available for irrigation. Politically astute, Alter concludes with a selective quotation from Willis L. Moore, chief of the Weather Bureau, implying that forests have no effect in flood control, then fires a final shot of his own at those who believed in preserving forests to conserve snow: "The claim that irrigation would be crippled by the removal of the forests or benefitted by reforesting large areas would also appear to be erroneous, as the late summer irrigator gets his water principally from the snow packs that lie in the open, solid from the effects of the sunshine, wind, and rain, the softer snow in the forests and other shady places having disappeared in April, May, or June."[31]

Four years later, two Forest Service employees came to the opposite conclusion in their study of the influence of a western yellow pine forest in Arizona on the accumulation and melting of snow. The winter of 1910–11 had little snow but frequent rains in that area. The winter of 1912–13 had a heavy snowfall. Two areas, similar in soil and altitude but one forested and one open except for grass and small plants, were studied. The conclusions were that the rate of melting during the winter was higher in the forest because minimum and mean temperatures were higher, but that in the spring the open areas melted much faster and caused floods because the ground was still frozen. In the forests, the snow melted more slowly and seeped into the unfrozen ground. Unable to accept these findings, both Marvin and Kadel wrote rebuttals. Marvin challenged the study on two grounds: the five-year period of observation was too short and

> presentation of the case permits the nontechnical reader to gain the unwarranted impression that the influence of the forest is general and fundamental rather than indirect and incidental. It is easily conceivable that the radiant heat over a park or open field could be partly cut off by the installation of artificial devices arranged to intercept solar radiation to almost any specified extent and thus artificially conserve the melting of the snow, much as the forest is found to do.[32]

Marvin the inventor was confident that he could build a better tree.

In the following years, the weathermen shifted their focus away from the influence of forests on snow melting toward inorganic factors, such as altitude, depth, density, temperature, and humidity. Church, no tree hugger in the present meaning of the term, offered a compromise that was acceptable then as it is now:

> The ideal forest, from the view point of conservation, is the one that can conserve the maximum amount of snow until the close of the season of melting. Such a forest should not be dense enough to prevent the snow from reaching the ground and yet should be sufficiently dense to afford ample shelter from sun and wind. The fir forest possessing a maximum number of glades or a forest of mountain hemlock meets these requirements both theoretically and practically. However, glades can be produced in any dense forest by the simple operation of cutting, the diameter of each glade being so proportioned to the height of the trees around it that the snow in early spring is effectually screened from the sun. Such a forest, when viewed from above, would resemble a gigantic honey comb, the glades of the forest being equivalent to the cells of the comb.[33]

Eighty years later, the complexity of forest-snowcover relationships is being studied with new methods and theories. Noting that "in the boreal forest, interception of snow can store 60 percent of the cumulative snowfall in mid-winter and sublimation of this snow can return over 30 percent of annual snowfall to the atmosphere as water vapour," J. W. Pomeroy of the National Hydrology Research Institute in Saskatoon, Saskatchewan, and R. A. Schmidt of the Rocky Mountain Forest and Range Experiment Station in Fort Collins, Colorado, have conducted studies of the interception, unloading, redistribution, melting, and sublimation of snow using both living and artificial trees whose branches were carefully weighed and photographed to determine amounts of snow intercepted, evaporated, melted, and unloaded. Using fractal geometry to model the surface area of intercepted snow patches, Pomeroy and Schmidt suggest that the processes of accumulation and sublimation are chaotic but not random. They write that since "interception of snow by vegetation strongly influences sub-canopy winter climate, snowcover chemical composition, snowcover physical characteristics and the snow hydrology of forested areas," understanding of the processes involved is vital. Add to this range of variables the differences in vegetative cover—for example, jack pine picks up very little snow, black spruce a great deal—and it becomes clear why it has taken so long to resolve Webster's dispute with Jefferson described in Chapter 1.[34]

Until the late 1920s, the Weather Bureau clearly held the upper hand in the bureaucratic and scientific disputes over snowfall measurement, ablation, and runoff, and the effects of snow on agriculture. The Forest Service came late to the study of snow, and the Bureau of Agricultural Engineering and the Bureau of Chemistry and Soils, both concerned with flood control and soil erosion, were not yet prepared to challenge the Weather Bureau's dominance. Marvin's interest in snow-measuring instruments, first as chief of the Instrument Division from 1888 to 1913, then as chief of the Weather Bureau from 1913 to 1934, meant that any new methods of measuring precipitation would challenge existing procedures and compete with his rain and snow gauges. Marvin had been quick to see the advantages of the Horton method of snow sampling and weighing and designed his own tube in 1910, but his promotion to chief effectively removed him from active involvement in the development of snow samplers. Meanwhile, Church and the Utah weathermen invented their own instruments in response to specific questions about the availability of water for irrigation. Although these were the prototypes for the apparatus adopted by the Soil Conservation Service in 1935, the Weather Bureau continued to experiment with other methods of snow surveying.

In the decade 1913–23, Benjamin C. Kadel assumed Marvin's role as definer of Weather Bureau orthodoxy on snow science. Writing from his position as local forecaster at the experiment station jointly maintained by the Forest Service and the Weather Bureau at Wagon Wheel Gap in southwestern Colorado, Kadel dismissed "such apparatus as gages, bins, and other devices for catching the snow as it falls" as "not only useless but misleading, first because of their well-known failure to make a correct catch, and second because they hold their contained snow up to the action of the sun and wind, thus accelerating the evaporation of their contents." Kadel created an iron sampling tube 15 centimeters (6 inches) in diameter and 1.5 meters (5 feet) long when two sections were joined. After the tube was pushed through the snow to the ground, an auger and measuring stick were screwed down through the tube, the contents removed, the snow placed in a bucket and weighed. The diameter of the tube was chosen not because it was convenient to insert in the snow but because 1 pound of snow would equal 1 inch of water. Despite the awkwardness and size of Kadel's equipment, A. J. Henry of the Rivers and Flood Division pronounced it superior to Church's.[35]

Kadel and Church were in agreement on one important element of the snow survey, however—using seasonal percentages rather than trying to determine the actual water content of a single watershed. Church also joined Kadel in judging snow gauges and snow stakes ineffective. By 1919 Kadel had recognized some of the shortcomings of his sampler and "improved" it by adopting the Marvin tube in lengths of 60 and 180 centimeters (2 and 6 feet). Although acknowledging that Church's sampler was capable of taking samples in much deeper snow, Kadel doubted that "greater lengths than 6 feet can be successfully handled in the field." His new tube could be weighed on a spring balance like Church's and the Utah sampler.[36]

Although Church's methods continued to receive greater popular attention, including a story in the December 11, 1920, *Scientific American,* illustrated with a dramatic color cover painted by Howard Brown, showing a lone snow surveyor taking a sample near the summit of a rugged mountain (Figure 4.3), the Weather Bureau stuck to its old ways. Kadel revised Circular E, *Measurement of Precipitation: Instructions for Measurement and Registration of Precipitation by Means of the Standard Instruments of the U.S. Weather Bureau,* in 1922. Following the pattern established by Marvin in 1893, the booklet describes the installation of the rain gauge and its use in measuring snowfall, the tipping bucket gauge, the Fergusson weighing rain and snow gauge, and other self-registering gauges. To these are added discussions of snow mats, snow stakes, and the Marvin snow sampler as modified by Kadel. By the end of the 1920s, so much data had been gathered on snow density that meteorologists were less concerned with how they were gathered than with how to interpret them. Moreover, the interest of state governments and private power companies in snow surveys was forcing federal agencies to expand the scope of their research.[37]

California was the first state to authorize funds for snow surveys. In 1917 Paul M. Norboe, chief assistant state engineer in the state's Department of Engineering, brought Church's work to the attention of the state legislature, which funded studies of the Sacramento River and San Joaquin Valley watersheds. Church helped Norboe find equipment, and cooperative studies were made along the California-Nevada border until 1923, when the California assembly failed to approve funding. The Nevada Cooperative Surveys continued from 1919, however, and the U.S. Reclamation Service began surveys at Jackson Lake, Wyoming, in the same year. Washington Water Power Company sponsored snow surveys in the Coeur d'Alene basin in 1920, the Utah Cooperative Snow Survey was organized in 1923, the Los Angeles Department of Water and Power started snow surveying in the Owens River basin in 1925, the Oregon Cooperative Snow Survey got under way in 1928, and California returned to snow surveying in 1929, stimulated by a study sponsored by the California Economic Research Council of the state Chamber of Commerce.[38]

The committee's report and the state legislature's action were announced at the June 1929 meeting of the American Meteorological Society

Figure 4.3. Cover of *Scientific American* (December 11, 1920), showing a snow surveyor taking a sample of the snowpack. The accompanying story celebrated Church's innovations, while Harold Brown's illustration depicts the snow surveyor as heroic.

in Berkeley and published in the *Monthly Weather Review* in October, in an article by Harlowe Stafford, a hydraulic engineer in the California Department of Public Works. Stafford, who played a significant role in promoting snow surveying for more than thirty years, concluded his remarks with a few words of caution. Snow surveys alone could not provide all the data necessary for understanding snowcover runoff. "In each quadrangle," he warned, "it will also probably be necessary to establish supplementary precipitation and temperature stations. In addition, at certain key stations it may be advisable to establish facilities for observation of more complete meteorological data, such as humidity, pressure, temperature, wind direction and velocity, etc." These words and the expansionist "etc." must have reassured the weathermen in the audience, but by the time Stafford's recommendations were printed, the stock market had crashed and all government programs faced reduced budgets and hard times.[39]

A NEW DEAL FOR SNOW

The election of Franklin Roosevelt in 1932 brought fresh opportunities for government-sponsored scientific research. In addressing the economic crisis, the federal government turned its attention to problems associated with soil and water conservation. As in the past, however, interdepartmental rivalries threatened to disrupt efforts to conduct research on flooding, irrigation, and reclamation. Secretary of Agriculture Henry Wallace and Secretary of the Interior Harold Ickes were both ambitious men, dedicated to improving the use of natural resources. Ickes acted first, creating a Soil Erosion Service on September 19, 1933, and luring two Department of Agriculture scientists, Hugh H. Bennett of the Bureau of Chemistry and Soils and W. C. Lowdermilk of the Forest Service, to his new agency. Eighteen months later, Wallace recovered his staff and the agency when Congress created the Soil Conservation Service and placed it in the Department of Agriculture. Although the Soil Conservation Service did not assume full responsibility for snow surveying until the late 1930s, Bennett and Lowdermilk had been directly involved in research on snow for more than a decade.[40]

Even before Roosevelt's election, the revival of the California surveys in 1929 gave Church the incentive he needed to promote snow surveying nationally and to expand the scope of the study of snow. In the spring of 1931, the Section of Hydrology of the American Geophysical Union authorized a Permanent Committee on the Hydrology of Snow with Church as chairman. The goals of the committee were cooperation among agencies studying snow, standardization of research methods, and publication of the results. With renewed energy the sixty-two-year-old Church began his most productive years. Seizing his opportunity to promote an arcadian vision of nature as a seamless web, Church reported to the union that "the field of the hydrology of snow has been broadened beyond the narrower limits of the economic aspects of snowfall and runoff to include the evolution of snow from its initial fall until its final emergence at the mouth of the stream." To achieve a balance between applied and theoretical approaches, Church selected three men to join him on the Committee on Snow—an engineer specializing in ice, H. T. Barnes of McGill University; Matthew Balls of the Shawinigan Power Company in Montreal, who represented the Canadian Committee on Geodesy and Geophysics; and Professor William Herbert Hobbs of the University of Michigan, who had employed Church as a meteorologist in Greenland in 1927–28. Later the committee was expanded to fifteen, most of whom, like E. S. Cullings, vice chairman of the Black River Regulating District, Watertown, New York, and George Dewey Clyde, professor of engineering at Utah State University, were leaders in the early Eastern and Western Snow Conferences.[41]

For the next two decades Church kept up correspondence with snow scientists in Europe, Asia, and South America, compiled bibliographies, and published reports. With Clyde's help he organized the Western Interstate Snow Survey Conference in 1933. In the same year the International Association of Scientific Hydrology, meeting in Lisbon, authorized an International Commission of Snow, the first meeting of which was held in Edinburgh, Scotland, in September 1936, with Church as chairman. In his address to the delegates, Church was characteristically ambitious:

> The Commission of Snow is novel in some respects. By generous consent of Dr. Mercanton, Secretary of the Commission of Glaciers and Provisional Secretary of the Commission of Snow, the latter commission was at once expanded to include all fields of snow and ice not specifically desired by the former. This plan was advantageous to both, for it gave the specialized Commission on Glaciers its preferred project of measuring the movement of glaciers and the broader Commission of Snow the opportunity to include all other snow and ice scientists in a single organization. This is an inevitable moment, for neither projects nor scientists can be divided on the basis of snow or ice. Rather the Commission of Snow should be designated the Commission of Snow and Ice with the Commission of Glaciers complementary to it. This would be the evolutionary view. Practically, the study of glaciers antedated the study of their parents, snow and ice, for the phenomenon of ice movement has always been fascinating. Now finally the process is being reversed.

To illustrate his concept of a holistic study of snow and ice, Church included a diagram consisting of a dozen intersecting circles labeled "Rain/Snow," "Glaciers," "Ice," "Forecasting Runoff," "Conservation of Water: Forest Cover," and other aspects of hydrology and oceanography. Significantly, this web of "various phases of interest of [*sic*] Commission of Snow and Ice," included the emerging field of snow ecology through its attention to "forest cover," and in 1938 winter recreation was also added.[42] Church was in control of regional, national, and international organizations for the study of snow.

The Western Snow Conference, which met for the first time on February 18, 1933, in Reno, Nevada, unified the Nevada and Utah surveys. With Church and Clyde providing leadership, the conference became the place where engineers, hydrologists, meteorologists, foresters, power and water company administrators, and snow surveyors met to discuss improvements in equipment, the design of snow survey courses, and the financing of research. At the first meeting, Joseph Kittredge, who would become the foremost authority on the effects of forests on climate, water, and soil, gave a paper on the relation of forests to snow, and W. C. Lowdermilk, who was about to begin his distinguished career with the Soil Conservation Service, presented his research on measurement of the percolation of water in snowpacks. The first conference was attended by forty delegates; ironically, the representatives from Southern California were prevented from coming by snow blocking the roads. Currently about a hundred attend the annual meeting, now held in April.[43]

The Western Snow Conference, as it has been called since 1943, not only is the oldest organization for snow science in the United States but represents, in microcosm, the history of snow in America. Attempts to define snow, manage it, and ultimately profit from it are chronicled in the proceedings of the Western Snow Conference, which were published in the *Transactions* of the American Geophysical Union from 1934 to 1944 and as separate volumes thereafter. During its first decade, members of the conference coordinated state and federal snow surveys, improved and standardized equipment, and settled bureaucratic and professional turf wars. Surveys were in progress in all the western states and parts of Canada by 1936. George Clyde and Ralph L. Parshall laid out snow survey course sites in Colorado and Wyoming; R. A. "Arch" Work, on the Rogue River drainage in southern Oregon (Figure 4.4); and James C. Marr, associate irrigation engineer with the Bureau of Agricultural Engineering, on the Snake River watershed in Idaho. Standardization of the sampler involved combining features of the Mount Rose and Utah apparatuses and improving them by adopting a "heavier alco-aluminum stock" for the tubes. Bright yellow signs made of porcelain enamel and bearing the words "Federal-State Cooperative Snow Surveys Snow Course Marker" were adopted (Figure 4.5). Standards for the location and maintenance of the courses were considered.[44]

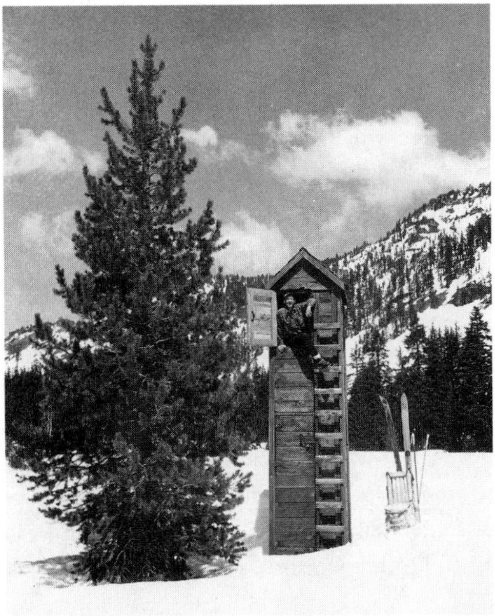

Figure 4.4. "Snow Surveyor R. A. Work looking pretty pleased with himself from the security of the Santy Claus chimney, which is the only entrance to the snow surveyors' shelter cabin during periods of heavy snowfall." Soil Conservation Service photo by Branstead. The original caption conveys some of the esprit of the Soil Conservation Service snow surveyors in the 1950s. The cabin is buried under more than 3 meters (10 feet) of snow. Photo courtesy of SCS, U.S. Department of Agriculture.

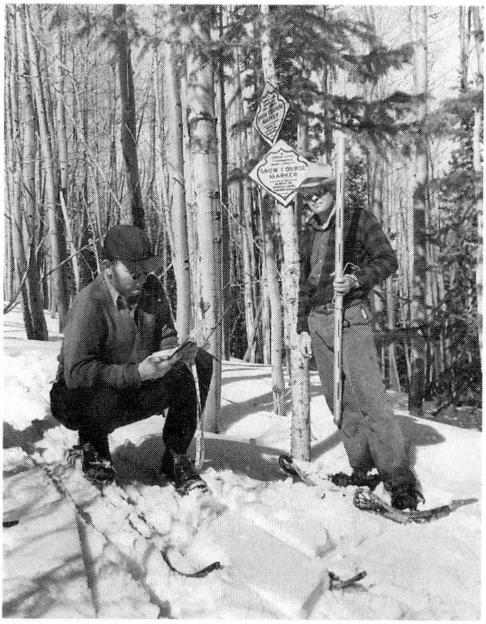

Figure 4.5. "U.S. forest rangers in the Sangre de Christo mountains above Penasco, New Mexico, weighing snow to predict the amount of irrigation for the coming year." Photo by John Collier, Jr., for the Office of War Information, successor to the Farm Security Administration photo project (ca. 1942). Prints and Photographs, Library of Congress. Note the Federal-State Cooperative Snow Surveys Snow Course Marker, meant to establish a scientific presence and warn backcountry skiers not to pack the snow.

From 1935 to 1939 the Division of Irrigation in the Bureau of Agricultural Engineering of the Department of Agriculture was responsible for supervising the Federal-State Cooperative Snow Surveys. In January 1939 Secretary of Agriculture Wallace directed Hugh Bennett of the Soil Conservation Service to take over the Division of Irrigation and Drainage, as it was then called, and the SCS became the agency primarily responsible for the surveys. Until then the SCS had been just one agency involved, along with the Forest Service, the Park Service, the Bureau of Reclamation, the Corps of Engineers, the Geological Survey, and the Weather Bureau. Most of the agencies were represented on Church's enlarged American Geophysical Union Committee on Snow in 1936.[45]

After five years of expanding snow surveys and improving the samplers and scales, Church, the Committee on Snow, and the state and federal agencies cooperating in the study of snow were shifting to other problems and concerns, such as total annual precipitation, soil types, climate trends, and winter recreation. Under the leadership of Merrill Bernard, the Weather Bureau reasserted itself as an important agency in snow study. Conceding the snow surveys to the Soil Conservation Service, the bureau revitalized its snowfall stations to provide additional information on precipitation, radiation, convection, evaporation, and condensation. In January 1941 Bernard reconceptualized the "Snow—Heat—Run-Off Balance" cycle for those attending the Ninth Western Interstate Snow-Survey Conference in Sacramento, California. As he conceived it, data from hydrothermographs, electric psychrometers, recording anemometers, radiometers, snow gauges, rain gauges, depth gauges, density tubes (snow samplers), calorimeters, soil thermographs, and streamflow gauges would be gathered, analyzed, and used to make accurate predictions of runoff.[46]

Thus, on the eve of World War II, snow hydrology was conceived as the field in which the work of snow surveyors, forecasters, meteorologists, power company and irrigation district engineers, and others should integrate their knowledge. The war years stimulated an interest in snow because of the strategic importance of the Arctic and Greenland, and the Army Corps of Engineers became a major institution in the expanding network of agencies and bureaus dealing with snow. Although Church remained as chairman of the International Commission of Snow through the war, the old order was changing. A new generation of snow scientists was emerging, and the success of the Western Interstate Snow-Survey Conference and the International Commission stimulated competition. An Eastern Snow Conference was held in Boston, October 8, 1937, with the support of the Weather Bureau and the American Society of Civil Engineers, then refounded in Cambridge, Massachusetts, in September 1940. The Eastern Snow Conference met irregularly until after the war, but it represented the reemergence of the Weather Bureau as the dominant government agency in snow science. As Merrill Bernard and Ashton Codd reported to the Western Conference in January 1940, the Weather Bureau was modernizing its procedures for measuring snowfall in the mountains of the West. Sampling was adequate for predicting the percentage of deviation from the normal water supply, but their goal was accurate hourly measurement of snowfall in hundreds of locations.[47]

Many meteorologists contributed to the bureau's work. Cecil Alter, who had risen from observer to official in charge of the U.S. Weather Bureau in Salt Lake City, began a series of experiments in 1935 to improve the accuracy of the Marvin snow and rain gauges. His most lasting improvement was the invention of a shield for the gauge to prevent loss of snow from the wind. After testing several types of shields, he determined that a lath fence, suspended several centimeters above the top of the gauge, with the bottom of the laths flared inward and attached to wires that allowed each lath to sway in the wind, provided the best protection. Alter shields were installed on the gauges at all ninety-two mountain snowfall stations in eleven western states by 1941.[48]

A Central Snow Conference was organized in Detroit in December 1940. Its second meeting, in East Lansing, Michigan, took place in the anxious days following Pearl Harbor. Ironically, the work of the great Japanese meteorologist Ukichiro Nakaya on the formation of snow crystals had just been introduced to American scientists at the Western Snow Conference in 1940. As the organization of regional conferences suggests, however, American interest in snow was stimulated by specific domestic concerns. Thirty-one papers were presented at the second Central Snow Conference meeting, on topics as diverse as snow melting; highway snow removal; the effects of snow on recreation, wildlife, forestry, and agriculture; and "cryologic hydrology." Robert Horton proposed a separate branch of hydrology, "cryology," to deal "with natural occurrences and distribution of water in the form of ice and the physical processes involved in its course through the hydrologic cycle." James Church reported on his experiments with melting snow. Lieutenant W. F. Johnson of the U.S. Army gave a brief introductory talk titled "Snow as It Effects [*sic*] Military Operations." Although the January 16, 1942, meeting of the Western Snow Conference in Pasadena, California, drew only thirty-four participants, because another meeting was planned for Salt Lake City later in the year, it was here that Fred Paget of the California Division of Water Resources and chairman of the Executive Committee proposed adoption of the term *niphometrology* for the science of snow measurement.[49]

NIPHOMETROLOGY AND THE WAR YEARS

Although the neologism *niphometrology* did not find favor, it indicates that snow surveying had reached its own watershed. On one side were those who recognized snow as a resource to be conserved. They had proven the importance of snow and established institutions to promote its study. But their success created new divisions, regional and topical. The war accelerated government support and scientific specialization and disrupted international cooperation. The military importance of Greenland and the Arctic shifted research away from snow as a source of water toward problems of airstrip and road building. Ultimately, larger questions about snowcover and climate, specifically surface-energy exchange, emerged from research at both poles. New institutions and new patterns of research evolved. The 1942 Report of the Committee on Snow of the American Geophysical Union gathered regional reports from more than twenty watersheds in North America, including Canada and Mexico. Church's report that year emphasized studies on the thermal quality of snow, while others commented extensively on new safety bindings for skis. A motion picture on avalanche control with trench mortars was provided by the Swiss Snow and Avalanche Commission. Interest in avalanche control was, as seen in the last chapter, closely linked to the growth of recreational skiing and winter sports, which, in turn, was linked to national defense by representatives of tourist-related businesses who anticipated restrictions on civilian travel.[50]

By the following year, the Committee on Snow had shifted to wartime concerns. With the support of the Weather Bureau, a snow study facility was opened at Soda Springs, California, near Donner Summit. This became the Central Sierra Snow Laboratory. Church continued experiments on snow melt and refreezing, but studies of the relation of forests to snow accumulation and melting at the Rocky Mountain Forest and Range Experiment Station in Colorado had to be curtailed for lack of staff. "Hydrology's part in the war effort," as one observer noted, was vital but as yet unappreciated. Yet a former engineer from the Los Angeles Water and Power Department employed the metaphor of war to pay tribute to "Water Works Ski-Troopers." Even before the end of the war, there were signs of tension between the snow surveyors and other snow scientists. The 1943 meeting, at which the Western Interstate Snow-Survey Conference voted to rename itself the Western Snow Conference, was marked by a debate over "bona fide" membership. To the charge that only 176 of 600 reported members had paid dues, Church replied that "it had been the policy of the organization that anyone attending a meeting was automatically a member."[51]

Such unbusinesslike inclusiveness was at odds with the approach of J. A. Fleming, general secretary of the American Geophysical Union, who was anxious to raise revenues for publications and to strengthen the organization's professional standards. The AGU was trying to convert its annual *Transactions* to a bimonthly journal. Fleming was, therefore, unwilling to continue publishing the proceedings of the Western Snow Conference. The union's position was that some of the papers presented at the snow conference were not directly related to scientific hydrology. Fleming sent a letter to the 1946 snow conference meeting that sparked a lengthy discussion revealing the many changes taking place in snow surveying. Arch Work of the Soil Conservation Service in Medford, Oregon, and chairman of the conference, read Fleming's letter and explained that the changes meant that snow surveyors would need to buy four issues of the *Transactions* annually, in addition to paying dues to the AGU.

Moreover, Fleming was unwilling to publish "notes" from field researchers under a separate heading to be called "Snow Surveyors Forum" unless more members of the Western Snow Conference joined the American Geophysical Union. After debate the members decided to publish their own annual proceedings, as well as "some kind of mimeographed sheet to send out to linemen, forest rangers, fellows in the field." Fred Paget summed up the sense of the meeting:

> Well, Mr. Work, we have always thought of the Western Snow Conference as developing aims and ideals a little apart from those of the A.G.U., and to some extent along different lines.
>
> We came into being just a few struggling people here and there. So few that for a while we despaired of ever getting very far. Dr. Church

was the only one who had faith in us; sometimes when we were almost ready to sink he threw out a life buoy in the form of a few hundred dollars and held us up awhile. We kept struggling along, and eventually got to the stage where we are today.

But I feel our progress was largely due to the fact we had the annual "Transactions" to send out, printed in a single volume, along with the annual meeting on Hydrology here; that we could have something in one unit, that would come out once a year, which had an appeal to the rank and file who are less technical-minded than most of us who attend these meetings.

Church added his imprimatur to the new publication. Deploring the over-specialization of the AGU *Transactions,* he concluded, "I should like therefore to offer 'popular' themes in the hope that science will be benefitted by them." Later in 1946 the first issue of the *Snow Surveyors' Forum* appeared, with cartoons, humorous verse, personal experiences of snow surveyors, and technical information on snow-ice cutters for samplers, mechanized over-snow transportation, and ski repair.[52]

The *Forum* was published fifteen times between 1946 and 1964, when it ceased publication. Its existence coincides roughly with what one observer calls "the period of 'expanding interest and systems improvement'" in snow surveying.[53] By the mid-1960s remote-sensing systems such as airborne and satellite snowcover measurements, radio telemetry from isotopic snow gauges, and pressure snow pillows were supplementing manual snow surveys, and computers were being used to analyze the data. The pace of snow surveying was accelerating, and the playfulness of the early years became more ritualized. One ceremony that survives to the present is the El Farsante [Faker] Award, presented annually since 1953 to a member of the Western Snow Conference who has made an embarrassing mistake during the year, such as grossly misforecasting the spring runoff or missing Kim Novak while photographing scenery at Squaw Valley. The Eastern Snow Conference bestows a similar dubious award, the Sno-Foo, on its errant members. The function of these awards and of the *Forum* is clearly to create a work culture in which fellowship compensates for some of the hardships of the occupation and humanizes science to encourage a creative tension between abstract theory and common sense.

Snow scientists, like others struggling to understand the vast and complex realms of nature, know that their data, their theories, and their definitions of reality undergo constant modification. Church bid his colleagues farewell in 1949 with the reminder that science is nothing more than "man's interpretation of his experiences," a definition that covers art and religion as well. In the worklore of snow surveyors, there is an explicit recognition of the absurdity of their quest. Humor often reveals deep feelings. One *Forum* contributor, playing on the words *snow,* meaning "to defeat by an overwhelming aggregate of tallies," and *surveyor,* "an officer who ascertains the content of casks . . . of dutiable liquor," defined *snow surveyor* as someone overwhelmed by drink. Pushing his analogy, he defined *forecast* as "the guess, hunch, or conjectured opinion uttered when overwhelmed. This is changed continuously as runoff season progresses and as unoverwhelming takes place." *Accuracy* is defined as "a rather vague term difficult to explain to one not overwhelmed."[54]

SNOW TUBES, SNOTEL, AND AIRBORNE SNOW SURVEYS

Snow surveying in the late 1940s and early 1950s was concerned chiefly with improving forecasting techniques through the refinement of statistical models. More variables were included in the formulas, such as soil-moisture deficiency in the fall, precipitation during runoff, temperature changes during runoff, and water storage and transmission in the snowpack. A major technological change in snow surveying took place in 1947 with the U.S. Army Corps of Engineers' introduction of radioisotope-telemetering snow gauges in King's River basin in California. The radioisotope snow gauge was developed by R. W. Gerdel and B. L. Hansen at the Central Sierra Snow Lab.

The first models consisted of a pole about 6.0 meters (20 feet) high supporting a 4.5-meter (15-foot) arm at a right angle to the pole, on which a Geiger-Müller tube (an instrument similar to a Geiger counter, but with an amplifying system) was suspended. On the ground, directly under the Geiger-Müller counter, a lead cylinder containing two units of cobalt 60 (40 millicuries) was secured in a concrete slab. The gamma rays emitted by the cobalt decreased as the water content of the snowpack increased. The number of counts from the Geiger-Müller tube was transmitted by radio to a central receiving station by a telemetering system that involved electrically wound, spring-driven clocks and storage batteries capable of operating at low temperatures. All the elements were parts of wartime research by the Corps of Engineers. Although this model measured only the water content of the snow, it had several advantages over manual surveys. It permitted continuous monitoring of a site and did not destroy the sample by removal. It also enabled snow surveyors to gather data in places difficult or hazardous to reach on skis. Nevertheless, the accuracy of the radioactive snow gauges had to be checked against the Federal Snow Sampler.[55]

More significant than the technological innovation of remote sensing of snow was the institutional reorganization of the snow sciences. The Central Sierra Snow Laboratory, which had begun on a modest scale in 1943, was taken over by the Army Corps of Engineers in 1946, then transferred to the Forest Service in 1952. The corps had already established a Frost Effects

Laboratory in Boston in 1944 and a Permafrost Division in St. Paul in 1945, both to study the design and construction of roads and airfields in cold regions. The Western and Eastern Snow Conferences continued to grow, but the Central Snow Conference disbanded. The separation of the Western Snow Conference from the American Geophysical Union in 1948 was followed by the termination of the AGU's Committee on Snow. In 1949 the Corps of Engineers created SIPRE (Snow, Ice, and Permafrost Research Establishment) in Evanston, Illinois, to conduct basic research and to coordinate the work of the various laboratories. This group constructed the Keweenaw Field Station near Houghton and Handcock in Michigan's Upper Peninsula in 1953 before reorganizing again in 1961 as the U.S. Army Corps of Engineers Cold Regions Research and Engineering Laboratory (CRREL) and relocating in Hanover, New Hampshire.[56]

This was a crucial decade and a half for snow science. Although Church was invited to St. Paul to meet with the administrators of SIPRE and pass the torch of snow science, it was clear from the beginning that water management was only one of a number of issues with which the Corps of Engineers was concerned. At first the corps retained direct involvement in snow surveying through Cooperative Snow Investigations, a working agreement of the Department of Commerce and the Department of the Army that established a command channel and policy consultations between the Weather Bureau and the Corps of Engineers in the western states. Under this agreement, the Central Sierra Snow Lab, the Upper Columbia Snow Lab in Glacier Park, Montana, and the Willamette Basin Snow Lab near Eugene, Oregon, were jointly administered. Their investigations culminated in a 1956 report on snow hydrology that focused on the hydraulics of flow, the effect of geological formations on groundwater storage, the capacity of soils to transmit and store water, and the influence of plants and forests on the atmosphere and the snowpack. The Weather Bureau continued to emphasize the measurement of precipitation, and the Soil Conservation Service specialized in streamflow forecasting from snow surveys. All the agencies were improving their analyses by applying newly developed statistical methods and using digital computers. The publication of C. E. P. Brooks and N. Carruthers's *Handbook of Statistical Methods in Meteorology* in 1953 initiated an era of sophisticated mathematical modeling. At the same time, SIPRE commissioned translations of recent Russian work on snow.[57]

By 1960, with the installation of computers at central receiving stations and an expanding network of snow survey courses, radioactive snow gauges, and meteorological observatories, researchers began to write confidently about "snowpack management." A paper presented at the Western Snow Conference declared, "The snowpack zone in California is nearly perfect hydrologically; we will have to watch our step in trying to improve it." The *snow zone* was defined as the area where more than half of the streamflow comes from melting snowcover. In this zone only one year in ten was a "dry" year, in which there was less than half the normal streamflow. "By way of contrast," the author continued, "below the snow zone in the commercial timber belt *two to three years in ten* are 'dry,' and in the Sierra foothills not only do we get less water annually but it is more variable—*three to five years in ten* are 'dry years.'" The implications of this observation were that better management of the snow zone could offset drought in the rest of the state and that snowpack management is principally a concern of state and federal agencies. A conclusion of the California snow management study was a reaffirmation of Church's conclusion fifty years earlier that forests cut in small blocks allow snow to accumulate without increased melting.[58]

Snowpack management was theoretically possible in the eastern states, but a decade of study led to the conclusion that it was impractical because of patterns of land ownership and use. Whereas most of the snow zone in the Sierra was in public hands, most of the watersheds in the eastern mountains were privately owned. Moreover, the owners used their property in a range of agricultural and recreational ways. Snowpack managers in Arizona and New Mexico also confronted a range of "timber, forage, wildlife, and amenity values required by society" but remained optimistic because, like the rest of the West, most of the affected land was administered by the Forest Service or the Bureau of Land Management. One hydraulic engineer expressed the belief that the computer, with its potential to analyze so much new data, would actually humanize snow management by allowing scientists to ask why as well as how they develop their strategies. He reasoned that in the face of an overwhelming number of variables, scientists would need to establish priorities and rethink the purposes of their work.[59]

The 1960s and 1970s also brought two technological innovations that made major advances in snow surveying in the United States. James L. Smith, a Forest Service hydrologist, developed an isotopic snow gauge that not only measured total water content of the snowpack but provided a profile of the stratigraphy of the fallen snow using two tubes; one containing 10 millicuries of cesium 137 to emit gamma rays, the other containing a Geiger counter–like detector. A lift mechanism raised and lowered the detector and the radioactive source simultaneously in increments of 1.27 to 2.54 centimeters (0.5 to 1.0 inch), the water content of the snow being measured horizontally between the two tubes. The amount of radioactive material was later reduced and the strength of the instruments improved for field operations. A portable model was also developed. Profiling radioactive snow gauges proved valuable in avalanche studies, since continuous monitoring showed the breakup of the snowpack. More than twenty-five years of experiments and improvements have made isotopic snow measurement gauges a permanent, if limited, tool in snow surveying. Advocates of the instruments argue that they are accurate and reliable, while conceding that in their early years of use they had a high failure rate and were expensive. More esoteric problems included sun cupping around the poles in the melt season, which caused the gauge to read the air where the snow had melted away from the poles, and underestimation of the water content caused by the blind spots of the collimator, the tube containing the detector. The use of radioactive materials created problems of licensing by the Nuclear Regulatory Commission, but some observers felt the signs warning of radioactive materials discouraged vandalism. These disadvantages continue to restrict isotopic instruments to research rather than operational uses.[60]

A less expensive and more practical instrument was developed in 1965. Known as the Mt. Hood Pressure Snow Pillow and now called simply snow or pressure pillow, this device is similar to an air mattress filled with antifreeze. The fluid responds to the weight of snow as it accumulates, and this pressure is measured by a manometer or pressure transducer. Data from pressure pillows are easily relayed to central stations by the same systems used by isotopic instruments. Pillow snow gauges were the work of R. T. Beaumont of the Soil Conservation Service and his associates. Beginning in 1958, extensive testing of more than thirty-five types of snow sensors led to the conclusion that rubber pillows must be at least 3.6 meters (12 feet) in diameter to catch an adequate sample. Further experimentation developed a stainless steel pressure tank 1.2 meters (4 feet) square, in which a thin sheet of galvanized metal floats on a few centimeters of antifreeze. These metal floats proved less susceptible to leaks and damage from gnawing rodents than rubber pillows and they were less affected by wind and thermal disturbances. By the end of the 1980s, more than 500 snow pillows were in operation throughout the western states.[61]

The 1960s also witnessed improvements in an older instrument, the lysimeter, used to collect and measure the liquid water outflow from the bottom of a snowpack or some level within it. In the late 1930s, Gerald Seligman used an early model in attempting to measure the amount of water melting from an alpine glacier, and James Church had conducted meltwater experiments at the Central Sierra Snow Lab. The Corps of Engineers made extensive use of "snowmelt lysimeters," and by 1984 there were at least three dozen designs, ranging from a small dish-shaped collector placed at the bottom of a snowpack with a drain connected to a measuring device to a pair of collectors 6 square meters (64 square feet) in size surrounded by an aluminum shell sunk deep into the bedrock. The shell could be raised to a height of 4.4 meters (14 feet), in an effort to prevent water from flowing around the lysimeter. Lysimeters are used to measure the amount and timing of daily snowmelt, the energy balance as affected by solar radiation and albedo (reflection of light off the snow), water transmission through snow, snow chemistry, the influence of forests on the snowpack, and similar problems related to streamflow and flood forecasting, as well as to develop a basic understanding of the nature of snow. Snow is, of course, just one form of water in the hydrologic cycle. The difficulty of defining snow and of measuring it is partly a function of its constantly changing form. The lysimeter in a sense defines the end of snow, the return to liquid water, the passage from snowpack storage to daily use. With the melting of the last crystal of snow, the snow surveyor yields to the water commissioner.[62]

By the end of the 1960s, snow surveyors had four basic tools for measuring the water content of snow on the ground: (1) the Federal Snow Sampler and similar tubes, such as the Bowman, Adirondack, and Canadian samplers; (2) isotopic instruments; (3) snow pillows; and (4) aerial markers, measuring rods that can be read from low-flying aircraft. Efforts to improve each of these methods continue with the result that the standards of accuracy have risen considerably. Nevertheless, the variability of the environments in which snow is surveyed and the complexity of snowcover ablation and runoff make prediction a daunting task. After extensive testing of several models of snow samplers under widely different geographical and meteorological conditions, R. A. Work and his colleagues concluded that "the popular Federal snow sampler tends to over-measure the water equivalent of snow. The error ranges from an average of about seven percent in shallow, light-density Alaskan snow, to as much as ten to twelve percent in deep snow of high density." It is possible, however, to adjust for this overestimation in the statistical formulas used to estimate the water content, and given the difficulties of conducting the surveys and maintaining sensitive instruments in the field, Work reported "that the present Federal snow sampler equipment is as utilitarian as can be found for deep dense western snows, and even for shallow sub-Arctic snow."[63]

Snow surveys for different purposes were developed by meteorologists in the same period. The Weather Bureau's primary interest in predicting the amount of precipitation in a storm system led to improvements in rain and snow gauges. Beginning in the late 1940s, the bureau added radar to its repertory. Because rain, ice, and snow appear as bands of light on the radar screen, it is possible to distinguish types of precipitation by their refractive indices. Changes in particle size and shape may also result in changes in echo intensity. The use of parallel beams of light to measure falling snow by attenuation was also developed in the 1950s and 1960s. Like the isometric gauges developed by the Soil Conservation Service, the optical instruments provide real-time measurements of the rate of snowfall. At the same time other meteorologists searched for simpler and more practical ways to predict total snowfall. Stanley Wasserman and Daniel Monte, working with weather observers at La Guardia Airport near New York City, worked out a rough correlation between visibility and snow accumulation. Using the standard definitions of intensity of precipitation—light intensity allowing visibility of more than 1 kilometer (0.6 mile) and heavy intensity limiting visibility to 0.5 kilometer (0.3 mile) or less—Wasserman and Monte revised the bureau's rate of fall estimates upward. Light intensity yielded 0.5 centimeter (0.2 inch) of snow per hour; heavy intensity averaged 4.0 centimeters (1.6 inches) per hour. While admitting the limited nature of their experiment, the New York weather forecasters believed that it could prove a useful technique for estimating both total accumulation and water content.[64]

In recent years the rate of snowfall and the changes in snow crystal types in falling snow have been measured by video cameras. During the winters of 1980–81 and 1981–82, the Atmospheric Optics Branch of the Air Force Geophysics Laboratory developed three new instruments in field experiments called SNOW ONE-A (Scenario Normalization for Operations in Winter Observation and the National Environment), sponsored by CRREL. The fact that the experiments were conducted at Camp Ethan Allen near Jericho, Vermont, the home of "Snowflake" Bentley, was especially appropriate since two of them were essentially refinements of his photographic technique. According to the Air Force report, "The purpose of the Fall Velocity Indicator (FVI) is to record the physical characteristics of individual flakes of naturally falling snow and to provide a means of measuring their fall velocities."[65]

A second instrument, the Belt Reader, was "designed to record the character of snow crystal type on a continuous basis in order to relate changes in crystalline form to changes in the electromagnetic attenuation, snow density, fall velocities, etc." The FVI is a small video camera that tapes falling snow illuminated by strobe-light flashes. Multiple images of the same snowflake are recorded because the strobe rate exceeds the video scan rate. Although the type of crystal is often hard to identify, comparison of "top and frontal images can provide information on particle orientation, tumbling, or oscillation." Computer recognition and classification of snow crystal types was considered the next step. The Belt Reader transports snow crystals captured during a storm on a continuous rolling belt to a closed-circuit television camera that magnifies and records them for later identification.

A third instrument, the Snow Rate Meter, was designed to detect, resolve, and record variations in the snowfall rate of less than a minute. The Snow Rate Meter is an electronic scale capable of recording weight as low as 0.01 grams (0.00035 ounces) at intervals of about three seconds. Although all of these instruments are susceptible to disturbances from wind and are too delicate for operational use, the experiments indicate that future measurements of snow will be finer and more precise, although many scientists are skeptical about the contribution of new tools because they require "more knowledge of the system being studied."[66]

After the *Monthly Weather Review* discontinued its snow maps in 1961, there was discussion of the need for snowcover maps and techniques for making them more reliable. H. C. S. Thom of the Environmental Data

Service of the Environmental Science Services Administration, an agency created in 1965 that included the Weather Bureau, linked the need for snowcover maps to the need to know the weight or water content of snow. Thom also suggested the desirability of weekly snowcover maps for parts of the country during the fire season. Other researchers compiled snowcover maps of the entire Northern Hemisphere for studies of climate change. In 1972 G. A. McKay, of the Atmospheric Environment Service of the Canadian Department of the Environment, reviewed the uses and limitations of snowcover maps. Noting the importance of snow to all outdoor activities, especially in the management of renewable resources, transportation, construction, and recreation, McKay conceded that most of the maps lacked the precision needed to evaluate total water resources and make regional comparisons but concluded that the present network of manual and remote-sensing surveys combined with airborne and satellite photogrammetry could overcome the deficiencies in snowfall mapping resulting from present data inadequacies. "Maps are needed which show absolute values in sufficient detail to be of value in the computation of heat fluxes between the atmosphere and the soil," he wrote, "or to enable planners to select highway routes and ski slopes. Part of the present deficiency can be overcome by sound practices, but the major requirement is for superior networks and methods of observing snow, as well as more intensive knowledge on the relationships between land forms and snowfall and snow cover."[67]

Surveying the art of snow surveying in 1973, scientists gathered in Monterey, California, to call for coordination of the rapidly proliferating techniques and programs engaged in studying snow. At least a dozen federal, state, and private agencies were involved in western states snow survey telemetry alone, some for flood control, some for weather modification, others for irrigation, power, and mixed purposes. Technological changes had again outstripped administrative reform by the mid-1970s, when the Soil Conservation Service announced SNOTEL (Snow Telemetry), a data collection, transmission, and processing system utilizing meteor-burst telemetry to relay signals from a central station to snow pillows and other instruments at remote sites and back again. Since the system was designed to cover a vast area of the western United States, more than 3,108,000 square kilometers (1,200,000 square miles), meteor-burst telemetry techniques were superior to and less expensive then satellite transmission. Although as Manes Barton and Michael Burke reported, "the occurrence of a meteor trail in the right geometric orientation, and of sufficient duration to allow a particular remote station to read its probe and transmit its data, is a random event," experimentation with the system demonstrated that, at worst, the conditions appeared several times an hour. Master stations were built in Boise, Idaho, and Ogden, Utah. As the system grew in the 1980s to include more than 550 SNOTEL sites and 2,000 climatological stations, an automated centralized forecasting system was put into operation, obviating the need for the Soil Conservation Service to cooperate with other systems.[68]

In 1970 the Weather Bureau, together with the National Environmental Satellite, Data, and Information Service, and the National Climatic Data Center, became part of the newly created National Oceanic and Atmospheric Administration. As had its earlier moves, this transfer of the Weather Bureau reflected changing national priorities and technologies. The space program extended physical frontiers, and the environmental movement focused on holistic perspectives of the earth. Thus, the National Weather Service, following the initiative of Soviet scientists, began to develop techniques to measure the characteristics of snowcover (depth, density, water equivalent, and extent) from low-flying aircraft using natural terrestrial gamma radiation. The expansion of this technique over the past twenty years has added a fifth tool to the snow surveyor's kit. The integration of airborne and satellite snowcover measurements, administered by the Office of Hydrology of the National Weather Service with SNOTEL and other Soil Conservation Service systems, is a major goal of the twenty-first century. As in the past, the new techniques require some redefinition of terms and further contribute to the public demand for more accurate measurements and reliable predictions. For example, is the snowcover defined primarily by area or by depth? Satellite instruments record the albedo or surface reflection of snow rather than its depth. Airborne measurements of natural radiation sense ice on the ground as well as snow, yielding higher estimates of water equivalence of the snowpack than manual samplers. Discrepancies among the measurements of different instruments may lead to the development of new hydrologic models and more reliable predictions.[69]

In 1979 a young Weather Service hydrologist, Tom Carroll, with a degree in glaciology from the University of Edinburgh and a Ph.D. in hydrology from the University of Colorado, began to participate in an operational airborne gamma radiation snow survey of the Dakotas, western Minnesota, and a portion of Saskatchewan. Enthusiastic and extroverted, Carroll, like many other snow scientists, combines a love of the outdoors with scientific training and possesses a strong sense of past efforts to live with and understand snow. Some of his first scientific papers dealt with the measurement of streamflow from alpine snowpacks in Colorado, which provided him a ground-level view before he took flight.[70]

The original purpose of the airborne snow survey program was improvement of spring flood forecasts. Using a variety of radiation detection devices, the first flights measured the gamma radiation over flight lines 20 kilometers (12 miles) long and 300 meters (984 feet) wide. Flying in twin-engine Rockwell Aero Commanders at an altitude of 150 meters (500 feet), Carroll and his colleagues collected data that were then checked against measurements made on the ground with snow and soil samplers. As they gained experience, errors caused by atmospheric radiation and accumulated radon gas on the ground were discovered and the model adjusted. Within a few years, the airborne survey was able to update the snow runoff models used in predicting floods by adding real-time data as conditions changed. It was possible, for example, to measure soil moisture in the fall immediately before the freeze-up, an indication of the degree to which rapid snowmelt will combine with impermeable ground to contribute to spring flooding. When Bernard Shafer and his associates studied the 1983 snowmelt runoff that caused flooding along the Colorado River, they concluded that the snow line was an important index of total water content of the mountain snowpack. Aerial observation was the only way to obtain sufficient information on a large and remote area. Carroll and his co-workers were able to show that the airborne surveys were substantially less expensive than flood damage.[71]

Airborne snowcover measurements provide data on snow water equivalents and the moisture in the top few centimeters of the soil but cannot measure snow depth or the total moisture in the soil. For that information and for calibration of the measurements taken from the air, older, manual instruments, including snow samplers, are used. The necessity of checking the measurements of one set of instruments against another when both are suspected of errors is one of the enduring paradoxes of science. Carroll believes that "snow tube measurements tend to *systematically underestimate* true snow water equivalent because of the sampling difficulties associated with ice lenses, ground ice, and depth hoar. . . . (An alternative procedure to make snow water equivalent measurements is to melt the snow sample to derive the water equivalent—a technique encumbered with equal difficulty.) Additionally, one point sample tends not to be representative of an area." While Work's tests with the Federal Snow Sampler came to the opposite conclusion because the cutter point forces more snow into the tube than the inside diameter of the cutter would suggest, Carroll's position enjoys support from experienced surveyors who have found that the sampler does not catch all the snow in a sample because of clogging in icy snow and the difficulty of picking up light powder. Carroll's solution is longer flight lines to collect more data for averaging, thereby reducing statistical errors. This is, of course, what all snow surveyors do when they collect samples from several points on a snow course. The paradox remains, however. What is being measured? Snow and snow water equivalent are distinct only by the definitions of those doing the measuring, or, as Carroll puts it in a final moment of frustration: "A significant rub in this business comes in knowing the TRUE snow water equivalent with which to compare the airborne snow water equivalent values. Ground-based snow measurements do not provide accurate mean areal snow water equivalent estimates." The rivals are, for the moment, standing on opposite sides of the bank.[72]

The integration of satellite and airborne snowcover data into the Soil Conservation Service Centralized Forecast System will be one more step in the journey that began with James Church's first ascent of Mount Rose in 1895. The many uses of snow survey data—reservoir management for irrigation, flood control, and hydropower generation; snow load maps for building design purposes; precipitation maps for regional planning; drought monitoring; fish and game management; winter recreation; acid precipitation monitoring; and avalanche forecasting—and the rapidly growing demand for water in the West make the snow surveyor a crucial figure in the coming century. Although the Omnibus Western Water Law of 1992 has nothing specifically to say about snow, it allows California's Central Valley Project water users to sell their water allotments at market prices as long as the sales pose no adverse environmental effects. It also creates a Western Water Policy Review advisory commission to study water resource and storage problems. Water stored as snow using snow fences or selectively cut forests is an option, less expensive and more technologically feasible than the proposed subsea pipeline that would bring Alaskan water to California. Whatever solutions are recommended and adopted, snow and water will require more human intervention. The number of federal, state, local, and private employees involved in water planning and management will grow beyond the present estimate of 410,000. There will be no business like snow business.[73]

The Names of the Snows

5

And here comes snow, a language
in which no word is ever repeated.
WILLIAM MATTHEWS, "SPRING SNOW"

QANIK TO SLUSH: FROM TWO KINDS OF SNOW TO TWELVE

Among the many problems of science, none is more fundamental than description. Description requires symbols, verbal or mathematical, which may or may not convey all the relevant details of the thing described but which take on a reality of their own. Thus it is with the descriptions and definitions of snow. The *Oxford English Dictionary* defines *snow* as "the partially frozen vapor of the atmosphere falling in flakes characterized by their whiteness and lightness; the fall of these flakes, or the layer formed by them on the surface of the ground." It traces the origin of the word to "Common Teutonic."[1] For more than a thousand years the word *snow* has been used by English speakers to describe a kind of frozen water, both in the air and on the ground. The words *hail* and *sleet* have been in use for almost as long, indicating a classification based on appearance and perhaps weather conditions. Meteorology, hydrology, glaciology, avalanche control, ecology, together with snow removal and skiing, have, in the twentieth century, redefined and reclassified traditional concepts of snow and have added hundreds of new names for types of snow. This is the subject of this chapter.

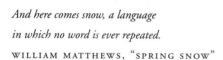

But don't Eskimos already have a dozen, 40, 200 words for snow? The reason for this frequently asked question is part of the history of snow and worth examining. "The Eskimo are among the best known of all native people of the North," write the anthropologists Nelson Graburn and Stephen Strong, "partly because of their unique and exotic way of life, and partly because their traditional lands extend over five thousand miles across the circumpolar region, embracing political domains of four nations—Russia, Alaska (USA), Canada, and Greenland (Denmark)." Although Eskimos, or Inuit as they are called, may be divided into as many as twenty-two regional groups displaying a great variety of adaptations to local conditions, they also exhibit considerable cultural and linguistic homogeneity, which is taken as evidence for their relatively recent (10,000 years ago, but thousands of years after the first migrations across the Bering Strait) arrival in and rapid dispersal across Alaska, Canada, and Greenland. Their remarkable adaptations to the Arctic have been studied for centuries by Europeans, with the result, as Graburn and Strong put it, that "the literature on the Eskimo and their land is more extensive than that of any other 'primitive' peoples in the world, in both professional and lay publications. For this reason we not only know more about past and present Eskimo life than about other peoples, but we hold more stereotypes about them which have to be modified."[2]

Chief among the stereotypes is that Eskimos have many words for snow while Americans have but one. The number attributed to the Eskimo has increased steadily over the past century and a half, from four in John Washington's *Eskimaux and English Vocabulary for the Use of Arctic Expeditions* (London, 1850) to "200 or some other large number," according to Elliott Norse in 1990. Moreover, the multiplicity of Eskimo words for snow has become a standard of comparison for those wishing to make a point about language and culture. Tom Horton, describing crabbing in the Chesapeake Bay, writes, "As Eskimos have many words for *snow,* and Arab Bedouins for *camel,* so do Smith Islanders for *soft crabs.*" A reporter commenting on types of crime in the 1990s, remarks: "Violence to the American has become as varied and omnipresent as snow to the Inuit."[3]

As early as 1890, the *English-Eskimo and Eskimo-English Vocabularies* compiled by Roger Wells for the Bureau of American Ethnography contained six snow-related words: two for falling snow, two for fallen, and one each for snowstorm and snowdrift. While it is obvious that the ability to discriminate among a variety of snow conditions is important for Eskimo, it is equally obvious that speakers of English and other languages can make fine distinctions when needed. The Russian naturalist A. N. Formozov lists at least two dozen from vernacular Russian and various Siberian languages, commenting:

> The word *nast,* usually little-known to city people, and which means a thickened crust on the surface of a mature snow cover, has a whole se-

> ries of synonyms. Thus, in Western Siberia, instead of *nast,* the word *charym* has been used; in Eastern Siberia—*chyr.* From the last root the verb *zachirat,* which means to cover with crust, has been derived. A thin, icy crust that breaks noisily under the weight of a man has been called by the hunter-traders of the northern part of Gor'ki Oblast' *skovoroda* (frying-pan) or *shorokh* (rustle).

The complicated snow terminology in Rumanian, resulting from borrowings from Slavic languages, has been explored by Yakov Malkiel and Pavel Sigalov. They found that Rumanians use a Slavic word for "snow" but a Latin-derived word for "to snow." They further trace the use of the Slavic form to an earlier, more limited meaning of "covered with snow" and conclude that Rumanian borrowings from Russian allow a more diversified description of snow conditions. Finally, American skiers have created more than seventy words for snow, as will be discussed in this chapter.[4]

The motives for attributing greater linguistic creativity to the Eskimo stem from a desire to criticize modern indifference to nature as well as a respect for the Eskimos' ability to survive in their harsh land. The impact of the Eskimo on American popular culture, from the motion picture *Nanook* in 1922 to the hundreds of cartoons depicting smiling hunters by snow igloos, deserves a book of its own, but the Eskimo vocabulary hoax, as one linguist terms it, can be explained briefly. Thanks to the research of the anthropologist Laura Martin, it is possible to trace the origins of the belief that Eskimos have multiple words for snow and to understand the confusion. The various dialects of the Eskimo language family have two distinct roots rather than words for snow. "In West Greenlandic," Martin writes, "these roots are *qanik* 'snow in the air; snowflake' and *aput* 'snow (on the ground).'" The distinction between root and word is important because Eskimo words "are the products of an extremely synthetic morphology in which all word-building is accomplished by multiple suffixation. . . . The structure of Eskimo grammar means that the number of 'words' for snow is literally incalculable, a conclusion that is inescapable for any other root as well."[5]

Martin traces the origins of the misunderstanding of the nature of Eskimo words to Franz Boas, who, in his introduction to *The Handbook of North American Indians* in 1911, cited four lexically unrelated words for snow—*aput,* snow on the ground; *qana,* falling snow; *piqsirpoq,* drifting snow; and *qimuqsuq,* snowdrift—and pointed out that Eskimo, like English and other languages, has some words that are independent and others that are derived from a single root. The matter rested there until 1940, when the Hartford Fire Insurance Company executive and linguist Benjamin Whorf published an essay asserting: "We have the same word for falling snow, snow on the ground, snow packed hard like ice, slushy snow, wind-driven flying snow—whatever the situation may be. To an Eskimo, this all-inclusive world would be almost unthinkable; he would say that falling snow, slushy snow, and so on, are sensuously and operationally different, different things to contend with; he uses different words for them and for other kinds of snow."[6]

The essay was reprinted the following year in S. I. Hayakawa's popular book *Language in Action* and continued to be reprinted and cited through the 1950s and 1960s. An excerpt from Whorf's collected papers appeared in the *Saturday Review*, February 4, 1961, under the title "You Can't Snow an Eskimo: In His Language, Slush Is Always, Irrevocably Slush." That slush is always *slush* in English seems to have been lost in the rush to prove that language limits perceptions of the world. In the context of postwar conflicts between colonizers and colonized, the idea that our perceptions of reality are limited by inherited categories was eagerly embraced by those who thought that it would be easy to broaden the categories and reduce cultural tensions. The Eskimo snow vocabulary example was, perhaps, a poor one, but it was used for the best of political reasons. Linguists will continue to be vexed by the popular notion that Eskimos have hundreds of distinct words for snow, but this belief is likely to persist.[7]

Eskimo vocabularies were not widely known when the International Meteorological Congress met in Vienna in 1873, however, and were not adopted in the set of terms and symbols for meteorological conditions that the U.S. Weather Bureau used. This classification included eight terms for frozen precipitation, each one further modified by symbols for small or large quantity. In addition to "snowfall," "hailstones," and "sleet," the Weather Bureau circular of 1894 identified "silver frost," an accumulation of snow and sleet on the limbs of trees; "glazed frost," an accumulation of snow and ice on trees in which ice exceeds snow so the appearance is smooth and transparent; "ice-needles," not yet defined by international usage; "drifting snow"; and "snow-covering." Responding to new concerns about snow conditions among railroad and highway maintenance personnel, and among those who maintain telegraph, telephone, and electric power lines, snow scientists began expanding snow terminology.[8]

The impact of new technology on the problem of defining winter precipitation, of separating sleet from snow, is illustrated in an article in *Monthly Weather Review* in May 1916, by Cleveland Abbe, Jr.:

> In undertaking to collect and discuss American statistics of the occurrence and the amount of ice coating or "glaze" (a term just adopted for the coating) deposited on electric transmission and other lines, the Weather Bureau had forced on its attention the prevailing diversity in the use of the terms "sleet," "ice storm," "glazed frost," "silver thaw," "glare ice," etc. As the phenomena bearing these names are all more or less of public interest, it is very necessary that our names for them shall be clearly defined and as specific as possible in application.

Abbe discusses the etymology of the word *sleet,* its early definitions, modern definitions, meteorological usage, and Weather Bureau usage. His primary concern is that sleet be restricted to mean "precipitation that occurs in the form of frozen or partly frozen rain" and distinguished from other forms of precipitation such as "soft hail" (the German, *graupel,* and French, *gresil*) and "glazed frost." This definition had apparently been adopted in 1897, changing an earlier definition that had equated sleet, soft hail, and graupel. Abbe's other concern is to distinguish the precipitation from the coating left by ice storms. Wires, he insists, are coated with "glaze," not sleet.[9]

Relying on the recently published *A New English Dictionary on Historical Principles* and some early American dictionaries for definitions, Abbe concedes that "the present divergent usages were thus early foreshadowed." These usages include snow and rain mixed together and a coating of ice on trees, sailing ships, and other objects. In response to a poll of forty professors, engineers, and meteorologists asking for definitions of *sleet,* Abbe was distressed to learn that engineers and managers in the wire-using industries "habitually call the ice-coating formed by cold rain, or snow, or rain and snow combined, or even 'small hailstones,' by the name of 'sleet.'" Usage among professors was even more heterodox. Robert DeC. Ward, professor of climatology at Harvard, preferred "'frozen rain' for what popular usage designates as hail, but which, occurring in winter, is really frozen rain drops without the concentric structure or origin of hail." Wilford M. Wilson, professor of meteorology and Weather Bureau official in Ithaca, New York, argued the need "to add to our meteorological vocabulary an expression to indicate the conditions when accumulations of ice occur on wires and other objects. It would seem that 'ice storm' is an appropriate expression and has the advantage of being already in use." Appropriately for a professor at the "E. A. Poe School of English" of the University of Virginia, C. Alphonso Smith wrote "that nothing is sleet that does not rattle on a tin roof or against the window pane. It seems to me that the element of sound is what differentiates 'sleet' from its winter congeners."[10]

Abbe goes on to review meteorological publications in several languages to demonstrate the confusion of terms, concluding that Arnold Guyot's instructions to Smithsonian weather observers in 1850 were correct in distinguishing between "small balls of snow, white and opaque, commonly without a crust of ice," and "little balls of transparent ice," but wrong in labeling the former "sleet." The official Weather Bureau definition of *sleet,* he reiterated, is "the precipitation that occurs in the form of frozen or partly frozen rain." A few months later, the *Monthly Weather Review* attempted to settle the issue by publishing a paper by the German authority Gustav Hellmann titled "Classification of the Hydrometeors." Following Abbe, Hellmann affirmed that *sleet* is the equivalent of the German *Eiskorner* and "consists of glass-hard, transparent spherules of ice that fall during the cold half of the year. The spherules strike the ground hard like bird shot, rebound elastically, and when they strike the dry Fall foliage make a rather loud noise [rattling] that at once attracts attention. Often the ice pellets are not round but rather angular or pointed." Sleet, Hellmann pointed out, is formed when raindrops fall from warmer to freezing air near the ground. Since it falls intermittently with rain, snow, and graupel, the atmospheric conditions in which sleet forms become an important part of its definition. Abbe's official definition removed sleet from the classification of snow and put it closer to rain in form as well as origin.[11]

Four more years of study by Charles F. Brooks and others restored snow as a possible source of sleet. Sleet is created by "partly melted snow or rain frozen while falling from a warm layer through a cold one," wrote Brooks, and differs from graupel, which forms and falls through freezing air, acquiring a coat of rime in the process, and from hail, which is an accretion of snow and rain formed by being carried up from a warm layer to a colder one. Although the issue seemed settled in 1920, the definition of *sleet* was more than a passing tempest. Thirty more years of conflicting definitions led to its rechristening as "ice pellets" by the International Commission on Snow and Ice in 1954.[12]

The 1920s and 1930s were years of rapid development in the study of snow.[13] Much of this work was summarized and codified by the English glaciologist Gerald Seligman in his 1936 book, *Snow Structure and Ski Fields.* This classic is still used, although some of Seligman's interpretations of the processes have been discarded. His contribution to the labeling and classification of snow is threefold: (1) close observation and photographs of snow forms; (2) a wide range of interests from meteorology and glaciology to skiing, and (3) a nomenclature based on theories of snowflake formation and the solidification of water. Early in his book, Seligman goes to the heart of the problem:

> There is a certain amount of indefiniteness in the nomenclature of snow and hoar deposits adopted by English physicists and meteorologists. I consider that the word "snow crystal," commonly used to describe snow, is not satisfactory, as it does not differentiate between falling snow and snow which has fallen to earth and lost its original structure. I therefore propose the term "snowflake" for all snow in the act of falling, whether it consists of bunches of crystals, single plate crystals, the more complicated dendritic structures, the beautiful "table" and "wheel" shapes, or the fine needle-like or prismatic forms. In order to differentiate the bunches of crystals, which stick together during their passage earthwards when the temperature is high, from other types, which are all composed of single crystals or single dendritic growths, it will be convenient to refer to the first-named as compound snowflakes and to call the remainder simple snowflakes or just snowflakes.

After a snowflake has reached the ground it quickly alters in character and to distinguish it from snowflakes on the one hand and from glacier ice on the other I suggest the group name "fallen snow." In this connection it is interesting to note that the Greenland Eskimos have two different roots for falling and fallen snow—*ganik* and *aput.*[14]

Seligman was less interested in falling than in fallen snow and offers no classification of snowflakes beyond the plates, columns, needles, and dendritic forms photographed by "Snowflake" Bentley. He recognized the role of temperature, humidity, wind, and nuclei in creating the various forms but left the falling crystals to cast his eye on the ground, where he found dozens of forms of solid water. Declaring that his categories are "not a classification but merely a list of all the ice forms met with in nature, roughly grouped in accordance with their more obvious characteristics and modes of formation," Seligman divided ice into four general categories: (1) ice formed by the freezing of water, (2) ice formed by the sublimation of water vapor, (3) ice formed by both freezing and sublimation, and (4) ice formed by changes in fallen snow.[15]

The fourth category, fallen snow, is the most complex, since snow on the ground is constantly changed by variations in temperature, wind, pressure, and moisture. For this category, Seligman constructed a branching diagram that divides fallen snow into powder snow and old snow, further dividing powder into new snow, settling snow, and settled snow, which, over time, become new firn snow and, ultimately, advanced firn snow, the two categories of old snow. New snow, "snow immediately after it has fallen and while it still possesses a fluffy, feathery or floury nature due to its not having changed materially from the flake condition," may be further subdivided into sand snow and wild snow. The former "is a type which has fallen at extremely low temperature so that neither ski nor sledge will glide in it," while the latter is a form "that has fallen in complete calm at low temperature and is immediately unstable." New snow and new firn snow may change into sun crust, wind crust, and rain crust as a result of melting and refreezing caused by those elements.[16]

Seligman went on to name another dozen forms of crust, each distinguished by its use or by the process of formation. Telemark crust "is the name given by skiers to any hard snow or crust thick enough not to be breakable, but sufficiently softened on top to permit turns being executed with the ski." Film crust is a layer of thin ice on top of hard snow but separated from it by a small air space. Perforated crust is a crust pitted by the sun's rays striking small depressions in the snow and causing the snow to melt and evaporate in the pit. Foam crust, *skavler,* and penitent snow are crusts carved out by the action of sun and wind. Obviously the latter forms represent the process of snowmelt in places where they are characteristic.

Foam crust, or plowshares, were identified by C. S. Wright and R. E. Priestly in Antarctica, *skavl* is found in the Nordic countries, while *nieve penitente,* pillars or columns of snow that remain standing as the snowcover melts, is a product of wind, sun, and terrain in the Andes. *Skavler* are similar to the Russian *sastrugi* and the Kobuk Valley Eskimo *kalutoqaniq.*[17]

The popularization of scientific interest in snow may be seen in two publications of the late 1930s. The Wisconsin novelist Frederic Prokosch's haunting novel of an imaginary journey across Central Asia, *The Seven Who Fled,* turns the process of firnification into a kind of poetry:

> He observed the soft slopes very gradually penetrated by the light and transformed from blue to silver. And as it grew more vivid and exact he could detect in the snow the most dazzling ornaments and devices, a million intricate shapes, a pure and crystal world unbelievably ornate and perfect. . . . A million crystals of infinite complexity, living for nothing else but the gradual destruction of their own perfect selves, growing slowly into each other, moving silkily downward during each moment of sunlight, motionless again at night, and then in the warm sun again becoming amorous and weak, like vast degenerate tribes drifting together, flowing away; demonstrating how close to one another were purity and decay, perfection and death.

The *Scientific Monthly* simplified Seligman's categories by announcing, "Twelve Kinds of Snow Recognized by Science." Reducing the number of Seligman's distinctions, the editors listed falling and fallen snow, powder, sand, wild, settled, new and old snow, sun crust, rain crust, firn ice, and glacier ice.[18] Seligman's "list" remained the standard in English until 1951, when the International Commission on Ice and Snow began to work on a new classification based on advances in snow science after World War II.

FROM SLUSH TO CRUD: SKIERS' CONTRIBUTIONS TO SNOW TERMINOLOGY

Selling Seligman's system to skiers was no schuss. As noted in Chapter 3, the Englishman W. R. Rickmers named twenty kinds of snow in his 1904 article on "ski-running," and his list was plagiarized in the first American ski book. Henry Baldwin, assistant forester for the state of New Hampshire, pointed out in a paper published in the 1938 that skiers had been developing their own descriptive terms for different snow conditions since the late nineteenth century. After reviewing the classifications of Fritz Huitfeld, Arnold Lunn, Vivian Caulfeild, C. G. Dyhlen, and others, Baldwin offered the system of Park Carpenter, editor of the *Ski Bulletin,* which had been published in Boston every Friday during the ski season since the early 1930s. Carpenter asked his volunteer observers in the mountains to report the depth of snow, the temperature, and the condition of the surface as of 8:00 A.M. Thursday. Surface conditions were divided into Soft Snow Conditions, Hard Snow Conditions (Crust), and Spring Conditions. Soft Snow was subdivided into eight types:

> *Fluffy.* Large, light feathery flakes. Scooped up by hand, it may be blown off like feathers. Even a tennis ball would drop out of sight in an accumulation of sufficient depth.
>
> *Powder.* Dry, cold, powdery snow; the perfect snow for sleighing and all winter sports. . . .
>
> *New.* "Fluffy" or "powder" snow that has fallen within 24 hours immediately preceding the observation should be described also as "new"; thus "new fluffy" or "new powder."
>
> *Heavy.* Occasionally a powder snow, long before changing to crust, becomes somewhat "packed," or "heavy," so that while skis sink in it readily, there is nevertheless a "toughness" to the surface which resists against lateral movement of the ski, causing slight difficulty in turning. . . .
>
> *Sticky.* Snow just beginning to melt. . . .
>
> *Wet.* A soggy state intermediate between "sticky" and "slush" reached after powder snow or wind crust has been thawing for some time. This is the "cotton batting" snow from which boys roll big balls, build forts and snow men.
>
> *Glazed.* A light, thin, transparent glaze frequently forms on any one of the above-described sorts of soft snow. It is the result of superficial melting by hot sun and subsequent freezing. . . .
>
> *Slush.* Requires no explanation.[19]

Carpenter remained closer to Seligman in his designations for Hard Snow, subdividing it into "Breakable," "Wind," "Common" (unbreakable), and "Icy." Spring conditions added only the term "Granular," for snow that had melted and refrozen night after night. In addition to being asked to name these conditions, the *Ski Bulletin*'s observers were told that it was better to underestimate the depth of the snow than to overestimate it. The reason for this advice, though unstated, is obvious. Skiers leaving Boston by train for New Hampshire on Friday afternoon arrived almost forty-eight hours after the ski report. Unless the forecast called for some accumulation of snow, they were likely to find even less than reported. As Baldwin and others knew, snow conditions change rapidly in response to local weather. Knowledgeable skiers knew their area and could make their own forecasts. In another paper presented to the Western Interstate Snow-Survey Conference in Davis, California, January 8, 1938, Baldwin elaborated on Carpenter's method of classification. Noting that the depth of snow varied greatly from open hillsides to the trails in sheltered woods, he recommended reporting for ski trails only. As for snow quality, Baldwin was critical of complicated classifications. "Bentley at Jericho, Vermont, did not see the snow for the crystals," he writes. "The opposite is true of most ski-runners." For this reason, Baldwin favored a simple "hard," "soft," and "breakable crust" reporting system. In a letter to Church, he went on to conclude that "a really sound classification of snow surfaces would have to be worked out by a committee representing the winter sport, highway, and water power interests in cooperation with physicists and meteorologists. I do not think I am competent to comment further."[20]

Baldwin and Carpenter were acutely aware that recreational skiing was attracting increasing numbers of participants and that these new weekend skiers were a powerful economic force in many small New England communities. Delays and inaccuracies in reporting snow conditions could result in disappointed, angry customers. Although ski clubs had been in existence for fifty years, the number of skiers remained small until railroad companies began luring city dwellers to the slopes with special ski trains in the 1930s. An estimated 2 million Americans skied in 1938, and the National Park Service, the Forest Service, and state agencies began to provide information and facilities for them. There was a growing recognition that snow conditions in the eastern United States differed from those in the Alps, where some Americans had learned to ski, as well as from those in the Midwest and the western United States.

Robert G. Stone, Harvard meteorologist and editor of the *Bulletin of the American Meteorological Society,* critiqued the situation in the same volume of the *Transactions of the American Geophysical Union* in which Baldwin's papers were published. *Glazed* is a poor term, Stone writes, because it might be confused with "true glaze, sleet, sun-crust, rain-crust, or incipient frozen granular, which are quite different, though all impracticable for skiing, but the description given for 'glazed' indicates it refers only to sun-crust, as it is known in Europe." The term *common crust* was "neither accurately descriptive nor in the vernacular," and the name *heavy* was too vague and overlapped with *wet* and *sticky.* The real problem, however, was not terminology but the different snow conditions caused by weather, topography, and season. The primary need was to educate American skiers in "snow-craft." Another factor causing a reevaluation of snow terms was the shift from Scandinavian cross-country skiing with some downhill runs to Alpine-style steep downhill runs, with rope tows and chairlifts. Moreover, the old descriptive terms, suitable for the skillful, were misleading and dangerous for the thousands of novices joining the ranks of skiers. As Stone puts it:

> It is also customary in the East to say only whether the skiing is good, fair, poor, etc. without giving any reason except the depth of snow-cover and some adjective such as slushy, hard crust, powder, etc. Now

different observers and skiers have different ideas of what is good or poor skiing, depending on their experience or training, as well as temperament. Often what looks like good skiing in these parts would be rated decidedly poor by many western skiers used to more alpine conditions. The possibility of standardizing the ratings is a knotty problem, and perhaps if science were able to do it, commercial interests would sabotage it, for obvious reasons.[21]

Democracy and capitalism hit the slopes.

Stone goes on to criticize even experienced skiers for misusing terms such as *corn-snow* to refer to all types of spring snow, rather than the specific dry granular firn with a layer of loose crystals on top that Seligman had labeled *Kornschnee.* "It is perhaps too much to expect that skiers turn amateur scientists, as they sometimes do in Europe, and answer the questions for us," he writes, "but it would be well to encourage any scientifically-minded skier to make a careful study of the friction of different snows on skis and waxes and of the relation of density and structure of the snow to the amount of speed and torque necessary to turn skis." Stone continues in this vein, urging skiers at least to understand the "evolution of snow from falling flake to eventual spring snow or firn ice, the diurnal effects of Sun and shade, and the ablational modifications from wind, Sun, rain, dust, temperature, and ground." The terms used, Stone writes, should be both descriptive of the appearance and structure of the snow, and grouped according to the genetic relations among them. Given the state of knowledge of snow metamorphism in 1938, a synthesis of morphological and process-oriented classifications was visionary. Sixty years later the problem is being resolved, but gaps in knowledge remain. The functions of snow typologies were also relatively simple in an era when a season's snows were sufficient for both recreation and irrigation. In conclusion, Stone offered a sage observation: "A multiplicity of terms in a classification need not be looked upon as a drawback in itself, since in practice very few of them are needed in any one section and time of the season, but the classification must at least cover all that are likely to occur in one region during any winter, which includes more than many people suspect."[22]

Discussion of snow terminology for skiers continued in the pages of the *Transactions of the American Geophysical Union* in 1939. Wayne Poulsen of the University of Nevada, reporting on a survey of skiing conditions on Mount Rose, near Lake Tahoe, for the purpose of locating a resort, observed that snow conditions in the Sierra depend on two factors: "(1) exposure and direction of the slope, and (2) elevation." With a few exceptions, west-facing slopes had the worst conditions; north, northeast, and east slopes the best; while south slopes had good conditions at higher elevations. At higher elevations the skiing was better in early winter and late spring. Since skiing before the development of lifts and runs was essentially a form of cross-country skiing, a trip over several kilometers meant that the skier encountered a great variety of snow conditions. With this in mind, Poulsen offered his own modification of the classification of snow.[23]

Poulsen's classification offers an interesting contrast to Seligman's and Carpenter's. Absent are the basic powder snow–old snow and soft snow–hard snow distinctions. Gone too is Carpenter's distinction between fluffy and powder. Carpenter had also added sticky and slush as snow conditions preceding and following Seligman's wet snow classification. Poulsen's definition of wet snow remains close to Seligman's and incorporates Carpenter's slushy. The simplification may reflect the fact that Poulsen, skiing at higher altitudes, in more variable conditions, and over a longer season, saw less reason to make fine distinctions among kinds of deteriorating snow, or it may result from Carpenter's attempt to educate New England skiers regarding the ablative process. While all the typologists agree with the Englishman Arnold Lunn that spring snow offers excellent skiing, Poulsen lumps several of Seligman's categories under this term. On crusts, too, there is general agreement, although West meets East on the term *glazed* in preference to Seligman's *Telemark snow* and Lunn's *Telemark crust.* Poulsen again follows Seligman rather than Carpenter in identifying pitted snow (penitente) and erosion forms, probably because snow conditions in the Sierra are closer to those of the Alps than to those of the mountains of New England. Finally, Poulsen recognizes the modification of snow by skiers themselves by creating the category of packed snow. Within a few years this would be the only type encountered by most weekend skiers.

Poulsen's system was simple compared with that developed by the Weather Bureau for the northeastern states in 1938. After dividing New England, New York, New Jersey, and Pennsylvania into seven districts, the bureau collected "Trail Observer's Reports" in each area every Thursday morning and disseminated the information from Boston on Fridays by radio, newspapers, and the wire services. The observers were given a form for each trail and asked to check off one of sixteen snow conditions. The list seems to combine the Carpenter and the Poulsen categories but changes "slush" to "thawing," perhaps, as Stone had remarked, to appease business interests whose customers might be driven away by unfavorable terminology. Observers were also asked to rate conditions for skiing, tobogganing, snowshoeing, skating, and iceboating on a scale from excellent to impossible.[24]

As the area of reporting expanded and as the federal government, through the Weather Bureau, National Park Service, and Forest Service, attempted to meet the needs of a variety of winter recreation interests, the naming of snow types became more complex and less specific. With hundreds of observers over thousands of square kilometers reporting for diverse purposes, standardization was impossible. Beginning in 1942 wartime travel restrictions curtailed the growth of skiing temporarily. In the postwar years changes in ski equipment, technique, resort management, and scientific knowledge obviated the old classifications and once again renamed the snows.

By 1989 the Aspen ski resorts advertised twenty-one kinds of snow with the comment "Eskimos may have more words for snow, but we have more lifts." Two years later *Snow Country* magazine came up with a list of thirty-two names for different snow conditions, while ignoring eight others listed in an advertising supplement in the same issue. Many of the terms used by Seligman, Carpenter, Baldwin, and others are omitted, but terms such as *ballroom, boilerplate, buffed snow, champagne powder, chowder, cold smoke, corduroy, crud, death cookies, mashed potatoes,* and *Sierra cement* demonstrate considerable powers of observation and imagination.[25]

SNOWFLAKES TO SNOW GRAINS: FROM A DOZEN TO 101 KINDS OF SNOW

It is a measure of the rapidity with which international science recovered from World War II that the Commission on Snow and Ice of the International Association for Scientific Hydrology, meeting in Oslo in 1948, set up a subcommittee on the classification of snow. The committee consisted of Vincent Schaefer of the United States, G. J. Klein of Canada, and Marcel de Quervain of Switzerland. Four years later their final report was accepted, and their classification remained in effect for the next thirty-seven years. As the committee was meeting, the National Research Council of Canada seized the initiative and produced two reports that bear on the history of snow nomenclature. The council invited de Quervain to tour North America and survey current research. His report, *Snow and Ice Problems in Canada and the U.S.A.,* published by the Division of Building Research in 1950, covers snow, glaciers, snowdrifts, avalanches, snow clearing, snow compaction, oversnow travel, river and lake ice, icing in the atmosphere, permafrost, snow as a water source, chemical compounds of snow, and winter sports. After describing the institutions for snow and ice research in North America, Europe, and Japan, de Quervain briefly discusses the methodology of snow classification. Later in the same year, G. J. Klein and his colleagues published a technical memorandum for the council called *Method of Measuring the Significant Characteristics of a Snow-Cover,* in which they also discussed some of the issues in snow classification. These two papers help put the final report of the international commission in perspective.[26]

De Quervain makes three central points. First, the classification should be useful to "any kind of snow observation and investigation . . . and allow refinements for special requirements." Second, "the snow properties should be defined by figures rather than by terms." Finally, definitions should be based on a minimum number of quantifiable properties. Klein's approach to nomenclature for snow appears to be more descriptive than process oriented. His classification follows Seligman closely, adding details of specific gravity to "dry new snow," "wet new snow," and various types of old snow. He frankly admits that the tests conducted by the National Research Council address problems associated with transportation, skiing, and snow removal and are concerned with falling snow "only because of its relation to snow in the surface layer."[27]

The classification developed by Schaefer, de Quervain, and Klein and approved by the Commission on Snow and Ice in 1952 is impressive in its comprehensiveness and ingenuity. Given the state of knowledge of snow metamorphism and crystal growth in 1952, it is somewhat surprising that it is descriptive rather than genetic, but its longevity attests to its usefulness. The classification has four sections: "solid precipitation," "deposited snow," "snow cover measurements," and "snow surface conditions." From the first paragraph, however, the authors qualify their categories: "When a sharp distinction is drawn between falling and deposited particles, the term applies to precipitation while it remains airborne; but in the classification presented here, 'solid precipitation' is taken to also include freshly deposited particles which have not undergone any perceptible transformation subsequent to being deposited on the earth. Neither English, nor Eskimo, can resolve the problem of the dichotomy between falling and fallen. When does a falling snowflake become fallen snow? Is there a single metamorphosis or two?"[28]

The classes of solid precipitation are essentially Schaefer's of 1945, rearranged, clarified, and given graphic symbols. Plate, stellar crystal, column, needle, spatial dendrite, capped column, irregular column, irregular crystal, graupel, ice pellet, and hail are the ten classes, which may be modified by four features: broken crystals; rime-coated particles not sufficiently coated to be classed as graupel; clusters, such as compound snowflakes, composed of several individual snow crystals; and wet or partly melted particles. Further, the classification calls for recording the size of each particle.

The second section of the commission's classification, "Deposited Snow," is based on the physical characteristics of "a mass of snow," which is composed of "ice, air, and water." The physical characteristics are essentially de Quervain's greatly elaborated, defined, and assigned alphabetic symbols. Specific gravity "is the ratio of the weight of any volume of a substance to the weight of an equal volume of water and is therefore non-dimensional. Density may be used as an alternative. It should, however, be expressed either in grams per cubic centimeter or in kilograms per cubic meter." Free water content, "expressed as a percentage per weight," is subdivided into dry, moist, wet, very wet, and slush. Impurities, such as dust, sand, organic material, and salt, become an important subclass if they influence the physical characteristics of snow.[29]

The next three primary features of deposited snow are structural—grain shape, size, and strength. Grain shape is arranged into five subclasses according to age on the ground and degree of melting, from freshly deposited snow to depth hoar. Each class is further distinguished by the metamorphosis of the original crystalline character of the snowflakes. By the third stage, or class, it has "lost all crystalline features and its grains become more or less rounded," the result of the temperature gradient. The final two classes have irregular grains with flat facets and hollow, cup-shaped crystals. Grain size is divided into five subclasses, from very fine (less than 0.5 millimeter) to very coarse (greater than 4.0 millimeter, about the thickness of a nickel). A five-part measure of strength is also used, with a sixth class for ice. The measurement is complicated, however, by the fact that strength can be determined by four different tests—compression, tension, shear, and hardness.

Snow is compressed by the weight of more snow falling on top of the snowpack, by wind, by the weight of skiers or vehicles, and by reduction of surface energy. Harder, denser snow supports more weight without collapsing. Snow is a complex material, responding to compression by becoming a solid mass of ice, by crumbling and melting, or by squeezing into new shapes. In certain conditions it will partially recover its original shape when the stress is removed. On pitched roofs and mountainsides a shear component comes into play. The steeper the pitch of the incline, the more the force of gravity acts to offset the compressive forces holding the snow to the surface. The pitch increases the tension and shear stress and raises the potential for the snow to slide. Just as the slightest bump on a roof will hold snow longer than a smooth surface, rocks, trees, and the rise and fall of a mountainside all act to cause variations in compression, tension, and shear.

There are several field tests for hardness, the simplest being the insertion of fist, fingers, pencil, or knife blade into the snow at 5.0 kilograms (11.0 pounds) of force. In North America avalanche workers apply 1.5 kilograms (3.3 pounds) of force, because softer layers may be missed using greater force. If you can push your fist into the snow, it is very soft; if it resists a fist but allows the insertion of four fingers, it is soft. If the snow resists four fingers but can be penetrated by one finger, it is medium hard. Snow too hard to penetrate with a finger but penetrable with a pencil is hard. Snow is called very hard if it can be penetrated only by a knife blade.[30]

Thus, the Commission on Snow and Ice provided a "basic framework which may be expanded or contracted to suit the needs of any particular group ranging from scientists to skiers." It also allowed for a combination of visual observations and testing by instruments. At its simplest, the commission's classification named ten kinds of falling solid precipitation, differentiated chiefly by shape. Shape also identified five basic kinds of deposited snow. The potential kinds of snow surfaces could be multiplied into the hundreds. As for snow in the atmosphere, the commission noted that its classification section 1, solid precipitation, "is not intended to replace that part of the International Meteorological Code which deals with snow, hail, etc., and care was taken to avoid any conflict between the two systems." This deference to the meteorologists indicates recognition of the important role their science played in early studies of snow, but it ignored existing differences among meteorologists in the naming and classification of snowstorms and related weather.[31]

While weather forecasters and applied meteorologists struggled to devise a vocabulary that would at once define the salient features of frozen precipitation and other weather phenomena and communicate to the public the information it needed to run farms, manage power plants, or plan trips, hydrologists, geophysicists, meteorologists, and glaciologists made new discoveries about the properties of snow that required new names and classifications. Ukichiro Nakaya, Choji Magono, and C. W. Lee in Japan, B. J. Mason in England, Lorne W. Gold in Canada, and R. A Sommerfeld, Edward LaChapelle, and others in the United States all contributed to our expanding knowledge of snow crystals. Nakaya, who began his work in Hokkaido in the 1930s, came up with seven principal types, which differed from the International Commission's ten types in several ways. Nakaya omitted hail. He gave needle crystals, columnar crystals, and rimed crystals such as graupel their own categories and created a separate category for "Columnar Crystals with extended side plates." Plate, stellar, and spatial dendrite crystals were placed in his "Plain Crystal" category. Capped columns were part of a "combination of Column and Plate Crystal" category. Irregular crystals and ice pellets appear in his "Irregular Snow Particle" category.[32]

When Magono and Lee published their classification in 1966, they identified eighty types. Although their system was based on falling snow, it was popularized by Edward LaChapelle in his *Field Guide to Snow Crystals,* which went through several printings after 1969. LaChapelle had been studying glaciers and avalanches since the early 1950s and had come to the conclusion that there was a relationship between avalanches and certain types of snow crystals. He was also attempting to bridge the gap between crystals observed under a microscope in the laboratory and what could be seen in the field.

As LaChapelle points out, Magono and Lee added subcategories to all seven of Nakaya's main types, and they introduced a new category, "germ" crystals, which are snow crystals in the earliest stage of formation. To Nakaya's three divisions of needle crystals, for example, they added two types of sheath crystals (hollow needles), a solid column, and three types of combinations of needles, sheaths, and columns. Columnar, plate, and combination crystals are similarly subdivided, as are rimed crystals. Nakaya's seventh category, irregular snow particles, which includes the catch-all subcategory "Miscellaneous," was refined by Magono and Lee to include "broken branch" and "rimed broken branch" types. LaChapelle considered this one of the greatest improvements on the older system because these types occur frequently.[33]

LaChapelle earned a degree in physics from the University (then College) of Puget Sound and did advanced studies at the Swiss Snow and Avalanche Research Station, where he continued the work of Henri Bader and others on snow metamorphism. In 1952 he began research on snowcover and avalanches in Alta, Utah, under the direction of Montgomery Atwater, of the U.S. Forest Service. In 1957 he joined the Faculty of Geophysics and Atmospheric Science at the University of Washington in Seattle. After fifteen years of research on problems such as ablation from settlement and subsurface melting, electrical forces in the development of snow cornices, density and tensile strength in snow, and deep slab avalanches by himself, John Montagne, Charles Bradley, Richard Sommerfeld, and others, LaChapelle was ready to offer a new classification of fallen snow. "The Classification of Snow Metamorphism," written with Sommerfeld, who had already published a classificatory outline of his own, appeared in the *Journal of Glaciology* in 1970.[34]

The authors begin by attempting to clarify the term *crystalline,* which refers to solids with atoms "arranged in an orderly way." Thus, snowflakes, metamorphosed snow, and ice are all crystalline. Since snowcover is an aggregate of a variety of crystals, it is polycrystalline. "The individual crystals in a polycrystalline material," they point out, "are called *grains* and are separated by *grain boundaries.*" Once snow begins to bond or pack, the grain boundaries are difficult to detect, especially under field conditions. In newly fallen snow, each grain may be a single crystal, but metamorphism soon turns the grains polycrystalline. Because of this relationship of grains and crystals, Sommerfeld and LaChapelle recommend using the term *grain* for both and describing a snow layer as "*small, medium,* or *large grained.*" From these assumptions, they offer a genetic rather than morphological classification of fallen snow.[35]

Their four-part system involves a first stage of unmetamorphosed snow and three kinds of metamorphism. Unmetamorphosed snow consists of three types: crystals newly fallen in still air; windblown shards and splinters of the original crystals; and surface hoar, or water vapor in the atmosphere deposited on the snow surface. Subclassifications of the first two types follow Magono and Lee. Sommerfeld and LaChapelle label the next stage "equi-temperature metamorphism" and divide it into "decreasing grain size" and "increasing grain size." Because equi-temperature metamorphism assumes a uniform, subfreezing, temperature in the air, throughout all depths of the snow, and at the surface of the soil, it is an ideal condition found in the laboratory rather than in nature. Its importance to a genetic classificatory system is that it involves a distinct physical process, a theoretical possibility in which a single crystal decreases or increases in size until it reaches a state of equilibrium. This process, sometimes called "destructive metamorphism," takes place because water vapor flows from small to large particles, from negative (concave) to positive (convex) surfaces, and from warm to cold surfaces.[36]

The third part of the Sommerfeld-LaChapelle classification and the second kind of metamorphism is called temperature-gradient or "constructive metamorphism." As with equi-temperature metamorphism, the key factors are time and temperature. The process is divided into "early" and "late" subclassifications, and the mechanism is the flow of heat and water vapor from the warmer (lower) layers of snow up toward the colder layers. As the warmer crystals evaporate and water vapor escapes upward, the colder grains grow and the facets of the crystals may become visible, eventually forming a layer of loose grains called depth hoar. Since all but wet snow undergoes a temperature-gradient metamorphism from the moment it strikes the ground, this process is a major factor in creating the conditions found in the field. It is as complex a process and creates as great a variety of snow crystals as the atmospheric formation of snow. There is, in fact, continuity between falling and fallen snow because the type of crystal that falls influences the type that forms on the surface.[37]

The final category in the Sommerfeld-LaChapelle system, firnification, actually consists of two processes, melt-freeze metamorphism and pressure metamorphism. Firnification is defined as a final stage of metamorphosis when the snow grains approach the density of ice. In melt-freeze metamorphosis, snow melts, and water is trapped between grains, refreezes, and adheres to them. The resultant mixture of ice and snow can be very dense and hard, but grains still retain their identity as single crystals. The density of the snowcover is also increased by pressure as annual accumulations of snow occur. This is the process of glacier formation. Sommerfeld and LaChapelle renamed the types of fallen snow identified by Seligman in 1936 based on new understandings of the physical process of metamorphism. In so doing, they shifted the focus from the surface of the snowpack toward the lower layers. This shift was appropriate for those interested in avalanche prediction and control and contributed to a better understanding of hydrologic processes as well.

By creating a genetic classification emphasizing the ongoing process of crystal transformation, Sommerfeld and LaChapelle attempted to bring to the study of fallen snow a theoretical and methodological sophistication equal to that reached by Nakaya, Schaefer, and others on the life history of the snowflake. Ironically, in offering their system as an extension of the International Classification of Snow, they also hardened the distinction between the systems. LaChapelle makes this explicit in his *Field Guide to Snow Crystals:*

> Classification of metamorphosed snow presents a problem different from that of precipitated snow. In one way the problem is simpler, for there are fewer distinctly different forms to consider. The crystals of precipitated snow are the end products of different growth regimes in the atmosphere, but the various types of metamorphosed snow are simply stages in ongoing processes. Classification thus has to take into account stages of evolution rather than end products. Glacier ice is the only true end product of snow metamorphism, and even glacier crystals continue to evolve.[38]

Crystals of precipitated snow are end products only if we choose to define them as such. The dichotomy between falling and fallen snow is an artifact of language and culture. There are, as LaChapelle acknowledges, other end products of snow metamorphism; they are called vapor and water. For students of snow, *melt* is a four-letter word. Eschatology is not one of the snow sciences. Nevertheless, all sciences limit their study to what is defined as real until forced to redefine those boundaries by the discovery of new data or the invention of new theories.[39]

Both things happened in the 1970s. New data on storms and snow crystal formation in the atmosphere were obtained from satellites and aircraft. Satellites and other remote-sensing devices collected data on snowcover in quantities that permitted complex mathematical modeling by computers. The use of computers to model everything from "cloud droplet accretion on snow crystals" to "freezing precipitation in winter storms" raised the possibility of new classifications of snow based on finer and finer distinctions.[40]

Research on snow in the past decade has led to a new classification for seasonal snow on the ground by Samuel Colbeck of the Cold Regions Research and Engineering Laboratory in Hanover, New Hampshire. Colbeck studied physics and petroleum engineering at the University of Pittsburgh and received his Ph.D. in geophysics from the University of Washington, where his research focused on the mechanics of glacier flow. Since that time he has done research on almost every aspect of the physical properties of snow.

In a 1982 paper, "An Overview of Seasonal Snow Metamorphism," written for a U.S.-Canadian Workshop on the Properties of Snow held at Snowbird, Utah, Colbeck proposed a relatively simple four-part classification based on the distinctions between falling and fallen snow and between dry and wet snow. The terms *dry* and *wet* are more technical than they first appear, since their definition depends on temperature as well as on the amount of liquid water, although wetness is clearly a result of temperature above freezing. While Colbeck's classification is, like Sommerfeld and LaChapelle's, process oriented rather than morphological, it focuses more closely on grains and grain particles recognizable with a 10x lens, thus making the system accessible to the field observer. By using water content and temperature as his chief determinants of classification, Colbeck emphasizes the bonding of crystals and consequently the permeability of snow and qualities such as stickiness and slipperiness, of practical interest to hydrologists, skiers, and road and power line maintenance personnel.[41]

Less than four years after the Snowbird workshop, Colbeck published "Classification of Seasonal Snow Cover Crystals," a paper explicitly challenging the older classificatory systems. In a clear opening statement, he reviewed why he thinks snow terminology and classification are necessary and why the task is so difficult:

> The terms used to describe snow are important to all aspects of snow studies and snow applications. The variety of interests and needs complicates the construction of a comprehensive classification system unless it is based solidly on the physical principles governing all aspects of snow metamorphism. For example, permeability is a basic parameter in many snow studies and is very sensitive to the type of crystals present. A permeability measurement should include a meaningful description of various snow parameters such as the crystal type to facilitate transfer of the permeability value to other situations.[42]

After disposing of the troubling distinction between *crystal* and *grain* by opting to use "'snow grains' to refer to the basic visible unit, whether it is a single crystal or not," Colbeck outlines the processes of snow metamorphism.

The review follows his 1982 "outline," adding detailed criticisms of the previous classifications. The International Commission "system does not classify snow crystals with sufficient detail to distinguish among the multitude of crystals that are observed in the seasonal snow cover. The processes leading to the observed shapes are barely described at all, although the system does describe many features of snow." Colbeck repeats his criticism of Sommerfeld and LaChapelle for giving equi-temperature metamorphism equal status with temperature-gradient metamorphism, since even in isolation a snowflake goes through phases that require temperature gradients. He also argues that a classification can be both process oriented and morphological because "more than just the shape is communicated by the label attached to snow grains."[43] Finally, he presents evidence that melt-freeze and pressure metamorphism do not account for all forms of wet snow.

Figure 5.1. Ice crystals in dry snow are usually well rounded and bonded together to form a lattice. These are 0.1 to 1.0 mm. Compare this representation of snow with the photographs of W. A. Bentley in Figure 1.9. Most snow crystals fall as irregular clumps, not symmetrical hexagonal flakes, and their metamorphosis in the snowcover produces new crystals of great variety and complexity. Photograph courtesy of Sam Colbeck, Cold Regions Research and Engineering Laboratory, Hanover, N.H.

Colbeck's 1986 classification, like earlier systems, is a combination of process and descriptive categories. Precipitation, dry snow, wet snow, and surface-generated features constitute his four basic forms, with dry snow subdivided into equilibrium (rounded) form, mixed rounded and faceted, and kinetic growth (faceted) forms. Wet snow is divided into pure grain clusters, melt-freeze particles, and slush (Figures 5.1 and 5.2). The three forms of dry snow are further divided according to processes that cause them to become rounded or faceted, solid or hollow, hexagonal plates or prisms. Surface-generated features include surface hoar, wind crust, melt-freeze layers, sun crust, and freezing rain crust. In moving beyond his earlier "outline," Colbeck makes two important innovations. First, he constructs a "generalized scheme of metamorphism of dry snow showing effects of temperature gradient and time," from a falling (precipitated) snowflake to a depth hoar crystal of mixed rounded and faceted form shaped by a decreasing temperature gradient. Second, he integrates the mixed snow particles or grain clusters into the dry snow metamorphic processes. Colbeck attempts two things in this classification: to make a system useful to both laboratory and field scientists; and to make the dry snow–wet snow distinctions consistent so that grain clusters that resemble one another, such as the melt-freeze particles of wet snow and the transitional mixed rounded

Figure 5.2. Ice crystals in wet snow form into clusters with liquid water in the crevices and pores in the interior. The individual crystals are 0.5 to 1.0 mm. Photograph courtesy of Sam Colbeck, Cold Regions Research and Engineering Laboratory, Hanover, N.H.

and faceted crystals found near the surface, where they are evaporating, will be adequately explained if the history of the snowcover is known or can be inferred.[44]

Colbeck's efforts contributed to the creation of a new committee of the International Commission on Snow and Ice of the International Association for Scientific Hydrology. After three years of study, the new classification for seasonal snow on the ground was published. It retains many of the features of the 1952 classification but expands it and synthesizes the morphological and the process-oriented classifications in a "framework which may be expanded or contracted to suit the needs of any particular group ranging from scientists to skiers."[45]

The primary features of deposited snow—density, liquid water content, impurities, grain shape, grain size, strength, hardness, and temperature—remain almost the same, although they are given some new symbols. Liquid water content is defined more clearly, from 0 percent for dry snow to more than 15 percent for slush. Grain size is also refined, reducing the size of very fine from less than 0.5 millimeter to 0.2 millimeter, of medium from 1.0 to 2.0 millimeter to 0.5 to 1.0 millimeter, and of very coarse from greater than 4.0 to greater than 2.0 millimeters. The most extensive changes are in grain shape. The new classification is divided into (1) morphological classification, (2) process classification, and (3) additional information on physical process and strength. Each of the twenty-nine subclasses of snow is given a graphic and alphabetic symbol. Process-oriented classification is divided into place of formation and process. Under additional information the new system provides details on the physical processes; dependence on most important parameters, such as temperature, wind, and humidity; and common effect on strength.

The complete classification of twenty-nine kinds of snow contains eight falling and twenty-one fallen. If these twenty-one are added to Magono and Lee's eighty kinds of falling snow crystals, we have 101 names for snow. It can be argued that these are merely grain shapes, some observable only in the laboratory and meaningful only to the initiated, that they lack the rich connotations of Eskimo words evoking the flowing, sculpted forms of drifting snow or the pendulous droop of a snow-covered pine branch. Yet the significance of the century-long orismology of snow is both scientific and poetic. The search for understanding, the desire to know and to name, to organize and to classify is based on the assumption of the orderliness of nature, and order has an aesthetic dimension. Order is both discovered and invented. This is why no system of classification—descriptive or genetic—can last forever. Names will change; even the definition of the reality for which the names are bestowed will change (Figures 5.3 and 5.4). The story of the names of the snows is familiar in many fields of study. It is fundamental to epistemology

Figure 5.3. Scanning electron microscope image of a radiating assemblage of hexagonal plates (collected at –5°C, 23°F). Photograph 1995 by Albert Rango, Agricultural Research Service, Beltsville, Md. Using a scanning electron microscope to produce three-dimensional images of snow crystals, scientists have observed dozens of previously unnamed forms.

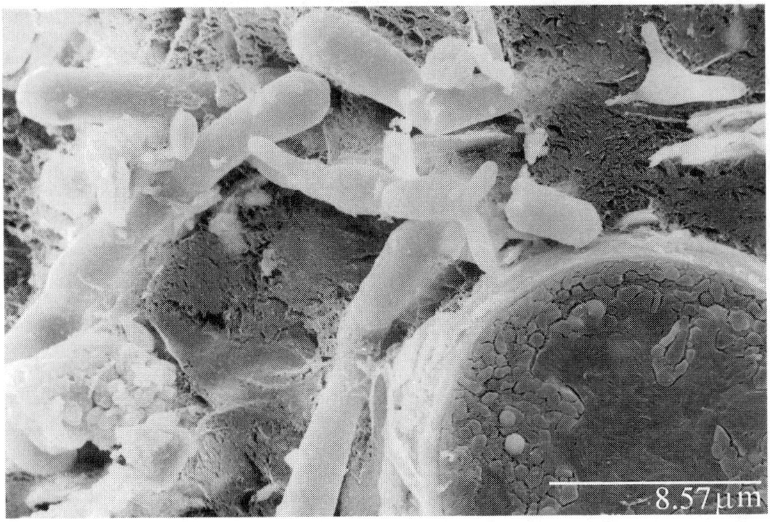

Figure 5.4. Scanning electron microscope image of "red snow" from a melting snowpack showing a portion of a fractured algal cell and other associated microbial forms (collected at a snowpack temperature of 0°C, 32°F, stored for three days in a dry shipping dewar). Photograph 1995 by Albert Rango, Agricultural Research Service, Beltsville, Md. This picture of the interior of a snow crystal was made at a magnification of 3500×. The arrangements of frozen molecules are perceived differently by different instruments.

and phenomenology. It continues to fascinate us because we find in lists some of the certainties we lack in daily life. The ability to recognize differences and likenesses allows us, as Barry Lopez has observed, to "talk clearly about the world."[46]

Snow, gentle, familiar, homey snow, whose very chaos stimulates the search for order, requires many names, a language, as the poet William Matthews says, "in which no word is ever repeated."[47] Naming the snows, whether by meteorologists, geophysicists, skiers, or poets, is a creative act, an affirmation of the cycles of snow accumulation and ablation that have shaped the earth for millions of years. Nature and culture exist in a moment between creation and destruction, for which one of the names is snow.

Ecology of the Snow Commons

6

The leading colonists of summer,
Carriers of what we called progress,
Uplift, or flat success,
Have gone south taking their plunder.
All crucial witnesses

Are safely hushed-up underground
Or live on on the season's scraps.
Thick snow blots out the maps;
The woods, the air, the memory's found
Compromised by gaps.

W. D. SNODGRASS, "SNOW SONGS"

NIVEAN FRONTIERS

Frederick Jackson Turner's frontier line of 1893 had two dimensions—one between civilization and wilderness and one that changed over time as the West was settled. The snow frontier line, first mapped in 1888, has a third—a vertical dimension within the snowcover. Like the old frontier, the nivean frontier is usually perceived as a commons, in the sense that Garrett Hardin popularized the term, a resource to be used for the benefit of all.[1] But does snow belong to the owner of the property on which it falls? Or is it a kind of unreal estate—seasonal, ephemeral, and transcendental? This question needs to be asked because we annually make and destroy snow without fully considering the claims of all its consumers. This chapter looks at the living snowpack—winter's deep freeze for the food chain—the efforts of ecologists to understand the role of snow in the biosphere, attempts to increase the snowpack by weather modification, and the pollution of snow from various sources.

Interest in snow expanded in the 1880s to include the microscopic, the algid animalcular as well as the chill embargo of blizzards. Wilson Bentley began making photomicrographs of snow crystals in 1884, and William Gibson and Frank Bolles wrote eloquently on weeds and insects in snow.

Alpinists had long noted the presence of red snow, and early arctic explorers brought back samples of colored snow. By the 1880s one of the common causes of red snow had been identified as a one-celled alga, named *Protococcus nivalis,* of the order Coccidia, a minute parasitic protozoa living in snow. Under microscopes other samples of snow revealed the presence of volcanic dust, vegetable mold, tree pollens, and various chemicals, yielding a rainbow of snow colors: black, brown, yellow, and green as well as red. During the drought years of the 1930s, brown snow fell in New England, colored by soil from the Dust Bowl of the Midwest. William Thomas, a research biologist at Scripps Institution of Oceanography, has found a potentially important relationship between the algae *Chlamydomonas nivalis,* which causes red snow at Tioga Pass in the Sierra Nevada, and bacteria that feed on organic molecules excreted by the algal spore. This tiny link in the food chain may, in turn, affect snow albedo (reflective power) and ultimately larger aspects of forest ecology. In the 1990s brown snow in the Central Canadian Arctic was found to contain hydrocarbons and pesticides from Eurasia.[2]

There was more to life on the nivean frontier than chemicals and cryoplankton, however. As early as 1847, Asa Fitch, who became the state entomologist of New York, identified and named a species of snow flea, *Podura nivicola,* but it was not until the end of the century that entomologists truly understood the rich insect life in the snowcover. In 1902 Justus Watson Folsom, an instructor in entomology at the University of Illinois, expanded Fitch's classification of the Collembola, *P. nivocola.* Folsom concluded that Fitch's description of the black, wingless insect (about the size of a period on this page) that is often seen in great numbers in the snow during winter and early spring thaws was too broad. Using time of appearance, abundance, habitat, and habits as well as anatomy, Folsom identified two new species of *nivicola* and renamed Fitch's snow flea, *Achorutes nivicola* Fitch. He also noted that the same species was known in Europe as *Achorutes socialis* Uzel and *A. spinifer* Schaffer. In Canada it was described in 1891 as *A. nivicola* MacGillivray, all of which illustrates the confusion that resulted from energetic, uncoordinated, and self-aggrandizing taxonomy.[3]

Taxonomy was also a major concern of the early zoologists and resulted in naming and renaming many species of smaller mammals, such as squirrels, shrews, and mice. Few of the early students of either plants or animals were specifically interested in their relation to snow, but a contribution to this topic was made by the popular writer and illustrator Ernest Thompson Seton in *Life-Histories of Northern Animals,* published in 1909. Although denounced by Theodore Roosevelt as a "nature faker" and damned with faint praise by the naturalist John Burroughs, Seton was a serious student of animals in the wild. Born in England in 1860, he grew up on a farm in Ontario, studied art in Canada, England, France, and New

York, and spent five years hunting, sketching, and collecting in Manitoba in the early 1880s. His drawings illustrated numerous ornithological books, and in the 1890s he began to publish the sentimental animal stories that earned him thousands of dollars and Roosevelt's scorn. He traveled extensively in the wildernesses of North America, finally settling near Santa Fe, New Mexico, in 1930 and becoming a citizen of the United States.[4]

The effects of snow and cold on each of the fifty-nine animals described in *Life-Histories* are considered, often illustrated, and sometimes supplemented by anecdotes gathered from travelers' books, friends, and personal experience. It is easy to see why Seton was popular, and his work must have heightened snow consciousness. Describing the habits of the common shrew (*Sorex personatus* St. Hilaire), he quotes the naturalist E. W. Nelson:

> After snowfalls they travel from place to place by forcing a passage under the snow, and frequently keep so near the surface that a slight ridge is left to mark their passage. On the ice of the Yukon [River] I have traced a ridge of this kind over a mile, and was repeatedly surprised to see what a direct course the Shrews could make for long distances under the surface. . . . These little adventurers sometimes tunnel far out on the sea-ice, and the Norton Sound Eskimo have a curious superstition connected with such stray individuals. They claim there is a kind of water Shrew living on the ice at sea which is exactly like a common land Shrew in appearance, but which is endowed with demoniac quickness and power to work harm. If one of them is disturbed by a person it darts at the intruder, and burrowing under his skin, works about inside at random, and finally enters the heart and kills him.

Seton also describes the snow-tunneling activities of the pocket gopher and the red squirrel. He writes of some squirrels he saw who had built a "snow-drift playground . . . a perfect labyrinth of galleries in a drift that was twenty feet long and six feet wide. This had ten entrances leading to chambers and passages innumerable, and on very cold days they evidently played tag here instead of in the tree tops. Around the entrances I found the remains of nuts and pine-cones, so maybe somewhere in the snow-drift was a feasting place—their winter palace was banquet hall as well as gymnasium; but I could not examine it fully without destroying it, so left it alone."[5]

Seton's paintings and drawings of animals in snow are often filled with detail. *Moose Family in Early Winter,* for example, shows a bull, a cow, and two calves in a grove of birch and pine (Figure 6.1). The cow's left foreleg is raised in the act of scraping snow from the ground for forage. A pair of jays, one perched, one flying, wait in the expectation that the moose will uncover something for them to eat as well. The most striking detail is the way the mooses' legs mimic the trunks of the saplings and underbrush, camouflaging the animals, especially the calves. Seton does not comment on camouflage, except to note seasonal color changes in some northern animals, but he may have been aware of the work of his contemporary the artist Abbott H. Thayer, whose *Concealing-Coloration in the Animal Kingdom* was also published in 1909. Thayer, who lived in New Hampshire, was a keen observer of snow who propounded a theory of natural selection based on the colors in an animal's habitat. He dismisses Darwin and other naturalists as deficient in understanding optical illusions and concludes his introduction with a challenge to the reader to conduct an experiment: "If, like a multitude of people, one cannot see that shadows on an open field of snow, or a white sheet, under a blue sky, are *bright blue* like the sky overhead, one will probably prove more or less defective in all color-perceptions. To prove that such shadows *are* colored, lay a colorless mirror on the snow in such a shadow—its reflected sky will match the surrounding snow."[6]

Thayer uses the blue jay as one of his illustrations, showing that the bird's blue, white, and gray coloring obliterates it among ever-changing shadows and sunlight on snow.

Figure 6.1. *Moose Family in Early Winter,* Ernest Thompson Seton. Illustration for *Life-Histories of Northern Animals,* 1909. Reproduced from the Collections of the Library of Congress. Seton comments on the complex interrelation of vegetation, mammals, and birds in the snow.

Seton's jays take on additional meaning when placed in the context of Thayer's work. They become a visual comment on the protective coloration of moose, a subject generally ignored in considering such species, which are assumed to depend on their size and strength, not camouflage, for survival. Thayer had other points to make about winter coloration. "Except when a wet snowstorm or an icestorm has plastered and veiled these twigs, the average northern landscape in winter is full of great masses of soft, purplish red, reaching here and there a brighter tint. Golden brown, varying to red and purple, is also the color of the cones of spruce and pine and fir trees. It is among these pink and bronzy twigs and buds, seed tassels and cones, that the northern grosbeaks, linnets, and crossbills get their food, and the red or reddish colors worn by many of them are therefore in full accord with their environment."[7] Seton and Thayer, like the Canadian painters discussed in the next chapter, discovered the chromaticity of snow, but unlike the landscape painters, they saw abundant animal life. Their contribution to the natural history of snow continues in the work of Robert Bateman and other wildlife artists.

Despite the interest of artists in the effects of snow on plants and animals, John Harshberger of the University of Pennsylvania wrote in 1929 that "little attention has been given to this phase of botanical investigation in America." In the 1920s, however, American botanists, entomologists, and zoologists began exploring the nivean frontier with the same enthusiasm as their colleagues in meteorology, hydrology, glaciology, and chemistry. It was no longer possible to treat snow as inorganic. Even snow crystals that did not form around organic nuclei in the atmosphere were soon infused on the ground with microscopic plants and animals, becoming an ice cream parlor of flavors in the food chain. From the beginning of the twentieth century, the biological sciences took an ecological approach to snow, looking at plant and animal communities.[8]

Harshberger may be the first American to have used the terms *chionophile* (snow-loving) and *chionophobe* (snow-hating). Drawing on the work of the Swiss ecologist Josias Braun-Blanquet, Harshberger summarized snow ecology in 1929:

> The beneficial effects of snow are found in the protection of plants from wintry winds and from the action of frost, the retention of heat by the soil, the protection of the early spring plants, the increase of air temperature by reflection of the heat from the snow surface, the fertilizing action of snow in bringing down to the ground dissolved gasses and floating particles of dust in the air, the accumulation of water, which is held in the form of snow as a reservoir, and the provision of smooth snow surfaces for the distribution of fruits and seeds. The destructive effects of snow comprise the rubbing action of small ice particles on exposed plants, the formation of a cold air layer on the upper snow sur-

face, the shortening of the growing period through the persistence of snow fields, mechanical injury to plants, abrasion of grass-covered surfaces; and its contribution to solifluction, or the downward flow of soils.[9]

Harshberger was part of the effort by biologists and ecologists to explain how plants and animals are distributed in regions of cold and snow. Their focus was also shifting from individual species to communities. They knew from the great surveys of the early botanists John Richardson and William Jackson Hooker that the boreal regions, the Arctic and Subarctic, had relatively few plant species, but that these species were widely dispersed. As geologists provided more information on the extent of Pleistocene glaciation, determining that the continental ice sheet had reached as far south as northern Pennsylvania, Ohio, Indiana, Illinois, and Iowa, several ecological questions arose. Chief among them were How and where did certain species persist despite glaciation? What are the natural limits of the tree line and where is that line now? And how do life forms adapt to snowy conditions?[10]

A century of botanical studies in the Arctic and Subarctic was nicely summarized in 1941 by Hugh Raup of Harvard's Arnold Arboretum. Asserting at the beginning of his review that work on the structure of plant communities in what he called "Boreal America" had "hardly more than begun," Raup discusses the extent of geological, climatological, and biological knowledge of the nearctic region, noting the contributions of Sir Joseph Dalton Hooker, son of William Jackson Hooker, who believed that arctic plants had retreated before the glaciers and returned to the Far North and to some mountaintops after the last ice age, and the Swedish botanist Herman Georg Simmons, who in 1913 identified the area west of the Mackenzie River, where glaciation was not widespread, as the refuge of many plants of the American Arctic. The work of Hooker and Simmons was modified by M. L. Fernald, who pointed to *nunataks,* small unglaciated areas within the ice sheet, as islands on which arctic and cordilleran species persisted. These species were thought to be "old," "conservative," and "non-aggressive," which explained their limited ability to spread to new areas now free from ice. Fernald's hypotheses were quickly challenged as simplistic, but they stimulated further research in the 1920s and 1930s, especially the work of Robert Griggs, Eric Hulten, and Raup himself.[11]

Griggs, a biologist at George Washington University specializing in the timberline, argued that the boreal forest was actively migrating into the Arctic at Kodiak, Alaska, but that similar forests seemed static in the northern Rocky Mountains. He advanced several reasons for this stability, including grazing by both domesticated and wild sheep, soil conditions, and wind-driven ice crystals. Identifying what he called "snow mats," the accumulation of falling and drifting snow on the largest lower branches of pines, he observed that branches and even the bark of the trunks above the snow mat are cut away on the windward side, leaving a bare trunk towering above the lower limbs that mark the depth of winter snows. Finding only young trees in hollows and ravines in contrast to mature trees on ridges, Griggs hypothesized that those in the hollows were uprooted or broken by avalanches and that "even the small creep of a heavy mass of snow which would not affect large trees will bear down saplings and so, coming every spring, shear trees off a slope." A half century after Griggs, ecologists are still studying the ways avalanches influence the patterns of trees and shrubs on mountains. A recent Canadian study concluded that "the shift from shrub- to tree-dominated growth habit down the avalanche path occurred when the average interval between avalanches was less than 15 to 20 years." Griggs discovered one other snow mechanism contributing to forest mortality—poor drainage. "Stagnant snow water rather than the snow itself" kills young trees, he reasoned, and a combination of several snow conditions was the determining cause of the timberline and timber atolls.[12]

Hulten, a Swedish botanist, mapped the ranges of hundreds of boreal species in northern Europe, Siberia, and Alaska. Reviving Fernald's idea of conservative and aggressive species, but discarding the terminology from the era of social Darwinism and Progressive politics, Hulten substituted terms more appropriate to Swedish democratic socialism and the decade of the New Deal: "plastic" and "rigid." Plastic species adapt to new environments, rigid ones cannot, and the difference is not environmentally induced but genetic senescence. Moreover, species with the widest distribution are those with the largest number of races. In Raup's opinion, it was the Danish ecologist Christen Raunkiaer who, working with the life zone classification of the American C. Hart Merriam, made the most important contribution to boreal ecology. Raunkiaer looked at the adjustment of plants to the least favorable season and used the position of perennating buds in relation to the soil surface for classification. He found that many arctic species survive by completing their life cycle quickly when conditions are favorable and remaining underground when conditions are unfavorable. Raup questions whether geographical terms such as *tundra* can be applied in a biological sense and stresses the ways arctic and subarctic soils are affected by frost. He is convinced that Chinook winds create distinctive microclimates in some areas and agrees with Griggs that vegetation is always in the process of adjustment to an environment and exists with only partial adjustment. At present the importance of the Raup and Griggs theory of plant migration in response to climate change lies in its relation to global warming. A 1994 report from Austria suggests that some plants are migrating upward at the rate of 1.0 to 3.6 meters (3 to 12 feet) a decade in response to global warming, leaving them with noplace to go by the end of the next century.[13]

Parallel to these developments in snow ecology, another movement took shape that influenced how the general public came to regard snow and its effects on plants and animals. Beginning in the late nineteenth century, concern over declining farm populations, especially in the state of New York, led a group of professors at Cornell University to establish the "nature study idea," as its chief proponent, Liberty Hyde Bailey, called it. Although *nature study* refers to a number of activities with sometimes conflicting purposes, its relation to winter ecology is relatively simple. One of the principal goals of nature study was to preserve rural life by instilling a love of nature in boys and girls. Curricula were organized and books written to help young people learn about science and, above all, to "increase the joy of living."[14]

Among the leaders in the movement were two remarkable women, Anna Botsford Comstock, scientific illustrator, author of *Handbook of Nature Study,* and wife of a Cornell entomologist, and Ann Haven Morgan, who received her Ph.D. in biology from Cornell in 1912, taught at Mount Holyoke College from 1906 to 1947, and wrote several books for the general reader, including *Field Book of Animals in Winter* in 1939. More than 200 pages of her 500-page handbook are devoted to insects and simple plantlike animals. Her chapter "The Snow Insects" discusses and illustrates four orders: Collembola, the springtails such as snow fleas; Plecoptera, the water-dwelling stone flies; Mecoptera, the scorpion flies; and Diptera, the wingless snow fly described by O. Lugger in 1896 and by T. W. Harris in 1862. A chart shows how several species of Plecoptera succeed one another from mid-November through July of each year. Since the majority of animals described by Morgan are dormant in winter, her book seems better suited to the armchair naturalist than the winter field enthusiast, but it must have encouraged many young people to dig beneath the snow in yards and woodlots to find the nests and cocoons she pictures in leaf mulch and rotting logs. Curiously, her chapter on mammals does not, as Seton did, include tracks made in the snow; it focuses instead on hibernation, as if the reader might prefer to emulate the sleeping species, since, as Morgan poetically states, "The snow cover is like a feather quilt spread over the earth, because air is held between snow crystals as it is among the delicate barbules of feathers."[15]

NIVECULTURE

The story of weather modification experiments by Irving Langmuir, Bernard Vonnegut, and Vincent Schaefer at the General Electric Research Laboratory in Schenectady, New York, is relatively well known. Beginning in 1943, with research on the effect of snow crystals on radio transmission from airplanes flying through snowstorms, Schaefer went on to study icing on aircraft wings. After conducting some winter experiments on Mt. Washington, New Hampshire, he became interested in re-creating snowstorms in his laboratory. In 1946, more than ten years after Ukichiro Nakaya had "grown" a snow crystal on a rabbit hair in his laboratory on Hokkaido, Schaefer created a snowfall in a freezer by adding a piece of solid carbon dioxide (dry ice). By November, Schaefer had successfully "seeded" a cloud above the Berkshires, and the era of weather modification began.[16]

Control of the weather is one of humanity's oldest dreams. Rainmakers followed the farmers onto the Great Plains in the nineteenth century. Tourists still flock to see Native American rain dances, and public relations firms create snow dances and snow-calling contests. The historian Clark Spence attributes the occasional success of "pluviculturalists" to individualism, belief in progress, and short memories. Schaefer's success in cloud seeding and the use of various snowmaking techniques for the past fifty years raise a number of questions about the role of optimism in American character and about humanity's relationship with nature, as well as about definitions of snow.[17]

Research on the physics of precipitation began about the same time as the snow surveys. The great German geophysicist Alfred Wegener, who died in Greenland while trying to prove his theory of continental drift, made the first important studies of the formation of rain and snow in the atmosphere in 1911. Tor Bergeron of Sweden and Walter Findeisen of Germany elaborated his theory in the 1920s and 1930s, hypothesizing that all precipitation is formed as ice particles in clouds where the temperature is below freezing. Whether the ice crystals reach the ground as rain or as snow depends, therefore, on other variables such as altitude. In the 1940s it was discovered that raindrops can form in warm clouds without freezing through a process called coalescence, the collision of tiny cloud droplets as they fall at different speeds because of thermal differences in the air. Nakaya's success in growing a snow crystal in a laboratory and Schaefer's discovery that dry ice would cause a supercooled cloud to produce snowflakes were the final steps needed to bring theory and practice together in a series of rain- and snowmaking experiments. The military, economic, social, legal, environmental, and cultural contexts in which weather modification appeared significantly shaped its development.[18]

The most immediate context was military. Following the first explosion of an atomic bomb by a little over a year, Schaefer's success at cloud seeding immediately evoked an analogy between atomic power and weather modification. Early advocates of weather modification even suggested using atomic weapons to dissipate hurricanes and other storms. A legacy of this association is the frequent use of nuclear explosions to measure storms. Thus a small thunderstorm is said to have the energy equivalent of thirteen Hiroshima-sized atomic bombs, or tornado damage is compared with a nuclear blast. The Pentagon, convinced that the next war would be fought in the Arctic and obsessed with the fear of Soviet scientific progress, was anxious to explore the strategic uses of weather modification. The controversial Project Cirrus, which by Schaefer's account involved 116 experimental flights to seed clouds with dry ice and silver iodide over the Southwest, was conducted in secrecy between 1947 and 1953. Concern about military modification of weather led to the Treaty on Weather and Environmental Warfare in 1977.[19]

Related to military involvement with weather modification was the dependence of big science on government support. Cloud seeding joined other technological innovations of the 1940s—radar, missiles, nuclear weapons and power—in competition for a share of the ever-increasing federal budget for research and development. Like other technologies, it required answers to questions such as Who determines public policy? What agency should coordinate its research, planning, operations, and regulation? But the forces unleashed by scientific research in World War II moved too swiftly for adequate review by either the legislative or the executive branch of government. Requests for funding were seldom denied.

In these circumstances it is not surprising that Langmuir, a Nobel Prize–winning chemist and senior research scientist at General Electric Labs, could convince many that it was scientifically feasible to regulate the weather of the entire United States, if not the world. The sky was no longer the limit. Enthusiasm for weather modification from 1947 to 1957 was a preview of the selling of the space program. As Walter McDougall points out in his history of the space age: "Through our technology, our ability to manipulate Nature, we are subcreators, a demiurge. Hence we have never, from Protagoras to Francis Bacon to Tsilkovsky, been able to separate our thinking about technology from teleology or eschatology. For reason cannot predict whether our tools and dreams, which together permit us to 'invent the future,' will lead us to perfection or annihilation or unending struggle against Nature and ourselves."[20]

Unusually heavy rains in the Midwest following cloud seeding in New Mexico in 1950 caused many people to fear that this latest advance in technology might lead to a catastrophe. Such fears were not necessarily antediluvian. General Electric, fearing lawsuits, opened its patents on cloud seeding to the public and waived all royalty rights, but questions of legal liability for damages claimed from cloud seeding continue to arise. In 1953 Congress authorized the Advisory Committee on Weather Control to review the scientific data and recommend future action. In the meantime, dozens of private companies sold farmers and state governments on the possibility of increasing precipitation through cloud seeding. By 1951, the year some of these firms founded the Weather Modification Association, more than one-third of the land west of the Mississippi River was under contract for weather modification activities. Although Schaefer regretted that the technique had been oversold, he remained enthusiastic about the potential for large-scale weather control as well as increasing precipitation through cloud seeding. Claims for dissipating hailstorms, hurricanes, and lightning were, he conceded, difficult to evaluate.[21]

The *Final Report of the Advisory Committee on Weather Control* was issued on the last day of 1957. The committee, chaired by Capt. Howard T. Orville, formerly of the U.S. Navy and technical consultant for Friez Instrument Division of Bendix Aviation, included Langmuir, Schaefer, and Vonnegut, and was decidedly pro-seeding. It told Congress that snowfall in the western mountains had been increased 10 to 15 percent. In a separate section of the report, entitled "The Future," Schaefer and his associates expressed their belief that it was possible to increase sunlight or cloud cover over large areas, to control water balance in the atmosphere by a combination of irrigation and vegetation, to control global weather by thermonuclear heat, and to control ocean currents. Blithely, they contended that laboratory experiments had shown that chemicals can determine the rate of growth of snow crystals. "It is conceivable that by applying this technique we may be able to exercise a control of the rate of growth of the ice phase in clouds and to make significant changes in the properties of snow." Designer snow was just becoming available at some eastern ski resorts, but Schaefer promised a more public demonstration.[22]

The immediate result of the advisory committee's report was to make the National Science Foundation the coordinating agency for weather modification research. The foundation was just beginning to establish itself as one of the major federal agencies involved in snow research. The International Geophysical Year, 1957, and the antarctic program soon gave the National Science Foundation virtual control of most aspects of snow research. In 1959 the House Committee on Science and Astronautics heard testimony from Robert Brode, associate director of the foundation, who outlined the more than forty grants awarded to universities and research institutions for the study of weather modification. Most of the projects related to cloud seeding and precipitation enhancement. Schaefer continued to beat the drum for rainmaking, telling the audience at the Western Snow Conference meeting in 1963 that the Japanese were reporting increases in rain of 150 percent downwind of the Sea of Japan, and he urged the United States to undertake large-scale projects. "A certain degree of gambling is necessary in any applied science development," he concluded. Public confidence in rain kings diminished, however, and gambling with nature was replaced by concern for environmental protection.[23]

Environmental damage was among the issues addressed by Howard Taubenfeld and his associates in a book published in 1970. One contributor raised the basic question of whether federal regulation should include both intentional and inadvertent weather modification, also noting that the definition of weather modification depended on its purpose. In 1968 Congress had removed weather modification from National Science Foundation control and placed coordination of such projects in the hands of the Interdepartmental Committee for Atmospheric Sciences, which resulted in a power struggle between the Departments of Commerce and Interior, with Agriculture skeptically observing and Defense simply ignoring the others. Ten functions of weather modification research were identified: data collection, research, monitoring, operations, coordination, comprehensive planning, project review, regulation, licensing, and indemnification. Congress failed to consider more than a few of these functions in any of the bills it took up in the late 1960s, but it passed the National Environmental Policy Act, which required environmental impact statements in connection with cloud-seeding projects. Public attitudes toward pollution, government science, even nature itself were changing rapidly. A 1966 report by the Committee on Water, Division of Earth Sciences, of the National Academy of Sciences–National Research Council, showed that Americans included recreation, environment, and aesthetics as desirable social objectives and that there was a willingness to pay for intangible benefits.[24]

It had taken almost twenty-five years to move from the belief that weather modification is an end in itself, analogous to splitting the atom, to seeing it as a tool in achieving the more specific goals of increasing food production or improving safety at airports. It had also required a quarter of a century to begin assessing the social impact of cloud seeding and hail suppression. In the early 1970s, under grants from the National Science Foundation, the National Oceanic and Atmospheric Administration, and the National Center for Atmospheric Research, J. Eugene Haas, professor of sociology at the University of Colorado, conducted public opinion polls with residents of the San Luis Valley, Colorado, central Florida, and South Dakota. Cloud seeding for hail suppression had been carried out in Colorado in the late 1960s, seeding for drought relief was attempted in Florida in 1971, and both hail suppression and rain augmentation had been going on in South Dakota since the mid-1950s. Haas claimed to be surprised by the extent to which scientists disagreed over the efficacy of cloud seeding and by the generally favorable orientation of the public, despite their professed ignorance of the techniques of weather modification. Certainly it is curious that 54 percent of those polled in Colorado thought that cloud seeding would upset the "balance of nature," while 75 percent supported experimenting with cloud seeding to find if it really works. Nor was the NIMBY (not in my backyard) factor significant, since 52 percent supported a cloud-seeding program in their own communities.[25]

Haas's report explored a number of other issues of democratic decision making and organized opposition to weather modification, and raised several important points regarding interest groups and the resolution of conflicts between weather modifiers and their opponents. In later studies Haas speculated on the effects of colder and wetter winters on the nuclear family (increased interaction among housebound family members), the journey to work (increased tardiness and absenteeism), education (more school closings, reduction in outdoor exercise for children), and leisure and recreation (sharp reduction in outdoor activity and increase in interrupted and canceled trips in winter), although he admitted that the decrease in outdoor activity might be offset by an increase in skiing, sledding, and other snow sports in some areas. Haas's attempt to apply social impact assessment techniques to snowpack augmentation is representative of the more thorough technological and social assessments carried out in the 1970s.[26]

Two of these studies, one of the residents of the Upper Colorado River basin and another of Northern California, provide answers to some questions about public attitudes toward cloud seeding and snow. Under a grant from the National Science Foundation, Leo Weisbecker and his associates examined the ecological, economic, and social impacts of what they called "snow enhancement" along the Upper Colorado in 1971. The report was specifically addressed to lawmakers in Washington, D.C., and the states of Colorado, Utah, and Wyoming. The Bureau of Reclamation was conducting a large-scale pilot project in the San Juan Mountains of southern Colorado to increase the snowpack by cloud seeding. Many residents in the affected areas were opposed to the project, and others were concerned about the potential damage to farming, summer recreation, and the environment caused by heavier snows and longer spring runoff—specifically, the impact of heavier snows on winter forage for both domestic livestock and wildlife, the increased likelihood of avalanches, and the probability of flooding. The reports attempt to address these fears by emphasizing the need to have public input in the policymaking process, by recommending programs to control environmental damage, and by supporting restitution to anyone claiming economic loss. Although Weisbecker clearly indicates that it is impossible to quantify any economic benefit to the region from cloud seeding, he acknowledges that increased runoff will cause major problems of silting and other environmental damage, and concludes: "The technology cannot be stopped from developing, even though certain applications in specific areas may be stopped. The important question for WOSA [Winter Orographic Snowpack Augmentation] generally is not whether or how it might be stopped, but how it should be operated."[27]

In ways snow surveyors never anticipated, the resource they had helped to manage for sixty years became a miscreation, synthetic snow that felt slipperier to farmers and that threatened to turn public opinion against its creators. Cloud seeding turned meteorologists into teratologists. The possibility of angry citizens attacking laboratories and pursuing professors and their snow monsters with blazing torches seems to have occurred to the technology assessors. Their recommendation was a public relations campaign to convince the public of the benefits of cloud seeding or at least guarantee prompt payoff for damages. Advocates of cloud seeding realized that opposition stemmed from fear of big government as well as from the augmented snow. Weisbecker recommended the establishment of a Federal Weather Modification Board, an independent agency responsible for licensing operators, collecting data, and ensuring that environmental standards were met. Only a national agency could properly serve the three objectives of the Water Resource Council: (1) to enhance national economic development; (2) to enhance the quality of the environment; and (3) to enhance regional development. Moreover, "the Colorado River Basin Project Act of 1968 (P.L. 90-537) declared that 'the satisfaction of the requirements of the Mexican Water Treaty from the Colorado River constitutes a national obligation which shall be the first obligation of any water augmentation project planned pursuant to . . . this Act and authorization by Congress.'" In other words, the United States, having already allocated to five western states more water than the river contained, was telling Mexico it would get its share as soon as it could be manufactured.[28]

Faced with the conflicting demands of this snow enhancement project and the primacy of economic interests over environmental and aesthetic ones, Weisbecker was optimistic but equivocal: "The multiple objectives are not mutually exclusive in that effects that appear as positive or negative in one category may appear in others as well." His optimism was based on an uncritical acceptance of estimated average increases in precipitation provided by the U.S. Geological Survey and the Weather Bureau, which ranged from 15 to 25 percent and predicted significant increases in the snowmelt runoff even in dry years. Few scientists outside the project believed in such high rates of success. Most wanted more data and would have agreed with Harold Klieforth, who wrote in 1969 that "while estimated increases in precipitation augmentation and expected runoff are unwarranted at present, [we] should concentrate . . . efforts on field studies and experimental projects directed toward answering the basic questions underlying weather modification possibilities."[29]

The Northern California study was conducted in August 1975 by Barbara Farhar for the Bureau of Reclamation. Farhar polled 200 citizens in thirty-one organizations in five counties. She found that most of the individuals and groups questioned felt that they lacked sufficient knowledge of cloud seeding and generally preferred water conservation and traditional methods of increasing water supply, such as reservoirs. Responses to her questions about the possible side effects of cloud seeding reveal a commonsense understanding of the consequences of climatic change: 28 percent of the 40 percent who thought seeding might damage the ecology of the area mentioned the strain of snow and cold on animal life, while another 15 percent mentioned damage to both flora and fauna; 13 percent feared chemical damage to plants and animals, presumably from the silver iodide used in seeding, and 10 percent were concerned about flooding. Others mentioned potential soil erosion; damage to buildings, power lines, and roads; and the negative effects of community controversy. Perhaps most interesting to the cultural historian are the responses indicating that 49 percent believed that humanity should take the weather as it comes and not try to alter it to suit our needs and that cloud seeding is likely to upset the balance of nature. Fewer than 30 percent disagreed with these feelings, while the remainder were unsure. These were the opinions of people who lived in the Sierra Nevada, who hunted and hiked, worked for the power company and plowed the highways. Their feelings were less religious—only 36 percent agreed with the statement that cloud seeding violated God's plans—than practical.[30]

Twenty years of research, experimentation, and public debate have changed little. Harland Cleveland, chairman of the Weather Modification Advisory Board, testified to the House Subcommittee on Science and Technology on October 26, 1977, that "extensive cloud-seeding will soon be a proven technology," but the best reasons he could offer were his observations that "a growing U.S. public concern . . . reflected in fewer jokes about the weather and more serious public discussion of meteorology" and that "in search of more year-round sunlight, Americans have moved by the millions to arid regions and weather-vulnerable coastal zones where water shortages and storm warnings become part of their daily life." "Public awareness of weather, in global perspective," he continued, "has been greatly increased by the nightly appearance on our TV screens of synoptic photographs of large weather systems, taken from orbiting space satellites." Synoptic photographs, colorful maps, radar screens, and earnest young weatherpersons notwithstanding, TV weather reports are seldom taken seriously. Access to meteorological information has not prevented residents from what one observer calls the "western oasis culture" from using three times the amount of water consumed in the eastern states. In an article titled "Cloud Seeding: One Success in 35 Years," *Science* magazine debunked claims for increased precipitation. The 1982 meeting of the North American Interstate Weather Modification Council, a small organization of commercial cloud seeders and others working in the field, reported on the failure of Senate and House bills that would have provided funding for weather modification and placed the authority to develop and regulate projects in the Department of the Interior. The following year funding for weather modification was virtually eliminated from the budget proposed by the Reagan administration. Everyone talked about weather modification, but no one did anything about it.[31]

Cloud seeding for snowpack augmentation continues in the 1990s, but too little is known about natural variation in precipitation from year to year to draw firm conclusions or make reliable predictions. The wide swings from above to below "normal" years of snowfall in California in the 1980s and '90s are a case in point. A 1988 article in the American Meteorological Society *Bulletin* compared two types of winter snowpack augmentation: the "statistical" method, which uses random seeding in a target area, and the "physical" method of monitoring storms and seeding clouds that show the highest probability of producing snow based on their distribution of supercooled liquid water. Both methods are currently in use. The latter depends on satellite and radar remote sensing to provide information on cloud formation several hours in advance of seeding, because "what is being modified in winter clouds is a rate process. That is, clouds contain insufficient nuclei, nucleation occurs too late, or nucleation is too slow to convert the available cloud condensate to precipitation-size particles before the condensate is lost to the lee of the barrier. Glaciogenic seeding is an attempt to convert the cloud condensate (small cloud droplets with low terminal velocities) before the particles pass over the crest." This method allows cloud seeding in dry months when storms are frequent but weak, because the weakening portion of the storms produces substantial periods of supercooled liquid water. The article concludes by conceding that even this potential may not be present in drought years.[32]

EATING YELLOW SNOW: THE NIVEAN FOOD CHAIN AND POLLUTION

While the physicists made snow fall, biologists were beginning complex studies of snow ecology. William O. Pruitt, Jr., a zoologist educated at the University of Michigan who has lived and taught in Canada for most of his career, combines fieldwork with an interest in the history of his discipline and an advocacy of native perspectives on snow. Pruitt brought his early research to a general audience in the January 1960 issue of *Scientific American,* where he discussed a variety of snowcover conditions and explained how hardness and density determine the distribution of caribou and other creatures in snowy regions.[33]

Later Pruitt published *Animals of the North,* reprinted in 1988 as *Wild Harmony,* an examination of life in the taiga that begins with an account of a giant spruce, its life cycle a lesson in subarctic ecology. As it grows from sapling to mature tree, it provides food and shelter for flocks of crossbills and packs of squirrels. The dead needles at its base form a circular mat that prevents the growth of mosses and lichens. Even after it is uprooted in a flood, its exposed roots and bark-stripped trunk become home for algae, plankton, and tiny crustaceans. Finally, beached in the treeless tundra, it becomes a scent post for arctic foxes, who are then caught by an Eskimo trapper. Other Eskimos later remove the trunk for wood to build a house and a boat. Ultimately, Pruitt concludes, the wood will be burned and the carbon, hydrogen, and other molecules will return to the atmosphere and be incorporated in another living spruce tree.[34]

Pruitt continues his survey of snow ecology with glimpses of the lives of red squirrels, voles, hares, lynx, wolves, caribou, moose, and humans. The sequence allows him to introduce typical snow conditions as they are encountered by arboreal, subnivean, and supranivean creatures. The squirrel dodges around the *qamaniq,* or bowl-shaped depression at the base of a tree, and is impeded by *qali,* or snow-covered branches. The voles struggle to survive as the temperature falls until 20 centimeters (8 inches) of snow covers the ground, establishing the "hiemal threshold," insulating the ground and allowing the voles to tunnel near the surface of the unfrozen earth. The hares, chionophiles, change pelage and behavior as the snow accumulates. Each layer of fresh snow allows the hares to nibble higher on the willow bark. Pruitt's lynx and wolves are as anthropomorphic as Seton's, characterized as artists of ambush and master hunters, but Pruitt provides more detail on their adaptation to snow. As the snow settles and hardens and a crust forms on the surface, the hares abandon their trails and the lynx catch them more easily on the open snowpack. In the late winter, as the air becomes colder and drier and transmits little of the odors of living creatures, the lynx's keen eyesight and hearing allow it to locate prey. The wolf, by contrast, is less well adapted to running in deep snow or on thin crust, but it locates its meals by scent and runs them down in the thick forests, where the snowcover is not as deep. Caribou and moose are, for Pruitt, the creatures most superbly adapted to their environment, combining efficient insulation, physiology, and migration patterns to utilize the annual cycle of brief summers and long, harsh winters.[35]

While admitting that the survival of the northern wilderness lies less in the hands of scientists than in those of politicians, Pruitt continued his effort to explain the complexities of snow ecology in a 1970 essay in which he divided snow into two types: taiga and tundra. The basic distinction, he argues, is modification of snowcover by wind. The snow of the subarctic, forested taiga of interior Alaska is "typical" in Pruitt's lexicon because it is little modified by either solar radiation or wind. The snow lies undisturbed on the branches of evergreens, and temperature gradient, the difference between the cold surface of the snowpack and the warmer, insulated earth below, accounts for most of the snow metamorphism. Pruitt is especially interested in *qali,* snow that collects on trees, because "it is one of the agents initiating forest succession." If a tree accumulates too much *qali,* it will break, setting off a chain reaction in adjacent trees until a glade is formed, which in turn allows the wind to blow excess *qali* off branches and end the cycle. In the glade, dead needles from the broken spruces choke out the feather mosses, allowing seeds from deciduous trees to germinate and grow to maturity. When these alders, birches, aspens, and willows die, spruces grow in their leaf litter and the cycle begins again. Although Pruitt's model assumes an equilibrium that many ecologists would say is never achieved in nature, it illustrates the crucial role of snow.[36]

Pruitt is also attracted to taiga snow because of his interest in animal adaptations to snow. He writes, "There are two main classes of animals in the taiga—those that are large and live above the snow cover and those that are small and forced to live beneath the protecting blanket. There is one animal, the red squirrel (*Tamiasciurus hudsonicus*), that lies just on the borderline between the two sized-groups of mammals. Red squirrels are active above snow most of the time, but when the air temperature falls below –30°C [–22°F], they vanish from the scene." Pruitt's admiration for the red squirrel, which he calls the "sentinel of the taiga," stems from its trickster-like ability to survive on the border between large and small mammals, and even mammals and birds.[37]

In a brilliant tour de frost, Pruitt synthesizes most of the known elements of boreal ecology—the importance of radiant energy and the angle of the sun; the interplay of snow, wind, soil, and permafrost; the supra- and subnivean bioclimates and their effects on living organisms; and the delicately balanced food web or chain that permits life to survive even in this harsh environment. Since "snow is one factor that unites the boreal regions," Pruitt expresses frustration with those who continue to ignore its importance in studying human adaptation to the North. Although he is capable of dismissing *qali* in regions of occasional snowfall as of only "transitory aesthetic interest," Pruitt never loses sight of his primary point, that the most important adaptations of mammals to snow have been behavioral, not physiological. As subsequent research confirmed, a holistic approach to snow is necessary because its apparent simplicity masks great complexity and because the boreal environment is so fragile.[38]

In 1984 Edmund Telfer and John Kelsall proposed a snow-coping index based on morphological and behavioral characteristics of the large ungulates of North America. Caribou and moose ranked first and second, bison and antelope at the bottom of the list. The hooves of the caribou provide better support relative to its weight than the hooves of the moose, but the moose's greater height allows it to browse higher on birch and willow branches. Caribou, however, can dig through the crust and forage near the ground more easily. Both leave food exposed for hares and other small animals.[39]

Paul Ehrlich of Stanford University has conducted population studies in various nivean environments. When a late January snowstorm struck in the vicinity of the Rocky Mountain Biological Laboratory near Gothic, Colorado, in 1969, Ehrlich and his colleagues were able to measure its effects on the plant, insect, and mammal species they had been investigating. Herbaceous perennials and bunch grasses, important food sources for insects and small mammals, were seriously damaged by the snow. The flowering of over 80 percent of a species of corn lily (*Veratrum californicum*) was totally destroyed, and only 3 percent of *Lupinus amplus* reached full maturity and fruiting after the storm. The destruction of these plants was a disaster for the butterfly *Glaucopsyche lygdamus,* which became extinct in the study area, and the populations of several other insects and mammals declined drastically.[40]

Ann Zwinger and Beatrice Willard echo Ehrlich's conclusion regarding the fragility of the food chain in regions of snow: "The simplicity of alpine ecosystems has its price. The food chain is short: alpine clover—meadow vole—weasel or hawk. When food chains are complex and elaborate, as in a lowland community, a shift in any one component is less felt, the change is easily absorbed in the interwoven web, and the community retains its essential stability. When the food chains are short, the alteration of a single member results in a prompt and drastic change in the rest of the chain." The inference here is that instability rather than stability or equilibrium is the distinguishing characteristic of the nivean frontier. Snow and its congeners, cold and wind, periodically disrupt the food chain by killing or retarding its fundamental building blocks—mosses, lichens, shrubs, and trees.[41]

The extent to which feeding patterns are adaptations to nivean conditions remains a mystery, but a 1987 study by Canadian foresters of the decomposition of leaf litter in a mixed conifer-hardwood forest in southeastern Ontario found that skeletonization of red maple leaves was significantly greater in plots where timber had been harvested and the ground was more open. Greater skeletonization was attributed to increased feeding by soil invertebrates active under the snowcover, although the amount of snow in the harvested and unharvested plots was not reported. C. W. Aitchison, a biologist who has specialized in subnivean food webs, offers evidence that tertiary consumers such as shrews feed not on hibernating and quiescent invertebrates but on winter-active species. All organisms show seasonal preferences in diet, and snow is an obvious factor in setting winter menus.[42]

A study of the effect of snowcover on the foraging of starlings in central New Jersey revealed that the birds shifted from invertebrates and grains on lawns and in fields to human-generated food scraps and garbage cans when the snow became deep. Moreover, the birds became more aggressive in defending these sites, presumably because they are more concentrated and easier to defend. Nicolas A. M. Verbeek hypothesizes that water pipits and horned larks forage on snowfields because the arthropods and insects they feed on are easily seen, but a test of this hypothesis concluded that the birds forage where the arthropods are largest, not on the snowcover but near it, suggesting that snowfields function more like tablecloths surrounded by crumbs than like buffet counters heaped with food. Studies of seed dispersal in species of dwarf willow in alpine areas of Scandinavia and of birches in northern Delaware found that windblown seeds are deposited in snow patches and that in some cases the longer-lasting snowcover promotes survival and growth of the species. Windblown seeds also travel farther over the smooth surface of snow, allowing the species to colonize.[43]

The range of responses to snowcover as an impediment to foraging is shown in studies of bighorn sheep in the Colorado Rockies and caribou in Labrador. Even small amounts of snow caused the bighorn to shift feeding sites and dietary composition, while woodland caribou apparently select "feeding sites based primarily on the expected presence of forage rather than on perceived or actual differences in snow conditions." They were observed digging feeding craters up to 123 centimeters (4 feet) in snow with a mean ram hardness of 222 to 536 kilograms (snow too hard to penetrate with a knife blade without considerable effort), that is, in snow much deeper and harder than previously recorded. By contrast, Elliott Norse cites a study that showed that "as little as three or four inches of new snow can drive Columbian black-tails from an area, and they begin to starve quickly when just a few inches cover evergreen trailing blackberries, one of their primary winter food plants."[44]

Snowcover serves zoologists with a litmus paper from which they can learn many things about the nutritional status of free-ranging animals. A recent investigation of deer populations in northern Minnesota found significant variation of nutritional status because of loss of food resulting from deep snow. Although this conclusion was not surprising, the testing of urine in snow for levels of urea nitrogen, sodium, potassium, calcium, and phosphorus provides one more tool for game management and a final twist to the warning "Don't eat yellow snow."[45]

Human activity that changes patterns of snowfall also affects wildlife. "Without question," Norse writes, "clearcutting can reduce forage availability. Snow accumulations are up to six times deeper in clearcuts than in forests with canopies having 70% coverage. Elk avoid clearcuts when they have difficulty plowing through snow and their food is covered." The widely publicized problem of ranchers killing protected buffalo that wander outside Yellowstone National Park is directly related to increased winter recreational use of the park. As use increased, more roads were plowed to facilitate traffic. The plowed roads provide easy corridors of travel for bison.[46]

A comprehensive study of snow ecology is under way at the Niwot Long-Term Ecology Research site in the Indian Peaks of the Colorado Front Range under the direction of D. A. "Skip" Walker of the Institute of Arctic and Alpine Research at the University of Colorado. The primary purpose of the project is "to understand how current snowpack distributions affect patterns of vegetation from species to regional scales." Behind this goal are, of course, current concerns about the effects of global warming and, more immediately, the impacts of mountain recreation and urban development. The project is unusual in its attention to both spatial and temporal dimensions of snow ecology. At the "plot" level, areas of 1 to 10 square meters (10.8 to 107.6 square feet) are studied for snow regime, nutrient cycling, primary plant production, animal disturbance, and microtopography. At the "landscape" level, areas of 100 square meters to 1 square kilometer (1076.4 square feet to 247.1 acres) are studied for the same factors, plus snow hydrology, geomorphology, water chemistry, and other large-scale phenomena. At the "regional" scale, areas 1 to 100 square kilometers (247.1 acres to 38.6 square miles), studies focus on elevation and temperature gradient, climate change, and geological factors. To overcome the limitations of human time scale, experiments are under way to simulate, through the use of snow fences, the effects of unusually heavy and unusually light snowfalls.[47]

Such studies help correct the impression expressed in W.D. Snodgrass's poem, that life in winter snow is diminished, existing "on the season's scraps." Though fragile, the snowscape teems with life. Some appreciation of this can be seen in the paintings of contemporary wildlife artists who carry on the tradition of Seton, A. B. Frost, and Thayer. In recent years Bonnie Marris, Sarah Woods, Clark Ostergaard, Pat Ruesch, and others have painted animals in the snow with almost photographic precision. Although their paintings are sometimes sentimental, with a nostalgia for a lost Eden and the amateur naturalist's passion for surface detail, they also confirm the ecologist's studies.[48]

No contemporary artist has painted or written more eloquently about snow and wildlife than the Canadian Robert Bateman. For more than twenty-five years, Bateman has recorded the effects of snow on plant and animal life in his native Ontario and in the far Canadian North. Some of his work echoes that of Seton and Thayer. *Window into Ontario,* for example, explores the concealment of the blue jay in the snowy landscape, but Bateman goes well beyond the earlier artists in both technique and ecological awareness of snow (Figure 6.2). Using acrylics to build up the backgrounds in his paintings, he provides a telephoto effect, a depth of focus that permits the viewer to study the animal and the larger environment in detail. In the painting *Awesome Land,* an elk is dwarfed by the mountains behind it, and neither seems as essential to the message conveyed by the title as the snow that covers both. Other pieces reveal the influence of Andrew Wyeth, for whom Bateman has expressed admiration.

In some paintings, Bateman goes beyond the superficial and explores the metaphorical and moral meanings of snow. In *Edge of Wood—White*

Figure 6.2. *Window into Ontario* (1984), Robert Bateman. Acrylic, 121.9 × 243.8 cm (48 × 96 in.), Copyright 1984 Robert Bateman, photograph courtesy of Mill Pond Press. Bateman's depiction of snow ecology is similar to Seton's in its use of the blue jay to echo the patterns of blue and white in the shadows on the snow, but it also includes the influence of humans in the split rail cedar fence that separates the maple forest from the cleared field.

Tail Deer, the placement of the trees, the decaying fence, and the shadows they cast not only camouflage but imprison the deer. The painting explores edges, margins, zones of transition between humanity and nature, and extremes of nature itself. Bateman's contrasts of light and shadow, woods and snow-covered meadows reveal diversity and complexity where others find simplicity and monotony. He provides visual evidence for the passionate beliefs of the boreal ecologists cited earlier: "The farther north one goes, the longer and more authentic winter becomes and the more pervasive and heroic nature's evolutionary adaptations and accommodations. Thus the natural heritage of winter is much richer and more fascinating in the North."[49]

Bateman's painting *Evening Snowfall—American Elk* is unusual not only because it shows snow falling but because it shows elk tracks in the snow that are lighter in the interior than on the surface. This is an optical effect explained by the meteorologist Craig Bohren, who has shown that when the snowcover is relatively thin, but thick enough to accumulate on boot soles, from which it drops onto the snow, it leaves a cast or "snow slab" of the foot, which, being optically thicker than the snowcover, reflects more light and appears to glow.[50] Bateman's careful observations link light and life, physics and ecology.

A patch of snow, in the wilderness or on a city street, is a laboratory for understanding ecology. Its lessons are simple. The living parts of our earth interact with the nonliving. What happens to both in the finite time that the snow remains is a microcosm of larger ecosystem dynamics. An experiment in this lab is merely a winter walk with our senses alert to the changes taking place in ourselves and in the snow.

The Modern Minds of Winter

7

One must have a mind of winter
To regard the frost and the boughs
Of the pinetrees crusted with snow;

And have been cold a long time
To behold the junipers shagged with ice,
The spruces rough in the distant glitter

Of the January sun; and not to think
Of any misery in the sound of the wind,
In the sound of a few leaves,

Which is the sound of the land
Full of the same wind
That is blowing in the same bare place

For the listener, who listens in the snow,
And, nothing himself, beholds
Nothing that is not there and the nothing that is.

WALLACE STEVENS, "THE SNOW MAN"

"NOTHING THAT IS NOT THERE . . ."

Wallace Stevens's poem, published in 1921, provides one of the essential texts for exploring the meaning of snow in the twentieth century. "The Snow Man" both sums up an earlier century of snow symbolism and sets an agenda for the next century. Stevens's poem shares Emersonian notions of natural order. Emerson's "The Snow-Storm" implies that nature mocks culture, that art, which laboriously tries to mimic nature, bases its order on apparent disorder (see Chapter 1). Stevens also theorizes that the wind mocks the "listener," who "beholds" "boughs/Of the pinetrees crusted with snow," "junipers shagged with ice," and "spruces rough in the distant glitter." This, literally, is the "nothing that is not there and the nothing that is." Yet, in the thousands of words written about this poem, the central visual image of snow on three species of trees, the effects of *qali* as William Pruitt would say, has been largely ignored. Snow intercepted by trees is lost as a resource, according to some scientists of Stevens's time. His language, like theirs, is utilitarian and stripped as bare as the place—trees are "crusted," "shagged," "rough," not hung with "Parian wreaths." Like other contemporary philosophers, Stevens wonders if there are limits to transcend, or if we create them by naming them. Nothing becomes something called "nothing."

The imagery of the poem is linked to the art as well as the science of the period. One artist, Rockwell Kent, painted a visual counterpoint of Stevens's poem in the same year it was written. In *The Trapper* (Figure 7.1), a solitary figure, holding a trap and a pelt (the end of game), stands on a hill crowned with five snow-covered stumps (all that remains of the forest and evocative of Emily Dickinson's snow-veiled "Stump, and Stack—and Stem"), silhouetted against a blue mountain. A branchless tree trunk or pole leans toward him from the right side of the painting, its shadow merging with others to form crosses on the snow, while he seems to "regard" or "behold" the light cast on the snow by the setting sun. A dog, traditionally a symbol of fidelity, disappears into the valley below. Kent's painting and Stevens's poem are, I believe, profound commentaries on America in the twentieth century.[1]

George Santayana, the Harvard philosopher with whom a young Stevens discussed poetry in 1899, provides the crucial linkage between ideas about American culture and character and the metaphor of snow. American life is free, Santayana wrote in 1920, because each individual acts alone and responds to the actions of others, and there is no predetermined pattern. "Any tremulous thought or playful experiment anywhere may be a first symptom of great changes, and may seem to precipitate the cataract in a new direction. Any snowflake in a boy's sky may become the centre for his *boule de neige,* his prodigious fortune; but the monster will melt as easily as it grew, and leaves nobody poorer for having existed." Seeing Americans as so many insignificant snowflakes, almost nothing, who together become an avalanche of nothings unable to control their future, but who believe in their common destiny, must have appealed to Stevens, who could easily roll a *boule de neige* into a *bonhomme de neige.* The snow man is a quintessential twentieth-century American, preoccupied with his own misery and vulnerability, yet keenly observant and superbly confident that the wind blows the same in inner and outer weather. As Stevens observed, his poem is about "the necessity of identifying oneself with reality in order to understand it and enjoy it." A trapper, whose way of life is constantly threatened by the forces of nature and human development, symbolizes American frontier individualism. Kent, best known for his book illustrations, was a very literary

Figure 7.1. *The Trapper* (1921), Rockwell Kent. Oil on canvas, 86.4 × 111.8 cm (34 × 44 in.), Collection of the Whitney Museum of American Art, New York City. This haunting painting captures the spirit of Wallace Stevens's poem "The Snow Man," published in the same year. Both raise questions about humanity's relationship to nature, metonymically represented by snow.

painter, who was less interested in the realistic depiction of snow than in snow as a symbol of isolation and introspection.[2]

Placed in the contexts of Kent's painting, other snow poems, and the reports of the emerging snow sciences, "The Snow Man" takes on added significance. Interpretations of the poem for the past seventy-five years have emphasized Stevens's playful treatment of epistemology: How do we know what we know? How can we free our minds from preconceptions and myths? What is appearance and what is reality? These are only a few of the questions that he grappled with in "The Snow Man." Is the snow man of the poem a Platonist or an Aristotelian? Is he Emerson's transparent eyeball ("I am nothing; I see all"), or has he achieved the freedom from self of an ancient Chinese philosopher? It is fashionable to read "The Snow Man" as a Nietzschean effort to create, with words, meaning out of nothing, and as an anti-Nietzschean belief in a higher order. My own interpretation is that the poem deals with the question of whether order is discovered or invented, and, in a very pragmatic American way, it accepts both positions. The snow man beholds both the "nothing that is not there and the nothing that is." Is this paradox so difficult? Recall the words of the meteorologist Napier Shaw, who wrote in 1913: "Every theory of the course of events in nature is necessarily based on some process of simplification of the phenomena and is to some extent therefore a fairy tale."[3]

In my reading, nature is the nothing that is there, while culture, art, and science are the nothing that is not—in the sense that it is a fairy tale. Nature and culture create the world we inhabit, an unknowable world and endless approximations of it. This interpretation makes sense in the context in which Stevens was educated and began writing—a world where snow, a central trope in American culture since Emerson, was valued as an ecological resource, cursed as urban refuse, and bandied as a symbol of change and uncertainty by writers and painters. The visual arts are a good place to begin an explication of modern philosophies of snow, "minds of winter," because Kent's *The Trapper* seems to be a perfect icon of it, because Stevens was influenced by Marcel Duchamp and other painters, and because the depiction of snow in U.S. and Canadian art in the period 1890–1940 is unusually rich.[4]

In the paintings of John Twachtman, for example, we see the influence of European Impressionism and Japanese calligraphic form, but we also see an American response to snow. Focusing on the 7 hectares (17 acres) of his home on Round Hill Road in southwestern Connecticut, Twachtman explored the optical effects of snow and created an ethereal, private world, where nature is almost nothing but is, nevertheless, there, visible in the outlines of trees, rocks, and, above all, unfrozen portions of brooks and streams, with their suggestion of change and their iconographic reference to life within the frozen snowscape. Twachtman's snow is foamy, appearing either fresh and dry or a degree away from melting; his snow softens and mutes the earth.

One of his contemporaries asserted that "he seldom attempts to render the delicacy of a fresh fall of snow, preferring, as a rule, to wait a day or two until the snow has begun to sink down between the rocks and to melt from the higher points." This may be true for the paintings *Winter Harmony* (Figure 7.2) and *February,* but it is clearly not the case in those titled *Winter, Snow,* and *Round Hill Road,* all painted in the 1890s. *Round Hill Road* anticipates the minimalist blank white canvases of the 1960s and could easily have inspired Stevens to think of nothingness. Moreover, Twachtman expressed his enthusiasm for new snow in a letter to a fellow artist in 1891: "We must have snow and lots of it. Never is nature more lovely than when it is snowing. Everything is so quiet and the whole earth seem [*sic*] wrapped

Figure 7.2. *Winter Harmony* (ca. 1890/1900), John Henry Twachtman. Oil on canvas, 65.3 × 81.2 cm (25¾ × 32 in.), National Gallery of Art, Washington, D.C., gift of the Avalon Foundation. Twachtman's foamy snow seems to float, cloudlike, above the banks of Horseneck Brook on the artist's Greenwich, Connecticut, property. The melting snow feeds the unfrozen stream in a life-renewing cycle that expresses more optimism than many late nineteenth-century paintings.

in a mantle. That feeling of quiet and all nature is hushed to silence."[5]

No wind blowing in the same bare place, few evergreens crusted or shagged with snow, just bare hemlocks quietly lining the brook, providing a simple, accessible nature. The philosophical distance between Twachtman and Stevens is a measure of the changes in American culture between the 1890s and the 1920s. Twachtman was part of a generation that came of age amid the first religious and scientific debates over Darwinism. Like many others, he chose to avoid conflict and seek solitude. Snowbound Round Hill Road was an ideal retreat, like his contemporary Robert Frost's in Franconia, New Hampshire. Frost's poems are generally bleaker than Twachtman's paintings, but they often express a similar concern for snow as a surface on which light plays its tricks. The speaker in Frost's "Afterflakes" sees his shadow "in the thick of a teeming snowfall" and wonders if it is cast by some internal darkness. But when he looks up to "where we still look to ask the why/Of everything below," he discovers:

The whole sky was blue;
And the thick flakes floating at a pause
Were but frost knots on an airy gauze,
With the sun shining through.

The critic Rachel Hadas explains this poem as an expression of Frost's search for answers to the questions "Is the world heliocentric or anthropocentric? Where are we to look for the explanation of what we perceive; and how far (or deep) can we perceive at all?"[6] Snow is a complex metaphor in Frost, but in general it conveys, as it does in Twachtman, the artist's attempt to escape from reality.

The art historian Deborah Chotner rightly contrasts Twachtman's snow scenes with those of other popular artists of his time, especially the New York painter Walter Launt Palmer, who was widely admired for his winter scenes and was thought to be one of the first American artists to paint colored shadows on the snow. Palmer's snow is more solid, frozen in place by the longer, harder winters of upstate New York. A typical Palmer snowscape, *Near Newcomb Lake, Adirondacks* (1898) or *The Glade* (ca. 1913), focuses on a snow-covered path through a forest of evergreens whose branches have shed their snow. The viewer's eye is led to a bright sky or snow patch; the mood is also sunny. Other New York painters, Birge Harrison, John F. Carlson, and Walter Koeniger, for example, similarly depict snow that has grown dense on the ground and looks to remain through the winter but is not threatening humans or nature. Regional differences in snowfall, all but ignored by art critics, became important in this period.[7]

The contrasts between U.S. and Canadian snow paintings were especially sharp in the years 1915–1930, when the Group of Seven, led by Lawren Harris and A. Y. Jackson, emerged. These artists were inspired to explore the meanings of winter landscape by the Finnish painter Akseli Gallen-Kallela and the Norwegian Edvard Munch, but the primary stimulus was love of their country. The young Canadians sought, in the words of Harris, to depict

> the charged air, the clarity and spaciousness of our north country. For it has in it a call from the clear, replenishing, Virgin north that must resound in the greater, freer depths of the soul or there can be no response . . . we are in the fringe of the great North and its living whiteness, its loneliness and replenishment, its resignations and release, its call and answer—its cleansing rhythms. It seems that the top of the continent is a source of spiritual flow that will ever shed clarity into the growing race of America, and we Canadians being closest to this source seem destined to produce an art somewhat different from our southern fellows—an art more spacious, of greater living quiet, perhaps of a more certain conviction of eternal values.[8]

At first, members of the Group focused on painting snow on the pines and tamaracks of Algonquin Provincial Park, north of Toronto; then they moved west to the Algoma region on the northern shore of Lake Huron, where they produced striking work such as Jackson's *First Snow, Algoma* (1920), in which hundreds of large snowflakes float above the red, green, and gold of the autumnal forest (Figure 7.3). Blackened stumps from an old burn provide an even clearer contrast for some of the snow crystals. The whole composition forces the viewer to think of the coming winter in terms of extravagant sensuality, not the stingy sound of the wind and a few leaves. Harris was the most restless of the Group, seeking "true north" in increasingly distant travels to the interior of Canada.[9] In *Mountains and Lake* (1929), painted after he visited the Arctic and the Canadian Rockies, a glacial Parthenon sits astride a notch in white and blue mountains above a green lake (Figure 7.4). Harris shared

Figure 7.3. *First Snow, Algoma* (1920), A. Y. Jackson. Oil on canvas, 108.0 × 128.2 cm (42½ × 50½ in.), McMichael Canadian Art Collection, Kleinburg, Ontario, in memory of Gertrude Wells Hilborn. Big snowflakes falling on an old burn in the red and yellow autumnal forests of the Algoma region of Ontario suggest the power of snow to bury the past and re-create the landscape. Nature is more active in the paintings of the Canadian artists of the early twentieth century than it is in those of their Yankee counterparts.

Figure 7.4. *Mountains and Lake* (1929), Lawren Harris. Oil on canvas, 91.9 × 114.4 cm (36¼ × 45 in.), McMichael Canadian Art Collection, Kleinburg, Ontario, gift of Mr. R. A. Laidlaw. In search of "true north" in the Canadian Rockies, Harris transformed nature into an imaginary temple, a Valhalla from which new heroes might emerge.

Kent's sense of awe when confronting glacial snows, but not until the paintings of the American Neil Welliver in the 1980s do we see a combination of bold abstract design and feeling for motion in nature, the constant metamorphism of snow.

Artists of the Pennsylvania School, loosely affiliated with Edward Willis Redfield of New Hope, combined impressionist techniques with realistic detail, "a direct, virile rendering," as the critic Eliot Clark put it in 1922. Following the conventions of Twachtman, almost all the New Hope painters incorporated a stream or river, framing its current with banks of snow broken by trees and their shadows. Buildings figure more prominently in their work than they do in that of other snow painters. Redfield's *New Hope* (1926), for example, is an almost Canadian scene of a village street, with cozy two-story houses and a sleigh pulling a boy on a sled. More impressive is his large canvas *River Hills* (1927), which depicts a steep, snow-covered hillside broken by yellow willows (Figure 7.5). The perspective combines close attention to snow conditions in the foreground and a more symbolic blanket of snow in the distance.[10]

Figure 7.5. *River Hills* (ca. 1927), Edward Willis Redfield. Oil on canvas, 127.0 × 142.4 cm (50 × 56 in.), Memorial Art Gallery, University of Rochester, gift of the Children of Daniel W. Powers. Continuing the realism of Moore, Redfield depicts the subtle effects of snow on trees, shrubs, and weeds that flourish just beyond the factories and towns that have transformed the riverbanks. Snow helps reduce nature to its essence and remind the viewer that there are many things beyond his understanding.

The reformer and journalist Benjamin O. Flower wrote that Redfield's "noblest work pictures the great Mother mantled in her shroud of snow or somber and silent in the recuperative sleep that so resembles death." The motif of renewal was not limited to the New Hope painters. Aldro Thompson Hibbard, the Vermont artist, Willard Metcalf of southern New Hampshire, and Canadian-born Ernest Lawson, who studied with Twachtman and painted in New York City, all did scenes of melting snow, expressing an optimism and faith in nature that artists such as Winslow Homer and Rockwell Kent had lost.[11] The regional artists' continued success after they were declared "sentimental" by avant-garde critics in the 1920s shows that loss of faith was not wide-

spread and that there was an intellectual toughness in their work that the critics missed.

The next generation of painters, contemporaries of Stevens and Kent, shared their skepticism about an underlying reality. Influenced by the bleak snow paintings of Winslow Homer—*Sleigh Ride* and *Fox Hunt* (both 1893)—George Bellows and John Sloan saw the traditional association of snow with seasonal renewal as anachronistic. Snow for Homer was a place of Darwinian struggle for survival, the last scene of the last act. For Bellows snow was also an experiment in urban aesthetics—the opening scene of a new drama in which the machines of city life, huge buildings, locomotives, tug boats, tested their strength against nature. The result could be striking, and, according to Bruce Robertson, Bellows's contemporaries recognized that his work offered an antidote to both Redfield's excessive optimism and Homer's unrelieved pessimism.[12] *North River* (1908), in which the white steam from a locomotive and three boats contrasts with the grayer tones of the snow, suggests both subtle harmony and conflict between humanity and nature (Figure 7.6). Snow rests on a park bench in sculpted forms, while bare trees seem artless.

Meanwhile, photographers of snowscapes were competing with painters for public support, but advice about snow pictures given by writers in both art and photography magazines tended to be similar. J. A. Blaikie, writing in the British publication *Art Journal* in 1885, advised photographers to avoid high contrast between sun and snow and seek the diffused light of overcast days. Eight years later *Photo-American* advised the same thing, telling aspiring winter photographers to "avoid a subject in which there is much dark material in the foreground, and equally avoid a subject which has no foreground at all." Whereas Blaikie encouraged pictures of "snowy tree-forms . . . [with] the exquisite delicacy of their tracery," the American critic condemned the "reticulation of strong dark against light, e.g., network of dark tree trunks, palings, open woodwork gates, etc. The effect of this is to irritate by a chessboard-like pattern of black and white patches." He recommended, "Wind-drifts and their exquisitely transparent shadows are of immense value in expressing the luminosity of snow." By the beginning of the twentieth century, artists, like scientists, were learning about the complexity of snow. Frank Townsend Hutchens refers artists to John Ruskin, encouraging them to remember that "every hour writes its history upon the surface of the snow which becomes pitted with little hollows and covered with a hard shiny crust which sinks down and breaks as the softer snow beneath melts away."[13]

The awareness of time, of snow metamorphism, and the reflecting power (albedo) of snow helped to create new standards for evaluating snow art. All commentators were especially anxious to dispel the belief that snow is purely white. Artists who failed to color their snow were judged naive. Increasingly they wanted to see evidence of the wind's effect on snow. Pho-

Figure 7.6. *North River* (1908), George Bellows. Oil on canvas, 83.0 × 109.2 cm (32⅞ × 43 in.), Courtesy of the Museum of American Art of the Pennsylvania Academy of the Fine Arts, Philadelphia, Joseph E. Temple Fund. Snow on the shores of Manhattan Island and New Jersey blends with the steam from boats and a locomotive to create a landscape that explores nature's interaction with culture. As the viewers' eyes travel from the park bench in the foreground to the distant shore, they journey to the unknown, leaving the security of steam-powered civilization frozen in the middle distance.

tographers, who were limited by the existing technology to black-and-white representations, learned to express both the psychological impression of a snow scene in the manner of the painters and the precise detail that the camera could capture. In 1893 Alfred Stieglitz, influenced by the Austrian physicist Ernst Mach's ideas about apprehending phenomena directly by focusing on underlying principles, and by the German Heinrich Hertz's "contention that science did not explain facts, but provided rational models of patterns of experiences," made his photograph *Winter—Fifth Avenue, New York,* in an effort to answer the question "Could what I was experiencing, seeing, be put down with the slow plates and lenses available?" He went on, "The light was dim. Knowing that where there is light one can photograph, I decided to make an exposure" (Figure 7.7).[14]

Figure 7.7. *Winter—Fifth Avenue, New York* (1893), Alfred Stieglitz. Carbon print 23.0 × 18.4 cm (9 1/16 × 7 1/4 in.), Alfred Stieglitz Collection, © 1996 Board of Trustees, National Gallery of Art, Washington, D.C. Stieglitz used a hand-held camera to capture a moment in a snowstorm in New York City. In the original photo three men are shoveling snow on the sidewalks. By cropping the image Stieglitz eliminated the organized struggle of humans and nature in cities and emphasized the more random effects of wheel ruts and hoofprints.

Thanks to the research of Geraldine Wojno Kiefer, we are better able to understand the interplay of science and art in the work of Stieglitz and his followers. In the photograph that became *Winter—Fifth Avenue,* Stieglitz used the snow's albedo to compensate for the lack of sunlight. The original photo contains a wide expanse of the avenue, including snow shovelers on the sidewalks. The horse-drawn carriage, which appears to be about 6 meters (20 feet) from the camera in the published version, was a much smaller element in the uncropped photograph. Cropped and enlarged, the version published in *Camera Work* in 1905 centers the carriage and encourages the viewer to speculate on the driver, whose outstretched right hand may be a signal to the sleigh behind him, a wave of greeting, or a warning to the photographer. I agree with Kiefer, however, that the focus of the photo remains "the paths of driving snow and the tracings formed around the approaching vehicle."[15] The slow film turns the falling snow into white blurs highlighted against the black horses and gray buildings, and the wheel-rutted snow in the street is a textbook illustration of snow metamorphism under compression, wind, and changing temperature.

Rudolf Eickemeyer, a commercially successful New York photographer, preferred rural to urban snow and was more sentimental toward nature than Stieglitz, but his work resembles the *Camera Work* photographers' in focusing on the snowy foreground. Eickemeyer's 1903 book, *Winter,* consists of several dozen photographs of snow scenes, each accompanied by a poem. Most of the photos resemble the paintings of Palmer or Twachtman. A Palmer-like path through a forest is captioned with a poem by the travel writer Bayard Taylor:

> The beech is bare, and bare the ash,
> The thickets white below;
> The fir-tree scowls with hoar mustache,
> He cannot sing for snow.

We are a long, cold journey away from "The Snow Man" and *The Trapper* here, but Eickemeyer's volume has a brief introduction by the bohemian poet and translator of Japanese literature Sadakichi Hartmann, which suggests that those who preferred more traditional nature photography were also becoming snow minded. "Winter is the most intellectual of all the seasons," writes Hartmann, who mentions that it is also the most honest. The book, he continues, "has . . . done us . . . [an] invaluable service by popularizing Nature pictorially and idealizing its more intimate realistic qualities." These qualities, Hartmann believed, were neglected by painters.[16]

Stieglitz and his generation sought to use their cameras as painters used their brushes—to capture the nothingness that is there, elusive but palpable. Younger photographers, born in the twentieth century, accepted the epistemology of science without agonizing about it and set to work recording in as much detail as their equipment allowed the aspects of nature that seemed to them to convey an ecological as well as an aesthetic message. Ansel Adams, whose long association with Yosemite, from 1920 to his death in 1984, has produced hundreds of beautiful snowscapes, is one of the most important of contemporary nature photographers. Adams took photographs of single pine branches, *Snow* (ca. 1932, Figure 7.8); portions of forests, *Winter* (ca. 1932); and large views of the whole park, *El Capitan, Sunrise, Winter, Yosemite Valley, California* (ca. 1968). Each conveys the feeling of what he called the "light heaviness" of new snow. Snow illuminates the dark, museum-like quality of the valley. Adams, playing the role of curator, invites us to look at but not touch nature's art.[17]

Adams and the snows of Yosemite are forever linked, but each snowy place has its photographic interpreter. Often a photographer was hired to promote a private ski development or winter tourism in general, but in the cases of Ray Atkeson on Mt. Hood and Ferenc Berko at Aspen, the results were happy combinations of art and commerce. Atkeson moved to Portland, Oregon, in 1928 because he had seen pictures of the mountains. A pioneer in ski photography, he also took photographs of trees twisted by snow and wind, frost crystals as large as lily pads on a pond at Sun Valley, and ski tracks winding sensuously across new snow. Atkeson is refreshingly unpretentious about his work: "Ideal photographic conditions have an annoying

Figure 7.8. *Pine Branch, Snow, Yosemite* (ca. 1936), Photograph by Ansel Adams, © 1996 by the Trustees of the Ansel Adams Publishing Rights Trust, All Rights Reserved. The close-up photo of *quali* highlights the way nature inspires an art that nudges the viewer back to a closer look at nature.

habit of synchronizing with superb ski conditions. Too often, I have succumbed to the lure of the physical pleasure and have missed many glorious opportunities to add to my photographic files." Atkeson's photos also reveal the difference in style between government-operated resorts, such as Mt. Hood (Figure 3.5), and the Disney-like developments of places like Aspen. In Walter Paepcke's vision of schuss and seminar, only a European photographer would be capable of recording the Aspen ideal. Ferenc Berko was hired in 1948 and remained to take artful photos of snow as well as portraits of the rich and famous. Fence posts, benches, sedges—all made abstractions in the snow that appealed to Berko's eye. Atkeson and Berko represent distinct points on a continuum of photographers associated with winter recreational areas. These men and women provide a visual record of American snow valuable as art and business.[18]

Photographers on other assignments have sometimes been diverted by the appeal of snow. Marion Post Wolcott, working for the Farm Security Administration in 1940, took a series of photographs in Woodstock, Vermont, and North Conway, New Hampshire, that are as aesthetically pleasing and culturally revealing as any by Ansel Adams. Her photograph of Woodstock at night, with the street light falling on the ridges of snow piled along the sidewalk, is singular in its focus on a familiar urban form of snow that passes almost unnoticed in everyday life unless we happen to stumble over it and break through its thin crust (Figure 7.9). The potential dangers of what might be called "curb moguls" are all but obscured by their beauty. Her photo of some cabins in North Conway, captioned "After a storm," combines the attention to village architecture and the delicate tracery of snow in the surrounding forest found in the great paintings of James Wilson Morrice and Lawren Harris. The wide expanse of even, white snow in the foreground invites contemplation of social frontiers and the boundaries of community.[19]

An interesting recent use of the camera to record snow conditions is Douglas Huebler's *Location Piece No. 5, Massachusetts–New Hampshire.* The photographer comments, "On February 7 [1969], ten photos of snow lying 12 feet from the edge of Interstate Highway 495 . . . [were] made at an interval of every 5 miles; or at every 5 yards; or of every 5 feet; or of a variable combination

Figure 7.9. *After a Blizzard, Center of Town, Woodstock, Vermont* (1940), Marion Post Wolcott. Prints and Photographs, Library of Congress. Taken while the photographer was on assignment for the Farm Security Administration, the photo explores the snow forms of the street, the wind-sculpted remains of snow removed by shovelers, the trampled sidewalk, the slow, viscous deformation of snow as it slides off awnings and the roofs of cars.

of all of those intervals. Ten photographs and this statement constitute the form of this piece."[20] Clearly this is in the tradition of Duchamp's "ready-mades," a seemingly random selection of snow forms created by highway plows and traffic. Dirty clumps of snow are far from Adams's aesthetics, yet the photos, arranged in two rows of five and looking at first glance like cloud formations, are wry comments on art and science. If Wilson Bentley could turn his snowflake photographs into an album of hexagonal designs used to illustrate meteorological textbooks and Christmas cards, then Huebler's interstate snow clumps might illustrate snow removal manuals and Groundhog Day greetings.

Humor and pathos pervade many of the depictions of snow in motion picture photography as well, from Charlie Chaplin's *Gold Rush* (1925) to Orson Welles's *Citizen Kane* (1941) and Frank Capra's *It's a Wonderful Life* (1946). Chaplin recalled in his autobiography that he was inspired to make a movie based on the Klondike gold rush because he had seen some stereograph photos of the event, and the opening scene of prospectors struggling up the Chilkoot Pass was an exact reconstruction of a widely reprinted news photograph. A visit to Donner Pass must have led him to read C. F. McGlashan's *History of the Donner Party,* because the scene in which Charlie eats his shoelaces is drawn from that book, in which the author writes, "So hungry were the poor people that some of them ate the strings of the snowshoes which part of the relief party company had brought along."[21] Chaplin's ability to build from one snow-related joke to another culminates in the snow-shoveling scene, in which Charlie, in an attempt to earn money to have a New Year's party for the dance hall girls, furiously covers one walk while clearing another. Although he doesn't break his arm, the resulting fiasco almost breaks his heart. Snow is absurdly funny, but it is also sensibly romantic.

Welles's use of a snow globe, which he had specially made in the RKO shops, to evoke Charles Foster Kane's memories of his childhood in New Salem, Colorado, was an inspired gimmick. As the story reels backward from Kane's death and the breaking of the globe, the audience learns that Kane had a sled on which were painted a flower and the word *Rosebud.* The sexual meaning of the name, as Robin Bates has explained so well, stems from a private joke between the scriptwriters, but it is meant to convey Kane's conflation of his love for his mother and his mistress. Welles plays with the American belief in the redemptive power of love. "Rosebud" does explain a man's life if it stands not just for the sled but for all Kane's memories evoked by snow. Snow symbolizes the inconstancy of his mother and mistress and Kane's futile attempt to stop time. The snowball thrown by young Kane strikes the sign of his mother's boardinghouse, making an emphatic period after the "Mrs." that precedes her name. By contrast, the snowball thrown by young George Bailey at his future wife, Mary, in Capra's movie is an oeillade, not an Oedipal glance. Capra uses snow more conventionally than Welles. As Charles Maland has observed, "Snowflakes signal George that he has returned to life," and Raymond Carney believes that Capra, like Robert Frost, used snow to explore a central tension in American art, the temptation to lose oneself in a vision of loneliness and the calling back to the world and society.[22] The fact that movie snow is often bleached cornflakes, soap flakes, gypsum, Foamite, and shaved ice merely underscores the paradoxes.

Finally, two American painters continue to make major contributions to the representation of snow—Andrew Wyeth and Neil Welliver. At eighty Wyeth is one of the few contemporary artists to receive both the praise of critics and the admiration of the general public. His realist style and recognizable rural American subjects make him immediately accessible. There is a strong narrative element in his paintings too; the same subjects recur and his human subjects age. A series of watercolors and tempera works in the 1940s depicts a freshly killed buck hanging in a barn after it has been gutted. Specks and clumps of snow dot the ground and buildings. The scene is desolate or haruspical, depending on one's feelings about hunting. For Wyeth's neighbor, a successful hunt meant food for the coming winter. It portended survival. Wyeth wrote to a friend, however, "I had my watercolors under my arm as usual in case something hits me. . . . I ended up in a barnyard where I found a beautiful young buck hanging from the top of an open barn door. . . . The sight of this beautiful young beast hung by its hind legs frozen stiff effected [*sic*] me deeply. I stood and watched it for some time as the snow sifted in through the open door on it. The result is a watercolor which I think is one of my very best."[23]

Although most of Wyeth's landscapes look bleak, his comments on his sketches and paintings sometimes reveal his love of snow, which he depicts almost one crystal at a time. In the painting *Snow Flurries* (1953), he has the snow outline the tiniest irregularities of a windswept hillside. His notes on this study mention the way the earth was soaking up the snow, the sound of crows in the distance, and the shadows on the clouds, which were changing too quickly to be captured in his sketches.[24] His techniques lend themselves to falling snow, and the drybrush work *First Snow* (1959) is highly successful in illustrating the effects of wind swirling around a barn and a tree.

Like most of the artists who have been attracted by snow, Wyeth focuses on specific places, Chadds Ford, Pennsylvania, and Kuerner's Farm, Maine, and provides a historical record of their winters, underscoring the close links of snow, place, and memory. In *Ring Road* (1985) tire tracks on a snow-covered road in Maine disappear at the crest of a hill; below are the collapsing walls of a snow-filled ruin, while a highway sign indicates a hidden curve (Figure 7.10). Wyeth has captured a tranquil moment after a snowstorm; the struggle between humans and nature has already occurred. Another paradox emerges—each snowstorm changes a place but also returns it to a place unchanged in memory.[25]

Neil Welliver is twelve years younger than Wyeth and more influenced by the Abstract Expressionist phase of American art, but his landscapes and

Figure 7.10. *Ring Road* (1985), Andrew Wyeth. Tempera, 42.9 × 101.0 cm (16⅞ × 39¾ in.), Private Collection, photograph courtesy of The Wyeth Collection. Wheel tracks on a snow-covered road at the crest of a hill and stone walls framing fields in the distance are geometric patterns that contrast with the wandering clumps of grass that have shed their snowcover, the leafless, branching tree, and even the ruins of a church and the sign warning of a curve. Nature and culture are mixed in form and substance.

snowscapes near his home in Lincolnville, Maine, are no less recognizable. The influences of modern art are in the size of his canvases—many of them are 2.4 meters (8 feet) square—and by the high gloss and hard edges of his trees, mountains, lakes, and snowfields. It is the paradoxical brightness of a winter day that appeals to Welliver. "Winter is just that white snow bouncing color in all directions," he wrote, "and everything is intensified terrifically. A lot of the brightest colors disappear because they are dead, they are dormant. But the color that is there, those brown violets, those blacks, those greens that are on very low life forms, they are phenomenally luminous. In the summer you cannot see them because they are completely overshadowed by the leaves and other organic forms." Nothingness becomes the colors that are and are not there. Both the critic Frank H. Goodyear, Jr., and the poet John Ashbery see in Welliver's snow scenes the celebration of confusion.[26] In *Snow Painting* (1974), for example, snow-covered underbrush and the trunks and branches of a dense forest of trees flecked with snow make a restless abstract pattern without a focal point. Another painting, *Old Avalanche* (1982), with its steep, rocky hillside overgrown with birch saplings, creates a houndstooth pattern of black and white, with hints of yellow in the birch trunks and blue in the shadows on the snow. In *Shadow* (1977) Welliver rejects the traditional penumbra of forest shade to polarize the snowy forest into bright branches and shaded ground (Figure 7.11). More than an illustration of the snowfall interception by trees controversy, this painting raises epistemological questions about what we see when nature plays with light.

In the paintings with falling snow, *Snow on Alden Brook* and *Storm's End and Sunlight* (both 1983), and the magnificent 2.4 by 3.0 meter (8 by 10 foot) panorama *Late Squall* (1984), Welliver gives us a glimpse of nothing and everything. As Ashbery writes:

> If a "slapdash" look were what he was after, that would be one thing, but the end result is as complex and unforgiving as mathematics, nowhere more than in the recent *Snow on Alden Brook,* with its thousands and thousands of snowflakes that threaten to obliterate the thousands of bare twigs on the trees that fade away into the heart of the snowstorm. Here is a major embodiment of one of the paradoxes of Welliver's art: the falling flakes are patterned as rigidly as God would have programmed them, nothing is left to chance, yet what emerges is a powerful, subliminal projection of an ephemeral moment: you can smell and taste the cold, and the damp penetrates your bones even as you wonder at the magisterially orchestrated and ordered precision of the complex surface, which would seem to leave no quarter to stray sensual impressions.[27]

It is an art that goes beyond Stevens and Kent to ask whether chaos, not order, is discovered or invented.

RHYME ON RIME

Snow and poetry share a balance of form and content. Like poetry, snow can fuse intellect and emotion, sense and intuition, by appealing to our eyes and to our memories. Poetry employs sight, sound, rhythm, and metaphor; snow draws its power from shape, color, texture, and temperature. Snow is to water what poetry is to prose, an elaboration and a source. "In prose," the Mexican poet Octavio Paz writes, "the word tends to be identified with one of its possible meanings at the expense of others . . . the poet, on the other hand, never assaults the ambiguity of the word." Poetry is to snow what prose is to rain, says Howard Nemerov, because "it flew instead of fell."[28]

Emerson, Dickinson, and Stevens discovered useful metaphors in snow. Poets of the late twentieth century elaborate theirs and propose others: (1) Snow as negation or nothingness, including isolation, silence, timelessness, death, and the unknown. (2) Snow as life, including love, purity, and fragility. (3) Snow as creativity, particularly the tension between intuition and imagination; and snow as memory, especially the recollection of

Figure 7.11. *Shadow* (1977), Neil Welliver. Oil on canvas, 243.8 × 243.8 cm (8 × 8 ft.), Museum of Modern Art, New York, gift of Katherine Lustman-Findling, Jeffrey Lustman, Susan Lustman Katz, and William Ritman in memory of Dr. Seymour Lustman. Photograph © 1996 The Museum of Modern Art, New York. Welliver's meticulous rendering of the effects of light and shadow on a birch grove in the snow comments on the delicate balance between fixity and change, the movement from present to past and future.

childhood playfulness. (4) Snow as order and apparent disorder, chaos and complexity. (5) Snow as word, the ultimate Logos, which, if it could be named, would become substance, resolving the old dualism of language and reality. The nothingness that is there.

Stevens's snow man sees beyond the nothingness suggested by snow, but thoughts of misery and death are present in the poem. Isolation, and the intimation of our ultimate separation from life, evoked when snow falls around the poet, is a traditional theme and not always a morbid one. Silence, an isolating condition that many poets have found in snow, is beautifully employed by Elinor Wylie in "Velvet Shoes." "Let us walk in the white snow," says the speaker in the poem, "In a soundless space." Using "veils of white lace," silk, wool, cow's milk, "the breast of a gull," "white down," and "silver fleece," as analogies of snow, Wylie concludes the short poem:

We shall walk in velvet shoes:
  Wherever we go
Silence will fall like dews
  On white silence below.
  We shall walk in the snow.

The image of velvet shoes is incongruous unless the reader sees the walking couple as snow men, in the sense that Stevens used the term. The figures, "shod in silk,/And . . . in wool," simultaneously suggest a nuptial and a funeral. The poem conveys emotions so pure that sound is superfluous.[29]

Snowfall-inspired timelessness is inherent in painting, a synchronic medium, as Anthony Sobin observes in his poem "Problems in Painting: Spring Landscape with Melting Snow." After commenting on the way artists depict rain in flight as either tiny globes suspended in air or thin lines, he writes:

And so the problem of snow
at the point of melting—
do we surround each clump of white

with a dozen parentheses—to indicate,
perhaps, a shuddering,
as a soul might shudder for an instant

before leaving its body? Or is it
to be, as we say, frozen in time?
Years later we hang the snowscape, unfinished

in the sun of a parlor in mid-July
and the light entering the windows is as
white as the lace curtains, which do not stir.

We await the thaw like a bulb.
We feel the green shoots stirring—
growing in our bodies like palm fronds,

ourselves the still life,
terrified of permanence
and too frightened to move.

Frozen by a fear of the permanence of death symbolized by the painted snow that cannot melt, we are left to contemplate the equally terrifying possibility of rebirth after a thaw, the transition from the nothingness we can behold to an as yet unimagined nothingness. Snow metamorphism illustrates the tension between fixity and flux, and suggests a distinction between timelessness and eternity.[30]

Death and snow are forever linked in the minds of readers of English by the final lines of James Joyce's *Dubliners:* "His soul swooned slowly as he heard the snow falling faintly through the universe and faintly falling, like the descent of their last end, upon all the living and the dead." American poets, from Robert Frost and Edna St. Vincent Millay to Donald Hall and Howard Nemerov, imbued with Joyce's mordant modernism, have given snow its familiar negative hue. First and last snowstorms are especially symbolic. In "First Snow," Howard Nemerov recalls:

> It used to be said that when the sun burns down,
> Being after all a mediocre star
> Of the main sequence, mortal as ourselves,
> One snow will seal the sleepy cities up,
> Filling their deep and canyoned avenues
>
> Forever. That will be the day. And for all
> I know it may be true; at least it was
> One vulgarized version of The Second Law
> A century ago, and almost all
> The celebrated authors did it up,
>
> A natural: "London, Peking, Moscow, Rome,
> Under their cerements of eternal snow,"
> And so on; writing was a powerful stuff
> Back then, and tales of entropy and The End
> Could always snow the middle class.

Nemerov and the naturalist-poet Loren Eiseley are eulogists of late twentieth-century civilization, and if there were no snow they would have had to invent it. In one of Nemerov's most powerful poems, "The View from an Attic Window," the speaker watches snow fall outside his childhood home and thinks of all those who have died and who will die. In "The Snowstorm," Eiseley rejects the notion of first and last snows to argue that "since snow/is always with us . . . can you say I am not composed of snowflakes?" and concludes:

> Do not bother to throw up the window,
> snow is already blowing
> the room is disassembled,
> our substance,
> the room's substance,
> is snowflakes;
> we are falling apart now,
> we have re-entered
> the eternal storm.

These musings are partly a product of the poets' postwar era, partly a reflection of their personalities, but mainly just another chapter in the constant tension between humanity's capacity to dream and our awareness of mortality, a condition made palpable by snow.[31]

The second theme of snow poetry—life—with the corollaries of love, purity, and fragility, was more typical of earlier centuries. In banishing sentiment, modern poets have banished love, at least love as pure as the wind-driven snow. Michael McFee, a poet and critic, manages in "Snow Goat" the difficult feat of writing an unsentimental love poem, in which snow masks desire:

> Stirred by the snow's rare erasure, my sister
> hikes across back yards quiet as a bedroom
> into the ruin-rich woods where Jack kissed her,
> then disappeared beyond the briers and gloom.
>
> *Who cares?* she says, skiing shallow ravines
> behind the vacant sausage plant. *What a bore.*
> *I hope he gets killed in Korea with the Marines*
> *or Army. Nothing could surprise me anymore.*
>
> Then snow at her elbow says *bah* and there
> he is, shocking as God, white, his slitted eye
> golden with curiosity. He tongues the bare
> salt lick of her hand, and she starts to cry,
> feeling his cloven hooves and her desire
> sharp as those horns whetted on the wire.

Robert Penn Warren invokes the purifying qualities of snow in "Function of Blizzard." Thinking of poverty-stricken ghettos, he asks God to redeem the Bronx by covering it with snow:

> Bless snow, and chains beating undersides of fenders.
> Bless insane sirens of the Fire Department
> And Christmas whirl of alarm lights. Bless even
> Three infants locked in a tenement in Harlem.
> God's bosom is broad. Snow soon will cover the anguished ruin.[32]

Memory triggered by snowfall, the third of the major poetic themes, usually returns to childhood, as in May Swenson's "Snow by Morning," which begins

> Some for everyone,
> plenty,
> and more coming—

And ends:

> By morning we'll be children
> feeding on manna,
> a new loaf on every doorsill.

Richard Hugo, whose lines introduce this book, is the great contemporary poet of bittersweet memory. The full text of "Snow Poem" is worth quoting:

> To write a snow poem you must ignore the snow
> falling outside your window.
>
> You must think snow, the word as a snotty owl
> high on the telephone pole
>
> glowering down and your forehead damp with fear
> under the glare
>
> of the owl who now is mating. On rare days
> we remember the toy
>
> owl we buried under the compost heap,
> white sky passing above, warm chirp
>
> of wren and the avenging hawk.
> That was summer. Let us go back
>
> to snow and forget that damn fool lecture
> I gave last winter.
>
> Well, then: here is your window.
> The storm outside. Outside, the dead dove drifting.

The sensibility here is as ancient and elevated as that of Sung dynasty poets. Hugo's poem also parodies Joyce Kilmer—fools can make only poems, not snow or trees. But the deeper meaning is articulated when, in another poem, Hugo advises us to "pray hard to weather, that lone surviving god,/ that in some sudden wisdom we surrender."[33]

Robert Bly is a poet who constantly raises the paradoxes of memory, whether it is retrieved or invented, and the related issue of the difference between intuition and imagination, in the sense that the former means becoming one with the world while the latter implies a more rational process of creation of order. Snow in Bly's 1962 book, *Silence in the Snowy Fields,* is a physical material to be used in inducing certain intense emotions, as in "Three Kinds of Pleasure":

> I
> Sometimes, riding in a car, in Wisconsin
> Or Illinois, you notice those dark telephone poles
> One by one lift themselves out of the fence line
> And slowly leap on the gray sky—
> And past them, the snowy fields.

II
The darkness drifts down like snow on the picked cornfields
In Wisconsin: and on these black trees
Scattered, one by one,
Through the winter fields—
We see stiff weeds and brownish stubble,
And white snow left now only in the wheeltracks of the combine.
III
It is a pleasure, also, to be driving
Toward Chicago, near dark,
And see the lights in the barns.
The bare trees more dignified than ever,
Like a fierce man on his deathbed,
And the ditches along the road half full of a private snow.

In this poem the amount of snow diminishes as the poet progresses toward a moment of religious ecstasy. A miracle occurs when the optical illusion that makes it seem as if telephone poles are leaping is replaced by an image of snow imprisoned in the wheel ruts of technology, only to become the pure sensory experience of a ditch "half full of a private snow." The snow is private, I think, because it belongs only to the poet, who links the leaping telephone poles, the dignified trees, and the dying man in a moment of transcendence made possible by the snow, which in its own natural cycle of melting contributes to regeneration. Making the snow in the half-full (not half-empty) ditch a private talisman confers pleasure.[34]

The fourth theme, snow as order amid chaos, appears on at least two levels. Jane Flanders's "The Students of Snow," a seemingly straightforward history of snow science, illustrates the first. After naming different types of snow crystals, she says:

The students of snow would never say
snowflakes are unique.
They would only admit they have never
seen two alike.
And no two students of snow are alike: Professor
Squinabol, Mrs. Chickering,
Herr Sigsun, and "Snowflake"
Bentley, the first microphotographer of snow,
bending almost invisibly
over his slide in a white smock.

Flanders concludes the poem with more scientific facts about snow and these haunting last lines:

And then they cease to speak, caught up
in the study of silence, transparency,
the radiant wedding of nebulae.

This simple but revealing history of snow science and pseudoscience in verse could be a coda for this book. Flanders suggests that some students of snow moved beyond taxonomy to teleology.[35]

The other approach to finding a sense of order in snow is that of A. R. Ammons. Ammons is a difficult and sometimes frustrating poet, always destroying the structures he creates, like a child with blocks. Even before his book *The Snow Poems* in 1977, he had discovered in snow a metaphor of his work:

Especially the fallen tree
the snow picks
out in the woods to show:

the snow means nothing by that,
no special emphasis: actually
snow picks nothing out:

but was it a failure, is it,
snow's responsible for
that the brittle upright black

shrubs and small trees
set off what caught the snow
in special light:

or there's some intention
behind the snow snow's too shallow
to reckon with: I take it on myself:

especially the fallen tree
the snow picks
out in the woods to show.[36]

Beginning with the wordplay of the title, Ammons looks to the fallen snow for "some intention/behind the snow," but punning again he finds the "snow's too shallow," leaving open the question of whether a deep snow might reveal a deeper order of things. Open too is the question of whether the meaning simply isn't there or whether the poet has just missed it. Ammons takes it on himself to provide an intention, but the final stanza merely repeats the observation of the first, the visual circularity of the poem is tautological. What Stephen Pyne says of Antarctic ice, Ammons suggests is true of both poetry and snow: they are solipsisms, about nothing but themselves.

With Ammons we begin to see a new sensibility and a different aesthetic, in which poets, searching for a higher order in the chaos of snow, work back through Stevens to Emerson, treating the snow itself as a poem, then begin tentatively to reexamine and rename snow in the context of late twentieth-century knowledge and experience. The renaming process is the key to revitalizing the symbol. Snow as word made substance.

Ammons makes a major contribution to this fifth category of snow poetry in his book *The Snow Poems,* written in 1975–76 and filled with references to an America obsessed by the late Cold War and the bicentennial. *The Snow Poems* is a long poem in the form of short verse meditations by a man living near Rochester, New York, in the winter before his fiftieth birthday. As Michael McFee points out, the very formlessness of the book is a key to understanding Ammons's choice of snow as a symbol for his life: "What he wants to offer . . . is provisional and piecemeal, not finished—a mishmash, a shindig, a high shimmy uncompletable, an infernal chaos rather than a unified paradise." Ammons has always maintained that he "prefer[s] confusion to over-simplified clarity, meaninglessness to neat, precise meaning, uselessness to over-directed usefulness." It is significant, in the context of this examination of snow, that Ammons's next long poem was about garbage, another exploration of the irony of refuse becoming resource.[37]

"Poetry is the smallest/trickle trinket/bauble burst/the lightest/windseed leaftip/snowdown," Ammons writes, and in another short verse he catalogs a day's snows:

> Snow showed a full range
> today, showers at six
> this morning with
> the temperature falling
> through sleet and grainy,
> gritty, and, now, dusty
> snow
>   a tying-off action with
> cold striking, congealing, the
> last skirts of action
>
> the lawn is whiter than green
> the hemlocks hold touchy sprays

Later in the book, after many snowfalls and meltings, Ammons sees spring returning and snow unable to tie off the cycle of life and death. He plays with the word, printing it backwards—"wons"—and partially upside down—"mons"—turning it as Bentley might have turned a snowflake on a glass plate, then ends his meditation with his now useless catalog:

grit, flakes, sleet, fluff
all day the snow snowed in
vain
nothing but green in the
grass nothing but leaves
in the trees

The next verse surprises, like snow itself, with a new beginning:

It snowed all night snow
like pear-petal snow and has
snowed all
morning, skimpy flakes,
solitary, wandering schools:

Winter is passing, the poet is irritable—"light showers soak my shoes/verse writers croak my nerves"—he has nearly finished his stormy weather report. Patches of words appear like ground breaking through melting snow. Do they contain messages? Earlier in the winter Ammons had observed:

I wrote
nothing: it is
the winter-deep, the
annual sink:
leave it unwritten,
as snow unwrites
the landscape

Ammons provides a dozen names for snow but leaves it to his readers to "make something of it:/think it over and out."[38] His contribution to nivean letters is the demonstration that snow is worth thinking about, that it still has the capacity to astonish and amuse. Centuries of snow poems have not exhausted its meanings.

Cathleen Calbert takes the notion of naming snow as a method of understanding it and its metaphorical meanings to the limit in "Lunatic Snow," in which each of the poem's fifty lines names a different kind of snow or snow condition. The last dozen lines are a good sample:

backsliding snow
snow that refuses to be known as snow
snow between toes
snow on clothes
light snow that is deep for walking
snow overhead and about to fall
avalanche

sex snow, or lying down with angels
mood snow, a special kind
laughing snow
lunatic snow, that steers you into questioning
snow that whispers the unknown names of snow.[39]

The artfulness of the selection and arrangement of names is deliberately concealed. The poem progresses from traditional, well-recognized forms—packed, powder, drifting snow—to more imaginative, personal creations, while retaining the semblance of an index or syntagma for an encyclopedia of snow.

John Frederick Nims makes an ingenious contribution to snow poetry as snow itself in his homage to Johannes Kepler, "The Six-Cornered Snowflake," a poem in twelve stanzas, each shaped like a hexagonal crystal, which tells the story of Kepler's *De Nive Sexangula,* the essay the great astronomer and mathematician wrote for his patron in 1610. As the stanzas progress, Nims describes Kepler walking in Prague in the snow, stopping to look at the city, and letting his mind wander as he thinks about the geometric perfection of snow crystals. Kepler played with the word *nix*—"snow" in Latin, "nothing" in German—and pondered folk and scientific explanations of snow. Other philosophers have sought the hidden meanings of forms, unwilling to accept snowflakes for their own beauty. In the final stanza, Nims expresses his own feeling:

*

He's
had no
truck with such
* "imagineless" commodities; he loves hang, heft, and edge, *
the five Platonic solids scaffolding the universe. All
hollow, yes, but hollow like calyxes for the essential Dream that seminates eternity, whose
faint bouquet is the "astounding beauty"
Harmonice Mundi raptures over still, now, in
our late December. Over Prague of the hundred
towers, jumbled roofs, the winter river, the reconciling
* bridge, down our endangered air, forgiving snow cajoles *
the earth in mu-
sical notes
yet

*

Nims, like Kepler, rejects hidden meanings and accepts snow as an irreducible thing, a material with which to contemplate an elusive harmony.[40]

If poetry is snow and prose is water, some novelists and essayists have managed to recrystallize snow in their writing. Ernest Hemingway is the American writer most obviously identified with snow. After reviewing Hemingway's use of snow and ice in dozens of short stories and the major novels, Bern Oldsey concludes that the author was constrained by the tenets of realism to find "practical," not "intellectual" symbols. Snow, for Hemingway, was simple and operated directly on the emotions. It could represent purity, the good life, even death, depending on context, or all three in "The Snows of Kilimanjaro." Another critic links his story "A Clean, Well-Lighted Place," with its existential refrain, "Our nada who art in nada," to Stevens's "The Snow Man."[41]

Conrad Aiken's frequently anthologized short story "Silent Snow, Secret Snow," first published in 1932, describes a twelve-year-old boy's withdrawal from reality symbolized by the snow falling in his imagination. The story begins with the boy dreaming of being a polar explorer during a geography lesson. On his way home from school, he imagines the street filled with snow and himself at home in bed. Alarmed by his behavior, his parents call a doctor, and they force the boy to tell them what he is thinking. "Nothing," he says, followed by the word "snow." The boy escapes to his room, where, in his mind, the snow promises to tell him the most beautiful story in the world, the story of everything growing smaller. Aiken's story ends with the snow whispering to the boy, "peace," "remoteness," "cold," and "sleep."[42]

My reading of the story focuses on three elements, the boy's desire to be a polar explorer, his evasive answers when asked what he is thinking about, and the final apocalyptic roar of the snow as it calls for peace, remoteness, cold, and sleep. Although Robert Peary, Robert Scott, and Ernest Shackleton are the only polar explorers mentioned in the story, Aiken, writing in the early 1930s, would also have been thinking of Richard Byrd and Vilhjalmur Stefansson and their withdrawal to the remote poles. Like the cartoon character Calvin, who regularly transports himself in his mind to the Jurassic, the boy in Aiken's story possesses an imagination capable of creating an arctic reality from the barest details of an autumn afternoon. The boy's explicit linking of nothing and snow is an allusion to Stevens's poem, which Aiken must have known. The boy has a mind of winter, which also explains the references to peace, remoteness, cold, and sleep. Going into the third winter of the Depression, on the eve of Franklin D. Roosevelt's election, the nation was psychologically disoriented. The boy's desire for escape was widely shared, and his journey home from school, which reviews the history of his town through trash in the gutter and marks on the sidewalk, becomes a quest for a new beginning. If snow, even imaginary or symbolic snow, could obliterate the past, perhaps the future would be better. All these interpretations are possible because the story is deliberately ambiguous, yet the use of snow in association with childhood, madness, and isolation confirms ancient myths and draws directly on our experiences with snow—childhood play, disorientation in a storm, being immobilized in traffic or cut off from communications. Snow plays its role whether mythologized or not.

Snow in recent short fiction usually functions, as it does in poetry, to frame the story with a symbol appropriate to the mood, as in Jeanne Schinto's "The Disappearance," in which an elderly man dreams of escaping from his tiresome wife to seek the solitude and freedom he has been denied by life's circumstances. The story begins with his memories of Whittier's "Snow-bound," as he hangs white sheets and underwear on a clothesline, and ends with his death in the snow a few weeks later, as he plans his getaway. Snow promises freedom, but only through death. The story is a variation of Aiken's themes.[43]

In Robert Stone's "Helping," snow symbolizes both the bleakness of his protagonist's soul and the promise of improvement. Elliot, a counselor in a Massachusetts veterans' clinic, is an alcoholic whose wife, an attorney, has just lost a child custody case to a violent, crazy biker. As they argue about Elliot's drinking, the biker calls, threatening them for causing him legal trouble. Elliot loads his shotgun and fantasizes about killing him. He drinks, waits, and watches the snow fall. Stone writes, "His eyes grew heavy as the snow came down. He felt as though he could be drawn up into the storm and he began to imagine that. He imagined his life with all its artifacts and appetites easing up the spout into white oblivion, everything obviated and foreclosed. . . . Snowflakes spun around his head like an excitement." During another argument his wife throws a glass sugar bowl at him, and it shatters on the wall, causing "a fine rain of crystals" to fall in his hair. Coated with the snowlike crystals, he walks in the real snow outside, "dazzled" by its whiteness. He removes his goggles to wave at his wife, realizing that his need for her is stronger than the intensity of the sunrise on the snow. The story gains its power from the naturalness of the snow metaphors. Elliot's need for violent action is driven by his inability to help the veterans who are suffering from the violence of war. His wife shares his sense of impotence because she has failed to protect a child from its violent father. Both feel threatened, a feeling that is at first heightened by the snow. When he is literally anointed by snowlike crystals from his angry wife, however, Elliot begins to realize that the snow is not malevolent, that, unlike the hate and insanity he and his wife experience in their work, snow can evoke more positive emotions and higher priorities. Snow can heal.[44]

As might be expected, nature writers are generally less portentous when describing snow than writers of fiction. Writing in the tradition of

Thoreau, Burroughs, and Muir, most twentieth-century naturalists are not professional scientists doing fieldwork, but serious amateurs, carrying on a nineteenth-century tradition. The most persistent cultural interpretation of snow by nature writers since World War II is its power to restore order. Henry Beston, whose *Northern Farm: A Chronicle of Maine* is justly cited as one of the essential nature studies since its publication in 1948, begins by remarking that the sense of direction had gone out of American life; Edwin Way Teale, whose volumes of travel through the four seasons conclude with *Wandering through Winter,* comments wistfully on changes in American society between 1945 and 1965; and Diane Kappel-Smith remarks in *Wintering* that the first snow on her Vermont farm makes her think of nuclear winter and the end of the world. All three look to snow to restore meaning in their lives.[45]

Several nature writers celebrate the changes in animal behavior that accompany snow. For Sigurd Olson, canoeing the wilderness of northern Minnesota, summer is too busy. "To me," he writes in *The Singing Wilderness,* "that is the real meaning of the first snowfall—not the cessation of effort, but a drawing of the curtain on so many of the warm-weather activities that consume so much time. The snow means a return to a world of order, peace, and simplicity. Those first drifting flakes are a benediction and the day on which they come is different from any other in the year." Another famous midwestern naturalist, Aldo Leopold, concurs: "January observations can be almost as simple and peaceful as snow. . . . time not only to see who has done what, but to speculate why."[46]

Annie Dillard, in *Pilgrim at Tinker Creek,* sees the first snowfall of her year in rural southern Virginia causing a reversal of order:

> The light is diffuse and hueless, like the light on paper inside a pewter bowl. The snow looks light and the sky dark, but in fact the sky is lighter than the snow. Obviously the thing illuminated cannot be lighter than its illuminator. The classical demonstration of this point involves simply laying a mirror flat on the snow so that it reflects in its surface the sky, and comparing by sight this value to that of the snow. This is all very well, even conclusive, but the illusion persists. The dark is overhead and the light at my feet; I'm walking upside-down in the sky.

Dillard's replication of Abbott Thayer's experiment seems more like a magic trick than science, not because she is less interested in the external world—in the passages that follow she tracks birds and animals in the snow—but because the popular meaning of science has changed in the sixty-five years between them. Optics takes on new meanings in an electronically mediated age.[47]

Snow, fact or nothingness? Order or chaos? The questions are as old as history, but the American experience adds considerably to the debate. The utopian spirit that motivated much of the settlement of North America, the immigrants' dream of a better life, the seeming abundance of natural resources, all forced an eventual confrontation with the harsher aspects of nature, including snow. Does real snow have a place in an imaginary Eden? More important, is there a place there for the paradox and ambiguity of snow? Can you laugh when you fall in the snow in paradise? The answer to that question is fundamental because snow is, if nothing else, a cosmic joke, even when the punch line is death. If you lack a sense of humor, you won't understand snow. Frozen precipitation, perhaps, but not snow. Snow in literature, as we have seen, is associated with memory, which is itself a form of play. It is also associated with covering up. Hide-and-seek is snow's game, a powerful metaphor for reality itself. Consciousness and unconsciousness, ignorance and knowledge. Knowledge, however, is not necessarily truth. This too is the meaning of snow.

The stories of snow told in this book are fragments of its natural and unnatural history. Frozen Charlottes and Durrie farmyards, cloud crystals and crosses of snow, malevolent snow bogies and jolly snowmen, snow makers and samplers, students of the snow flea and those who flee snow, these are important episodes, but there are others. Memories of first encounters and unusual experiences with snow that people shared with me in the course of my writing are filled with wonder and joy. A woman who grew up in Manhattan recalled the city reduced to a child's scale in the blizzard of 1948, allowing her to pretend that she was crossing the western plateaus of her geography book. Another remembered making houses for her dolls after an Iowa snowstorm, then losing the dolls for the winter when another snowfall buried them. Iowa was also the site of an incident in the life of Janice White, whose memory of leaping from her swing into soft new snow remains vivid after many years. "When I was a three-year-old in snow country," one correspondent wrote, "I was familiar with snow on the ground but not snowstorms. During the night snow fell good and strong. When I awoke to see the result, I exclaimed, 'Oh look Daddy! the snow climbed up the trees.'"[48]

The role of a parent in introducing children to snow and its possible meanings is also strong in the memory of Judyth Powers of New York City:

> In my first recollection of snow, my mother and I are at the kitchen window of our first house, looking into the back yard, she seated, me standing. A friend of mine had moved away the day before and, in such a short life, that felt like a terrible loss. I was four years old. In the morning my mother brought me to the window. The sun shown

> brightly on snow that had frozen over during the night. The crystalline snow sparkled on the trees and blanketed the back yard with glitter. My mother told me that those "jewels" were left for me by my little friend.[49]

Memory does not always turn snow into a symbol. Jerre Conder, who grew up in Colorado in the 1940s and 1950s, remembers walking in a large, open field where "there was, as usual, the faint white of reflected light. But over the entire area (I would guess perhaps thirty acres), there was—at ground level—a faint but undeniable present aura of blue." The blue was the color of sparks generated by static electricity and illuminating the scattered trees and bushes. Conder sat in the field for an hour. "It was then," he writes, "that I *heard* the snowflakes hitting the ground. It was an icy tinkling sound, but as best I remember, the snow itself was soft—not coming down as ice chips." The sound may have been the discharging static of each flake in the extremely dry air. In 1983 the writer Pat Winter observed a snowstorm with blue lightning in the Boston Mountains of northwestern Arkansas. A woman who grew up on the New York–Pennsylvania border remembered snow that "smelled" blue, while snow about to fall "smelled gray." Dan Graf, a meteorologist at Pennsylvania State University, believes that the type of agriculture, and industrial or residential pollution, in an area determines the smell of snowstorms. In the Northeast, storms coming from the South bring aromas from places where the vegetation is still green.[50]

The sight, sound, smell, touch, and taste of snow, the five senses, help us form perceptions; our minds work to create attitudes. As suggested in the Introduction, snow consciousness is part of everyday life. Even those who have never seen or felt snow, if they have heard of it, experience anticipation. Anticipation is keener for those who know snow will fall on them sometime during the year. The last snow of spring reminds us of the first snow of fall. Midsummer heat often stimulates longing for snow, and longing, as Susan Stewart cogently explains, leads to the desire to describe and explain.[51] Imagining a thing we have lost or have yet to attain is the beginning of a narrative that attempts to bring it to life. Stories of snow, filled with fact or fantasy, are a recognition of the impossibility of knowing the external world. Know it or not, our next step in the quotidian experience of snow, is to prepare for it. Clothes are purchased or unpacked, houses and cars are "winterized," vacations planned. The third part of the cycle of snow consciousness, the actual experience of snow, may last a few minutes or several months. It may involve pleasure and pain, joy and anger, play and work. The fourth stage involves recovery from the period of snow. Too much snow or too little can have serious consequences. Even norms and averages affect individuals differently. Finally, memory, which leads again to anticipation, but which forms our attitudes in all phases of consciousness, provides images of snow that help us hold on to special moments in life's inevitable decay.

The expansion of the nation turned snow from resource to refuse to resource again. As cities grew, snow literally became a kind of garbage, assigned to sanitation departments for removal from streets. Urban people, accustomed to daily mail, newspapers, and commercial services, resented disruptions caused by snow. Snow became "that fucking snow." Fresh snowfalls not only disrupted garbage collection but failed to cover the trash and grime of cities for long. Within hours, curbside snow looks like X rays of rotting lungs—carbon from car exhaust black on slushy tire tracks and footprints. Panic buying when weather forecasters predict snow has led supermarket managers in Washington, D.C., to dub the top-selling items "snowbread and snowmilk." Snack foods are the third most popular items for snowbound urban consumers, who have been known to walk kilometers in the snow to buy a bag of potato chips. Kitty litter, used to sand icy sidewalks, is the fourth biggest seller. Toilet paper and film for cameras are fifth and sixth in presnowstorm sales.[52] This list strongly suggests that snow in most cities is not a serious disruption of life but an excuse to go to the store, to socialize, and to record and later share memories of the latest "blizzard."

Snow as resource, principally for water, is important for cities as well as for agriculture. The role of snow in global climate regulation is even more significant. David A. Robinson's long-term studies show that snowcover in the Northern Hemisphere has been shrinking over the past twenty-one years. Whether this is natural variability or the result of global warming is uncertain. The causes of global warming themselves are unclear. What is clear is that changes in the extent and duration of snowcover will affect the weather in the temperate zone. Some regions will become drier, others may become wetter, many regions will be hotter and for longer periods. If the nivean frontier is permanently retreating, then this book might be called "The Significance of the Snow Frontier in American History," for the United States has been shaped by the line of seasonal snowcover as surely as it was by the advancing line of western settlement. The disappearance of the snow line will have consequences as far-reaching as the end of the frontier. Moreover, the atmospheric changes come just as the political climate of the Cold War thaws. The *Wall Street Journal,* forecaster of economic weather, noted in April 1993 that the military portion of the budget of the Institute for Snow Research in Houghton, Michigan, had shrunk from 75 percent to 20 percent in the past five years. Applied snow research continues on snow groomers for recreation and snow-melting compounds for highways, but basic research may change if snow becomes less important in U.S. foreign policy.[53]

Each month the Cold Regions Research and Engineering Laboratory puts out *Current Literature: Cold Regions Science and Technology,* a bibliography of articles published throughout the world. In 1996 there were more than 7,000 items. These articles represent attempts to behold snow as it is, to sharpen our perceptions of snow and how it behaves. As scientists look ever more closely at snow crystals, they make finer and finer distinctions, seeing what is there and what is not. Much snow research is about negative results. In July 1993, for example, James H. Cragin, Alan D. Hewitt, and Samuel Colbeck reported on an experiment designed to explain why 50 to 80 percent of the total solutes (chloride, nitrate, and sulfate) in snow are contained in the initial 30 percent of meltwater. This phenomenon, known as fractionation or the ionic pulse, contributes to the episodic acidification of stream water, known as acid flush. Two mechanisms had been proposed: that the snowpack acts as an ion chromatograph, with ice grains selectively retaining and retarding the elution (washing out) of certain ions over others, the selection based on both chemical and physical properties; or that during snow metamorphosis some individual snow grains grow larger than others and the concentration of certain ions at their surface excludes other ions. When the grains melt, ions are eluted in an order related to the way they were excluded during the recrystallization process.[54]

A series of carefully controlled laboratory experiments demonstrated that snow grains and ice crystals do not possess selective affinity for inorganic ions and cannot act as a chromatograph. Fractionation and the preferential elution of certain ions is caused by the way crystals grow—the mechanisms of snow metamorphism. The effort to see what is there and what is not is never conclusive. A complex experiment to rule out one mechanism does not necessarily mean that yet another unseen mechanism is not at work. Snow science creeps forward a crystal at a time. There is a classic distinction in the natural sciences (relevant to the study of snow) between splitters, who work to find the essential differences among types, and lumpers, who see the larger patterns of similarities. One researcher's snow grains are another's snowcover. Both perspectives are necessary, but are they enough?

"Snow is a living thing," announces a glossy brochure from the Snow Research Center of Japan, a "Snow-Think-Tank" founded in 1990 to sponsor research and provide information on snow (*yuki*), "overcoming the problems of snow, putting it to good use, and creating a sense of affinity for it." No American snow research institution has such a broad and ambitious mandate. Japan is no more snowy than the United States—an estimated 25 percent of its population lives in "Snow Country"—but it has a long tradition of snow study, dating from Suzuki Bokushi's 1835 book, *Hokuetsu Seppu,* translated as *Snow Country Tales: Life in the Other Japan.* Although Bokushi explicitly rejects the idea that snow is a living thing, the influence of Taoism leads him to conclude that it "possesses the vital force of activity; thus within many of the hexagonal-shaped yin flakes are rounded portions, after the yang principle." Snowflakes, snow fleas, snowshoes, snow shovels, snow games, all these things held the interest of this provincial businessman, an interest shared by some of the great Japanese artists of his time, including Utagawa Hiroshige.[55]

Snow offers every culture a choice of symbols and questions. Chinese interest in snow has been expressed in poetry and painting for more than a thousand years. It may be significant that *xue,* the Chinese word for snow, is used metaphorically in an expression meaning "to elucidate or make clear," the opposite of the meaning of "to snow someone" in English. Other Chinese snow proverbs are similar to European ones, however, as in "a fall of snow gives promise to a fruitful year" and "to bury a child in the snow," meaning what was hidden will come to light. "Smoke on snow is easy to look at in other people's pictures but difficult to put in your own" is an expression that sums up a central paradox of snow—that subtle differences in nature are easier to see than to duplicate or explain.[56] Both Japanese and Chinese sensibilities seem more positive toward snow than American. We give someone a snowball's chance and fear that things have gotten out of hand when they "snowball."

In the 1970s the sociologist Hervé Gumuchian conducted a study of winter in the northern French Alps which recognizes that attitudes toward snow differ according to place of residence, occupation, and age. Based on hundreds of interviews and economic and demographic statistics, Gumuchian's *La Neige dans les alpes françaises du nord* shows how the perception of snow as a "natural constraint" has changed as the mountain villages have been incorporated into the national economy by winter recreation and tourism. Heavy snowfalls that were once considered part of everyday life are now seen as unwelcome disruptions of business. Mountain residents and tourists look at the same snow and see different things and behave differently. Attitudes toward avalanches provide a good example. Residents have long regarded them with fear and superstition, accepting them as an inevitable hazard of life. Tourists demand preventive action and accountability. Now, however, residents realize that snow is a commodity to be sold, while for winter sports enthusiasts it is a product to be consumed.[57]

Of all comparisons of attitudes toward snow, perhaps the most instructive is the Canadian. Canadian painters looked beyond the nothingness of the snowy landscape to find a national identity in the seemingly endless snow of the North, as did Canadian writers, especially since 1965, the year Gilles Vigneault introduced his song "Mon Pays," with its defiant lines placing snow at the center of the French Canadian experience:

Mon chemin, ce n'est pas un chemin, c'est la neige;
  Mon pays, ce n'est pas un pays, c'est l'hiver.
(My road is not a road, it is the snow;
  My land is not a land, it is the winter.)[58]

Here snow is married to the wind in a "white ceremony," and a house with walls of ice is built to welcome all people, but because the "land" is a season, a time, and not a place, it must be "sung" into existence. Winter, with its paradoxes of hardship and beauty, hospitality and loneliness, must be made perpetual. For Canadians, not only Quebecois, who have felt exiled in the country of their birth because of English traditions, or the economic and cultural hegemony of the United States, snow is a useful symbol of both national distinctiveness and the psychological conflicts that a search for identity entails.

The key text for understanding snow in Canadian culture, the equivalent to Stevens's "The Snow Man," is Frederick Philip Grove's short story "Snow." First published in 1927, and revised and republished many times, it remains one of the most widely anthologized and studied pieces of Canadian literature. The story is deceptively simple in both plot and language. Mike Sobotski walks through heavy snow after an all-night blizzard on the Canadian prairies to tell his neighbor, Abe Carroll, and Abe's hired man, Bill, that another neighbor, Redcliff, failed to return to his wife and children after a trip to town. The woman had walked to Mike's for help after one of the horses from her husband's sleigh wandered into her yard. The three men ride to Redcliff's house to tell his shivering wife that they will search for him. They find the other horse frozen to death in Abe's yard. Abe reconstructs the path the horse took, and they discover Redcliff's sleigh wedged between charred stumps in a burned-over scrub forest. Redcliff's body is under the sleigh, where he crawled rather than attempt to save himself. The men take his body to Abe's. Mike goes to tell the widow, while Abe breaks the news to her parents, German immigrants who have failed at farming and who were counting heavily on their son-in-law for help. The bleak story ends with the mother sobbing, "God's will be done!"[59]

Snow in this story is dangerous; it is also a test. Those who understand it survive. Grove conveys the knowledge needed in two ways: by naming different kinds of snow to suggest its complexity, and by having Abe read the signs in the snow left by the wind. Five kinds of snow are described—deep, dry powdery, light and fine, snow slabs, and mealy—while shifts in the wind are noted four times—when it has stopped entirely, when a light northwest breeze chills the men as they ride to Redcliff's, when the smoke from Redcliff's chimney blows to the southeast, and finally when Abe shows Mike and Bill "a fold in the flank of the snow-drift which indicates that the present drift had been superimposed on a lower one whose longitudinal axis ran to the north-east." This information allows Abe to understand how the horse had been able to follow the scent of the yard, which in turn tells him to backtrack until he finds Redcliff's sleigh. Abe not only reads the signs in the snow but has the will to survive, unlike Redcliff, who did not even attempt to walk to safety. In contrast to Mike and Redcliff's father-in-law, he does not physically shrink in the face of the cold or bad news. Instead, there is a hint that Abe will assume responsibility for Redcliff's family. Life will go on despite the snow.[60]

Snow silences Grove's women, who communicate by shivering and by uttering platitudes, but a half century later the writer Margaret Atwood, one of the leading interpreters of Canadian landscape in both prose and verse, provides a strong voice for women and snow. For her, snow is not a paradox, two kinds of nothing, but an ephemeral moment, glimpsed when

> We have a minute, maybe two
>
> in which we're walking together
> towards the edge of that evergreen forest
> we'll never enter
>
> through the drifted snow
> which is of no colour,
> which has just fallen.

"Small Poems for the Winter Solstice," from which these lines are taken, consists of fourteen free-verse meditations of different lengths. The speaker in the poem addresses an absent lover, but not necessarily a specific person. The poem is about the meaning of absence in the abstract. Snow, which will inevitably melt after the solstice, when the earth tilts southward, is the central metaphor. The first lines, "A clean page: what/shines in you is not nothing," make the reader think of the albedo of snow and of Stevens's musings on nothing. What shines in the lovers walking in the snow is, near the end of the poem, a knowledge that things that can't be understood, like snow, are unstable. "I don't trust love," says the speaker, "because it's no shape or colour."[61]

Atwood's ideas about snow, nature, and Canadian identity are most fully developed in *Survival: A Thematic Guide to Canadian Literature,* published in 1972, a book that sets the tone for much of the subsequent discussion of these subjects. Canadians, she argues, distrust nature, fear it, and are unable to see it except as an enemy to be destroyed. Canadian writers, Atwood believes, are obsessed by death from water and snow, the dominant elements of the Canadian landscape. Nature can be actively hostile to humans, or merely indifferent, as in Grove's "Snow." Atwood recognizes, of course, that some writers deliberately reject this bleak view to maintain that nature is an "all-good Divine Mother" even as they are "being eaten by mosquitoes and falling into bogs." A third position declares war on nature, refusing to be victimized. A fourth perspective, which includes Atwood's own, is that nature is neither divine mother nor devil but a "living process which includes opposites: life and death, 'gentleness' and 'hostility.'" It is, she writes, in their attitudes toward snow "that Canadians reveal most fully their stance toward Nature—since . . . winter for us is the 'real' season."[62]

Love of "the North," fear of "wilderness," and ideas of environmental determinism are not unique to Canadians, although what might be called snowphilia, as expressed in the opening chapter of Farley Mowat's *The Snow Walker,* is seldom found south of the border. Snow as he envisages it "is the bleak reality of a stalled car spinning wheels impinging on the neat time schedule of our self-importance . . . the invitation that glows ephemeral on a woman's lashes on a winter night . . . the banality of a TV advertisement pimping Coca-Cola on a snowbank at Sun Valley . . . the gentility of utter silence in the muffled heart of a snow-clad forest."[63] Steadily snow has become a shibboleth distinguishing Canada from the United States.

In *Our Lady of the Snows,* the Canadian novelist Morley Callaghan took Rudyard Kipling's slightly derisive label for Canada and turned it into a multilayered story of hope and desire, loss and yearning. In Callaghan's skillful hands, Our Lady of the Snows is a beautiful Hungarian prostitute, Illona Tomory, who unexpectedly appears at Gilhooley's Bar in a Toronto hotel. She charms the regulars, from judges to gangsters, during a season—Christmas to spring—marked by heavy snows. Snow makes everything look better, the author writes on the opening page, and throughout the novel snow, like Illona, is associated with mystery and wisdom, possesses a hint of danger, is soft and peaceful, and is full of promise. After Illona leaves, one of the patrons longs to "see her in the snow. Always in the snow. The snow's all gone now, but it'll be back as she'll be back." But Illona does not return. At the end of the novel her heartbroken lovers at Gilhooley's learn that she married a Greek ship's captain. Illona is a siren retired to the sea. Callaghan's playful linkage of Canada as Our Lady of the Snows and Our Lady of the Snows as a meretricious ideal—desirable, yet unattainable—is a satisfying conclusion to Vigneault's journey on the road of snow. The snow covers the earth each year, promises an escape from the past, is all things to all people, but it is also just snow, inevitably disappearing into rivers and the sea.[64]

In the United States and Canada, in Russia and Japan, everywhere, snow is the nothing that is there, the something that is by turns valuable and worthless, resource and refuse. Snow has been named and renamed, but there are snows for which we still have no names. We have named imaginary snows, but in the long and continuing dialogue of humanity and nature, both symbolic snow men and real snow grains exist. Like snow itself, our mental images of snow crystallize, accumulate, and metamorphose under changing individual and historical conditions. Neither nature nor culture is fixed. This is the story and the meaning of snow.

Ablation

The snow is sick. The pure
page breaks and greys and
drools around the edges, sucks
at my snowshoes every heavy step saying
fuck it, just
fuck it, softly to itself.

DON McKAY, "MARCH SNOW"

Falling snow makes us think of beginnings, snowcover of ends. Both should make us think of life. The end of one cycle is the beginning of another. Snowflakes contain the history of individual crystals. Snowcover represents the history of a storm, a winter, an age. As soon as it stops falling, each snowflake begins to change, subject to many variables—its original shape, air and surface temperature, pressure and wind. The metamorphism of seasonal snow on the ground is as complex as the creation and growth of snow crystals in the air. Whether snow is sticky or slippery, whether it makes good snowballs or poor ones, whether it glides downhill like syrup or breaks suddenly into an avalanche depends on its changing crystal structure.

Is the snow sticking? This is the first question asked when a storm begins. Topography and temperature, wind velocity, extent and type of vegetation all affect snow on the ground. This is one reason why it is so difficult to measure the depth of a snowfall to everyone's satisfaction. Nor does total depth tell us much about the nature of the snow. Are the crystals dry and feathery, or are they wet and heavy? Will 25 centimeters of snow make more than 3 centimeters of water or less? Will melting snow flow safely into rivers and drains, or flood fields and basements? Will the snow slide harmlessly from roofs or cause them to collapse? These are questions that can only be answered by peeling back the layers of the snowcover.

Snow crystals that fall as columns, needles, plates, stellar dendrites, or irregular crystals disintegrate at different rates, the stars collapsing first because of their larger surface to volume ratio and because the elaborate branching of these crystals creates differences in vapor pressure and temperature that cause them to implode. In this process the larger, central part of the crystal sucks water vapor from the warmer, smaller extensions, even when the surrounding temperature remains well below 0°C.

The beautiful six-sided stellar dendrites become rounded snow grains, often made up of more than one crystal. This is called destructive metamorphism by some scientists, who differentiate it from temperature-gradient metamorphism, which makes crystals lose their distinctive forms because of sometimes extreme differences of temperature at the top and bottom of a snowpack. This difference can be 10°C (18°F) per meter in low-density snow and 20°C (36°F) in high-density, wet snow. Even in relatively shallow snow, the temperature near the ground can remain about 0°C (32°F), while the surface just 20 to 30 centimeters (8 to 12 inches) above can fall to –35°C (–31°F) or lower.

Such differences in temperature cause several things to happen to snow crystals. On the surface, where wind and temperature fluctuations are at work, the crystals may remain fragmented, irregular for a while, but eventually they become rounded grains, packed tightly by the wind before they melt into warmer grains below. Large grains grow as the temperature gradient increases, paradoxically changing faster as the surface temperature falls. Water vapor from the lowest layers of the snowcover rises toward the surface, where it freezes into larger and larger grains. This process helps to form a hard crust. Below the upper, denser, stronger crust lie faceted crystals, often hexagonal prisms, which become larger and looser as they lose water vapor to the rounded grains above. Below the faceted crystals are cup-shaped, hollow crystals, which, when they become large and loose enough, are called depth hoar. This fragile layer is often the cause of avalanches because the denser, heavier layers above begin to roll along its surface like a plank on marbles.

Marbles and planks are only two of the many building materials of snowcover. It is a Lego box of forms. Wet snow can turn to slush, refreeze, and become rounded granules. Faceted particles and columns are scattered through the layers from surface to depth hoar. Melt-freeze metamorphism, occurring when rainwater or surface melting percolates down through the snowpack before refreezing, makes a jigsaw puzzle of ice fragments. Feathery crystals may form in cold surface snow or in cavities near tree trunks and buried bushes. Thin, breakable crusts on the surface are faceted like pieces of glass in a kaleidoscope. Rime, moisture deposited by fog; rain crust, a glazed surface that freezes without penetrating; and wind crust, refrozen meltwater, all wrap the snowcover in thin, glistening sheets.

The wind-driven snow may be pure, but it is far from uniform, as you know if you have hiked, or skied, or photographed winter snow. Surface is what most of us know best. Wind and changing temperature can cause snow grains to take on fantastic forms—snow rollers, the tubelike snowballs that blow across hills and fields; *skavler* or *sastrugi*, the irregular ridges formed by wind erosion and deposition; plowshares, formed when the sun melts a pit in the snow and the wind builds up the sides; cones of snow that form around small trees or bushes; snow garlands that hang from roofs and trees when wet snow slides down an incline without breaking apart; and *nieve penitente* of the Andes, where the ablating snow leaves pillars bowed like penitents on their way to mass.

A pit dug in the snow, like an archaeologist's trench, reveals the history of the snowcover. Near the bottom of the pit are the fragments of snow crystals that began forming months before and thousands of meters above the earth. Metamorphosed beyond recognition, these snow grains are to the original crystals what butterflies are to caterpillars. Like butterflies, these buried snowflakes sustain life. They insulate the ground, saturate the soil, and transport water to streams and aquifers. Near the top of the pit, newer snow resists melting. New snow reflects up to 100 percent of the sun's rays and blinds unprotected eyes. Reflection, or albedo, is also a function of snow grain size, surface roughness, and impurities. Thus, some light penetrates and triggers responses from plants, animals, and insects in the subnivean environment.

The next time the snow slides off a roof down your collar, or you break through the crust in a pile of snow and freeze your feet, or your snowshoes suck slush, think of all that has happened to cause your discomfort. The millions of little pieces that had to fall and thaw and freeze their way into place. Once I was privileged to be walking across a snow-covered field when a moderate earthquake struck. With an audible sibilance, clumps of snow sprang up, then fell back. Astonished, I knew that such events are rare. This book is my attempt to make the snow sing and leap again.[1]

Notes

INTRODUCTION

1. On history as stories, see William Cronon, "A Place for Stories: Nature, History, and Narrative," *Journal of American History* 78 (March 1992), 1347–76.
2. Measurements of total amounts of snow are notoriously unreliable because of the difference between the terms *snowfall* and *snow depth. Snowfall* is the total amount that would have accumulated if the snow was unaffected by wind or compaction on the ground; it is an estimate based on several variables. *Depth* is the amount of snowcover measured with a ruler. See Thomas Schlatter, "Weather Queries," *Weatherwise* (December 1994–January 1995), 42–43.
3. G. A. McKay and D. M. Gray, "The Distribution of Snowcover," in *Handbook of Snow: Principles, Processes, Management, and Use,* ed. D. M. Gray and D. H. Male (New York: Pergamon Press, 1981), 170; J. F. Rooney, "Let's Be Objective about Snow and Ice Control," *American City* (October 1969), 106; W. O. Pruitt, Jr., *Boreal Ecology* (London: E. Arnold, 1978), 26; Peter J. Marchand, *Life in the Cold: An Introduction to Winter Ecology* (Hanover, N.H.: University Press of New England, 1987), 10.
4. *New York Times Book Review* (September 11, 1988), 18. I thank all the *Times* readers and those who heard me on National Public Radio and shared their snow experiences with me, especially Hiram Rodriguez-Mora, N. Polinsky, and Myrtle Simon, who remembered Doña Fela. Others are named in the notes and in the acknowledgments. Carlos Bulosan, "History of the Heart,"

Philippine Writing, ed. T. D. Agcaoili (Manila: Archipelago Publishing House, 1953), 280.

5. Carolyn Merchant, *Ecological Revolutions: Nature, Gender, and Science in New England* (Chapel Hill: University of North Carolina Press, 1989), 29.
6. Corydon Bell, *The Wonder of Snow* (New York: Hill & Wang, 1957); Ruth Kirk, *Snow* (New York: William Morrow, 1977); Douglas Helms, *Readings in the History of the Soil Conservation Service,* NHQ, historical notes no. 1 (Washington, D.C.: U.S. Department of Agriculture, Soil Conservation Service, Economics and Social Sciences Division, 1992); Samuel Colbeck, "History of Snow-Cover Research," *Journal of Glaciology* (Special Issue, 1987), 60–65.

   Books on the history of water in California that ignore snow include John Walton, *Western Times and Water Wars: State, Culture, and Rebellion in California* (Berkeley and Los Angeles: University of California Press, 1992); Norris Hundley, Jr., *The Great Thirst: Californians and Water, 1770s–1990s* (Berkeley and Los Angeles: University of California Press, 1992).
7. Stephen Pyne, *Burning Bush: A Fire History of Australia* (New York: Holt, 1991), xii; Donald Worster, "Transformation of the Earth: Toward an Agroecological Perspective in History," *Journal of American History* 76 (March 1990), 1090–91. See also Pyne, *Fire in America: A Cultural History of Wildland and Rural Fire* (Princeton, N.J.: Princeton University Press, 1982), and Donald Worster, *Rivers of Empire: Water, Aridity, and the Growth of the American West* (New York: Pantheon Books, 1985; reprint, New York: Oxford University Press, 1992).
8. John Haines, *The Stars, the Snow, the Fire* (New York: Washington Square Press, 1992), 5.
9. Vilhjalmur Stefansson, *The Northward Course of Empire* (New York: Macmillan, 1922); Wallace Stevens, "The Snow Man," in *The Collected Poems* (New York: Alfred A. Knopf, 1967), 10.
10. On explanation and truth in history, see Murray G. Murphey, *Philosophical Foundations of Historical Knowledge* (Albany: State University of New York Press, 1994).

ACCUMULATION

1. The most useful explanations of snow formation and precipitation are R. S. Schemenauer, M. O. Berry, and J. B. Maxwell, "Snowfall Formation," in *Handbook of Snow: Principles, Processes, Management, and Use,* ed. D. M. Gray and D. H. Male (New York: Pergamon Press, 1981), 129–52; John Hallett, "How Snow Crystals Grow," *American Scientist* 72 (November–December 1984), 582–89; Nancy C. Knight, "No Two Alike?" *Bulletin of the American Meteorological Society* 69 (May 1988), 496; B. J. Mason, *Clouds, Rain, and Rainmaking,* 2d ed. (New York: Cambridge University Press, 1975); and Vincent J. Schaefer and John A. Day, *A Field Guide to the Atmosphere* (Boston: Houghton Mifflin, 1981).

1. Thomas Jefferson, *Notes on the State of Virginia,* ed. William Peden (Chapel Hill: University of North Carolina Press, 1955), 80.
2. Noah Webster, *A Collection of Papers on Political, Literary, and Moral Subjects* (New York: Webster & Clark, 1843; facsimile ed., New York: Burt Franklin, 1968), 148. The exact relationship between forests and snowcover remains unclear, as does the extent of the impact of humankind on climate and the meaning of terms such as *weather* and *climate.* See G. A. McKay and D. M. Gray, "The Distribution of Snowcover," in *Handbook of Snow: Principles, Processes, Management, and Use,* ed. D. M. Gray and D. H. Male (New York: Pergamon Press, 1981), 158–59; and H. H. Lamb, *Climate, History and the Modern World* (London: Methuen, 1982). "By climate," Lamb writes, "we mean the total experience of the weather at any place over some specific period of time. By international convention the period to which climate statistics relate is now normally thirty years . . . although we shall see arguments for preferring different periods for different purposes" (8).
3. Noah Webster, *An American Dictionary of the English Language* (New York: S. Converse, 1828), n.p.
4. Quentin Hope, "Snow Imagery in Love Poetry," *Arcadia: Zeitschrift für vergleichende Literaturwissenschaft* 13 (1978), 1.
5. See Quentin Hope, "Lovers in the Snow," *College English* 4 (Winter 1977), 1–20; "Winter Pastoral and Winter Reverie," *Comparative Literature Studies* 15 (1978), 284–304; and "Snow as Deformity, Decoration, and Disguise," *Orbis Litterarum* 36 (1981), 37–52.

   The idea that lovers may leave messages in the snow is expressed in the bawdy story of the father who recognizes his daughter's handwriting in the love note written with her boyfriend's urine in the snow. See *Pissing in the Snow and Other Ozark Folktales,* comp. Vance Randolph (Urbana: University of Illinois Press, 1976), 5. In the libidinous 1980s, the association of love with snow assumed cruder forms. A Valentine's Day card carried the message "Sex is like snow. You never know how many inches you'll get or how long it will last." Original design by Game Plan, Recycled Paper Products, Chicago, Ill., 1988.
6. Hope, "Lovers in the Snow," 5–6.
7. Johannes Kepler, *A New Year's Gift or On the Six-Cornered Snowflake* (Oxford: Clarendon Press, 1966). This edition contains the original Latin and a modern English translation, with extensive notes on the similarity of *nihil, neige,* and *nix.*
8. Karen Ordahl Kupperman, "Climate and the Mastery of the Wilderness in Seventeenth-Century New England," in *Seventeenth-Century New England,* ed. David Hall and D. G. Allen (Boston: Colonial Society of Massachusetts, 1984), 3–37.
9. Karen Ordahl Kupperman, "The Puzzle of the American Climate in the Early Colonial Period," *American Historical Review* 87 (1982), 1285–86. For further examples of Puritan attitudes toward snow, see *The Diary of Samuel Sewall,*

1674–1729, ed. M. Halsey Thomas (New York: Farrar, Straus & Giroux, 1973), vol. 2, 847–48; *Diary of Cotton Mather, 1709–1724,* Massachusetts Historical Society Collections, 7th ser. (Boston: Massachusetts Historical Society, 1912), vol. 8, 439–40 and 506; Increase Mather, "The Uses of Snow," in Frances Chickering, ed., *Cloud Crystals; A Snow-Flake Album* (New York: D. Appleton, 1864), 127–29; and especially Cotton Mather, *The Christian Philosopher: A Collection of the Best Discoveries in Nature with Religious Improvements* (London: Emanuel Matthews, 1721), 59–60; reprinted and edited with notes by Winton U. Solberg (Urbana: University of Illinois Press, 1994), 67–69. Solberg notes that Mather's entire essay is taken from Nehemiah Grew, "Some Observations Touching the Nature of Snow," *Philosophical Transactions* 8 (March 25, 1673), 5193–96.

10. St. John de Crèvecoeur, *Sketches of Eighteenth Century America,* ed. Henri L. Bourdin, Ralph H. Gabriel, and Stanley T. Williams (New York: Benjamin Blom, 1972), 39–50. The first publication of the essay was in 1787.
11. Ibid., 49.
12. *The People's Almanac* (Boston: Willard Felt and Charles Ellms, 1833), 12–15.
13. *The People's Almanac,* 1840, n.p.
14. *Vance Randolph Ozark Folksongs,* ed. Norm Cohen (Urbana: University of Illinois Press, 1982), 528–32. My thanks to Alan Jabbour for this reference. The popularity of "Young Charlotte" survives into the present, and porcelain dolls whose legs are one immobile piece are sometimes called Frozen Charlottes. See the poem "Frozen Charlottes" by Susan Prosper, *New Yorker* (September 17, 1990), 44. On "She Perished in the Snow," see *Folk Songs of the Catskills,* ed. Norman Cazden, Herbert Haufrecht, and Norman Studer (Albany: State University of New York Press, 1982), 255–58. "The Mother's Sacrifice" is reprinted in *The Library of Poetry and Song,* ed. William Cullen Bryant (Garden City, N.Y.: Doubleday, Page, 1925), vol. 1, 86. Sally Webster, *William Morris Hunt, 1824–1879* (New York: Cambridge University Press, 1991), 62.
15. C. F. McGlashan, *History of the Donner Party* (Truckee, Calif.: Crowley & McGlashan, 1879; 2d ed., rev., San Francisco: Bancroft, 1880), 71, 129, 164, 204. For a more recent history of the Donner party, see George R. Stewart, *Ordeal by Hunger* (Boston: Houghton Mifflin, 1960; reprint, Lincoln: University of Nebraska Press, 1986).
16. Joy Kasson, *Marble Queens and Captives: Women in Nineteenth-Century American Sculpture* (New Haven, Conn.: Yale University Press, 1990), 101.

    This may be the place to consider the popularity of Hans Christian Andersen in America. His stories "The Snow Queen" (1845) and "The Ice Maiden" (1861) have been read by and to generations of children. A recent psychological interpretation of these tales, Wolfgang Lederer, *The Kiss of the Snow Queen: Hans Christian Andersen and Man's Redemption by Women* (Berkeley and Los Angeles: University of California Press, 1986), argues that they are expressions of Andersen's personal sexual problems as well as the more universal psychological dependence of men on women. Without Gerda's love Kay could not escape from the Snow Queen's palace and become human again. The frozen woman is sometimes victim, sometimes superhero, sometimes supervillain.

    David Cameron Miller, *Dark Eden: The Swamp in Nineteenth-Century*

American Culture (New York: Cambridge University Press, 1989), makes a case for reading eighteenth- and nineteenth-century literary and artistic depictions of swamps as metaphors of the primitive, the "other," the victim, and therefore of women and sexual passion. Snow, however, transforms the familiar landscape into "another," more primitive environment in which liberated women might be a danger to themselves as well as to others.

17. Nathaniel Hawthorne, "Snow-flakes," in *Twice-Told Tales* (Columbus: Ohio State University Press, 1974), 343–49; Arden Reed, *Romantic Weather: The Climate of Coleridge and Baudelaire* (Hanover, N.H.: University Press of New England), 1983.
18. Henry David Thoreau, "A Winter Walk," *Dial* (October 1843), 233; *The Variorum Walden,* ed. Walter Harding (New York: Washington Square Press, 1963), 224; *Henry David Thoreau Journal,* ed. John C. Broderick (Princeton: Princeton University Press, 1981), vol. 1, 207.
19. *Thoreau Journal,* vol. 1, 210; *Variorum Walden,* 194.
20. Ralph Waldo Emerson, "Nature," "An Address," and "The Snow-Storm," in *The Complete Essays and Other Writings of Ralph Waldo Emerson* (New York: Modern Library, 1950), 6, 76–77, 768–69. For an extensive analysis of Emerson's use of weather metaphors in discussing nature, ethics, politics, rhetoric, and similar matters, see Eduardo Cadava, "Emerson and the Climates of Political History," *Boundary 2* 21 (Summer 1994), 179–219, and his book, *Emerson and the Climates of History* (Palo Alto: Stanford University Press, 1997). Wallace Stevens, "The Snow Man," in *The Collected Poems* (New York: Alfred A. Knopf, 1967), 9–10.

    On chaos theory, see James P. Crutchfield, J. Doyne Farmer, and Norman H. Packard, "Chaos," *Scientific American* (December 1986), 46–57; and James Gleick, *Chaos: Making a New Science* (New York: Viking Penguin, 1987).
21. James Rodger Fleming, *Meteorology in America, 1800–1870* (Baltimore: Johns Hopkins University Press, 1990). For European backgrounds, see Theodore Feldman, "The History of Meteorology, 1750–1800: A Study in the Quantification of Experimental Physics" (Ph.D. diss., University of California, Berkeley, 1983). Other useful histories of meteorology in the United States are Eric R. Miller, "The Evolution of Meteorological Institutions in the United States," *Monthly Weather Review* 59 (January 1931), 1–6; Donald Whitnah, *A History of the United States Weather Bureau* (Urbana: University of Illinois Press, 1961); and Keith Thompson, "The Question of Climatic Stability in America before 1900," *Climatic Change* 3 (1981), 227–41.
22. James Pollard Espy, *The Philosophy of Storms* (Boston: Little, Brown, 1841). Biographical information on Espy is from *Dictionary of American Biography* (New York: Charles Scribner's Sons, 1960), vol. 6, 185–86.
23. Miller, "Evolution of Meteorological Institutions," 3; Smithsonian Miscellaneous Collections, no. 148, *Directions for Meteorological Observations and the Registry of Periodical Phenomena* (Washington, D.C.: GPO, 1872). Whitnah, *History of the U.S. Weather Bureau,* notes that no common time for snowfall measurements was instigated until after 1900 (73).
24. Elias Loomis, *A Treatise on Meteorology* (New York: Harper & Bros., 1868), 122–28.

25. "Harmony," "Talks with You," *Ladies' Repository* 13 (January 1853), 32–34. Compare Mary Noel M'Donald's untitled poem in *Graham's Magazine* 42 (January 1853) for the analogy between the hidden human spirit and seeds under the snow.
26. Fred. B. Adelson, "An American Snowfall: Early Winter Scenes by Alvan Fisher," *Arts in Virginia* 24 (1983–84), 3–9. My thanks to Professor Adelson for sharing his work with me. *Farmer in a Pung* and three other paintings by Alvan Fisher, as well as Francis Guy's *Winter Scene in Brooklyn* and Thomas Doughty's *Winter Landscape,* are reproduced in Adelson's essay.

    On the concept of *plain painters* as opposed to the frequently used term *folk painters,* see John Michael Vlach, *Plain Painters: Making Sense of American Folk Art* (Washington, D.C.: Smithsonian Institution Press, 1988). Vlach's point is that folk artists must come out of a specific ethnic or regional tradition. Most painters of snow scenes, such as Grandma Moses in the twentieth century, were influenced by reproductions of popular and studio art.
27. John Ruskin, *Modern Painters* (New York: Lovell, Coryell, 1846), vol. 2, 38–41.
28. Many of the Currier & Ives lithographs are reproduced in *Currier and Ives,* ed. J. L. Pratt (Maplewood, N.J.: Hammond, 1968). Tait's work can best be viewed in the Adirondack Museum, Blue Mountain Lake, N.Y. On Durrie's career, see Martha Hutson, "The American Winter Landscape, 1830–1870," *American Art Review* 2 (January–February 1975), 64; and *George Henry Durrie, American Winter Landscapist: Renowned through Currier and Ives* (Santa Barbara, Calif.: Santa Barbara Museum of Art, 1977). A copy of Durrie's diary is in the Archives of American Art, Smithsonian Institution, Washington, D.C.
29. J. Russell Harper, *Krieghoff* (Toronto: University of Toronto Press, 1979); Peter Mellen, *Landmarks of Canadian Art* (Toronto: McClelland & Stewart, 1978). The idea that Durrie's snowscapes represented New England's superior virtues was first suggested by the art historian Alan Gowans in his chapter in *Arts in America: The Nineteenth Century* (New York: Charles Scribner's Sons, 1969), 228.
30. "On the Crystals of Snow," *Art Journal* (1857), 73–76, 125–28. A booklet advertising Warren, Fuller and Co., Artistic Wall-Papers containing an example of the Tiffany ceiling paper is in the library of the National Oceanic and Atmospheric Administration, Rockville, Md. The poet James Whitcomb Riley was one customer. See Historic American Buildings Survey, *Historic America* (Washington, D.C.: Library of Congress, 1983), 154. My thanks to Judy Grogg for this reference.
31. Tim Armstrong, "'A Good Word for Winter': The Poetics of a Season," *New England Quarterly* 60 (December 1987), 568. My thanks to Professor Armstrong for sending me a copy of his article.
32. James W. Watson, "Beautiful Snow," in *The Humbler Poets: A Collection of Newspaper and Periodical Verse, 1870–1885,* ed. Slason Thompson (Chicago: McClurg, 1913), 371–73. Thompson prints a poem of the same title by "Major Sigourney" with the note that it was written in 1852, implying that Watson's poem is a parody. Sigourney's poem celebrates snow and ends with a plea to remember the homeless; it lacks the ironies of Watson's verse. Hazel Felleman's *Best Loved Poems of the American People* (Garden City, N.Y.: Garden City Publishing, 1936), 188–90, reprints Watson's poem without the last stanza.

33. John Greenleaf Whittier, *Snow-Bound, A Winter Idyl* (Boston: Ticknor & Fields, 1866). This poem was reprinted many times and went through twenty-one editions before 1930. For a thoughtful reading of "Snow-Bound," see George Arms, *The Fields Were Green: A New View of Bryant, Whittier, Holmes, Lowell, and Longfellow with a Selection of Their Poems* (Stanford: Stanford University Press, 1953), 44–47.
34. L. Edwin Folsom, "'The Souls That Snow': Winter in the Poetry of Emily Dickinson," *American Literature* 47 (1975), 363, 367; Martin Bickman, "'The Snow That Never Drifts': Dickinson's Slant of Language," *College Literature* 10 (Spring 1983), 142; *The Complete Poems of Emily Dickinson,* ed. Thomas H. Johnson (Boston: Little, Brown, 1960), no. 285, p. 132; no. 311, p. 146; no. 1133, p. 508; no. 927, p. 436.
35. Dickinson, *Complete Poems,* no. 525, pp. 256–57.

A case may be made for considering Edgar Allan Poe's *Narrative of A. Gordon Pym* (1838), with its use of the ice of the South Pole as a symbol for death and the unknown, and Herman Melville's *Moby-Dick* (1851), especially the chapter "The Whiteness of the Whale," with its discourse on white as a symbol of evil and terror, in the context of a philosophy of snow, but these examples seem too specific to their authors' concerns with death and evil to contribute to an understanding of snow in American culture. Nevertheless, Melville's conclusion is relevant: "Is it, that as in essence whiteness is not so much a color as the visible absence of color; and at the same time the concrete of all colors; is it for these reasons that there is such a dumb blankness, full of meaning, in a wide landscape of snows—a colorless, all-color of atheism from which we shrink?" *Moby-Dick* (New York: Modern Library, 1950), 194–95.

36. *Snow-Flakes: A Chapter from the Book of Nature* (Boston: American Tract Society, 1863), 5–8, 13–20.
37. Frances Chickering, ed., *Cloud Crystals; A Snow-Flake Album* (New York: D. Appleton, 1864), 10–12. Biographical information on Frances E. Knowlton Chickering is scant. She was the wife of John White Chickering (1808–1888), who graduated from Andover Theological Seminary in 1829 and was a minister in Portland, Maine, from 1835 to 1865. Their son, John White Chickering (1831–1913), was a professor of science at Gallaudet College in Washington, D.C., in the 1870s. See *Twentieth Century Biographical Dictionary of Notable Americans* (Boston: Biographical Society, 1904) and *Who Was Who* (Chicago: A. N. Marquis, 1942), vol. 1, 216.
38. Chickering, *Cloud Crystals,* 15–19. On Smallwood, see Suzanne Zeller, *Inventing Canada: Early Victorian Science and the Idea of a Transcontinental Nation* (Toronto: University of Toronto Press, 1987), 168. On electrical theories of precipitation, see W. E. Knowles Middleton, *A History of the Theories of Rain and Other Forms of Precipitation* (New York: Franklin Watts, 1966), 111–15.
39. Chickering, *Cloud Crystals,* 13–14, 93.
40. Ibid., 156, 86, 119–24, 133, 157–58.
41. Oliver Wendell Holmes, "Nearing the Snow-Line," *Atlantic Monthly* (January 1870), 86.
42. Jana Bara, "Through the Frosty Lens: William Notman and His Studio Props, 1861–1876," *History of Photography* 12 (January–March 1988), 25, 29. My thanks

to Julie K. Brown for this reference. Newton MacTavish, "Some Canadian Painters of Snow," *International Studio* 66 (January 1919), 78; Brian Osborne, "The Iconography of Nationhood in Canadian Art," in *The Iconography of Landscape,* ed. Denis Cosgrove and Stephen Daniel (London: Cambridge University Press, 1989), 166; Peter Mellen, *Landmarks of Canadian Art* (Toronto: McClelland & Stewart, 1978), 158–63; Michael Tooby, "Orienting the True North," in *The True North: Canadian Landscape Painting, 1896–1939,* ed. Michael Tooby (London: Lund Humphries in association with Barbican Art Gallery, 1991), 26.

43. James Russell Lowell, *Fireside Travels* (Boston: Houghton Mifflin, 1904), 374; *Winter Poems by Favorite American Poets* (Boston: Fields, Osgood, 1871); Clifton Johnson, "The Winter of the New England Poets," *New England Magazine* 41 (December 1909), 393–403, illustrated by eight photographs in a section called "Beautiful New England."

44. William Henry Jackson, *Time Exposure: The Autobiography of William Henry Jackson* (New York: Van Rees Press, 1940; reprint, Albuquerque: University of New Mexico Press, 1986); Peter B. Hales, *William Henry Jackson and the Transformation of the American Landscape* (Philadelphia: Temple University Press, 1988); Thurman Wilkins, *Thomas Moran, Artist of the Mountains* (Norman: University of Oklahoma Press, 1966); Joni Louise Kinsey, *Thomas Moran and the Surveying of the American West* (Washington, D.C.: Smithsonian Institution Press, 1992), 140–76. William Cullen Bryant, ed., *Picturesque America* (2 vols., New York: D. Appleton, 1872–1874; 4 vols., 1881–1885). Henry Wadsworth Longfellow, "The Cross of Snow," in *The Complete Poetical Works* (Boston: Houghton Mifflin, 1893), 323.

45. John La Farge, *Snow Field, Morning, Roxbury,* in the Art Institute of Chicago. My thanks to Bill and Carol Pollak for this reference. See also Kenneth Myers and Margaret Favretti, "'In Most Extreme Need': Correspondence of C. H. Moore with J. F. Kensett," *Archives of American Art Journal* 26 (1986), 14. Russell Sturgis, untitled review in the *Nation* 1 (November 23, 1865), 663, quoted in Myers and Favretti, "In Most Extreme Need," 14.

46. For an interesting discussion of the metaphor "reading the book of nature" and the use of the miniature as a way of containing experience and protecting it from contamination, see Susan Stewart, *On Longing: Narratives of the Miniature, the Gigantic, the Souvenir, the Collection* (Baltimore: Johns Hopkins University Press, 1984; reprint, Durham, N.C.: Duke University Press, 1993), 69.

Critics gave advice on how to paint snow. See J. A. Blaikie, "The Mask of Silence," *Art Journal* (January 1885), 1–5. This was, of course, the period in which Impressionism began in France, and painters immediately recognized the appropriateness of the style for depicting snow. Claude Monet, Auguste Renoir, Alfred Sisley, Paul Cézanne, Camille Pissarro, Paul Gauguin, and others painted innumerable street and road snow scenes. Their focus on muddy, rutted snow expresses an attitude that needs further study. See John Rewald, *The History of Impressionism,* 4th ed. (Greenwich, Conn.: New York Graphic Society, 1973), and *Couleurs de neige* (Geneva: Skira, 1992), a catalog for an exhibition of snow in European art shown at the Musée savoisien, Chambéry, France.

In Russia the village snow scenes of A. K. Savrassov and Vasily Perov and the brooding landscapes of F. A. Vasiliev are very similar to their American contemporaries. See Dmitri V. Sarabianov, *Russian Art: From Neoclassicism to the Avant-Garde, 1880–1917* (New York: Harry N. Abrams, 1990). In Finland the work of Magnus von Wright in the late 1860s resembles that of Durrie. The serene snow-covered farmlands of Fanny Churberg resemble Church's Olana, while her more impressionistic pieces, such as *Winter Landscape* (1880), are forerunners of Nordic Mysticism that influenced the early twentieth-century Canadian painters. Some of the early paintings by Akseli Gallen-Kallela, such as *Path on the Ice* (1887), resemble work by La Farge, Church, and Homer. See *Ateneum Guide* (Keuruu, Finland: Otava Publishing, 1987).

47. Wilson Alwyn Bentley and W. J. Humphreys, *Snow Crystals* (New York: McGraw-Hill, 1931; reprint, New York: Dover Books, 1962); Bentley, "Twenty Years' Study of Snow Crystals," *Monthly Weather Review* (May 1901), 212–14. One scientist who tried to build on Bentley's work was John Shedd; see "The Evolution of the Snow Crystal," *Monthly Weather Review* (October 1919), 691–94. A juvenile biography of Bentley by Gloria May Stoddard, *Snowflake Bentley: Man of Science, Man of God* (Shelburne, Vt.: New England Press, 1985), is the only biography I found. Bentley's photos are used to illustrate Rupert Sheldrake's *Presence of the Past: Morphic Resonance and the Habits of Nature* (New York: Times Books, 1988).
48. On Frost, see Henry M. Reed, *The World of A. B. Frost: His Family and Their Circle* (Montclair, N.J.: Montclair Art Museum, 1983). John Burroughs, *Winter Sunshine* (Boston: Houghton Mifflin, 1875), 41–42, 66. On Burroughs, see Clara Barrus, *The Life and Letters of John Burroughs,* 2 vols. (New York: Russell and Russell, 1968); and Perry D. Westbrook, *John Burroughs* (New York: Twayne, 1974).
49. William Gibson, "A Winter Walk," *Harper's* (December 1885), 70, 74.
50. Frank Bolles, *Land of the Lingering Snow: Chronicles of a Stroller in New England from January to June* (Boston: Houghton Mifflin, 1891), 4–5, 19, 24. Asa Fitch, in "Winter Insects of Eastern New York," *American Journal of Science and Agriculture* 5 (1847), 274–84, may be the first American to have described snow fleas. See also O. Lugger, "The Snow-fly," University of Minnesota Agricultural Experiment Station, bulletin 48 (1896), 256–57; and J. W. Folsom, "The Identity of the Snow Flea (*Achorutes nivicola* Fitch)," *Psyche* 9 (1902), 315–21. Biological study of snow is discussed further in Chapter 6.
51. John Muir, *The Mountains of California* (1894; reprint, New York: Doubleday, Anchor Books, 1961), 28–37.
52. Clark Orton, "A Boy's Life on the Frontier," unpublished MS, Minnesota Historical Society.
53. Mitford Matthews, ed., *A Dictionary of Americanisms on Historical Principles* (Chicago: Free Press, 1951), 133–34; Allen Walker Read, "The Word Blizzard," *American Speech* 3 (February 1928), 191–227; Walter S. Avis, ed., *A Dictionary of Canadianisms on Historical Principles* (Toronto: W. J. Gage, 1967), 56; Frederic G. Cassidy, ed., *Dictionary of American Regional English,* vol. 1 (Cambridge, Mass.: Harvard University Press, 1985), 286; Russell Tabbert, *Dictionary of Alaskan English* (Juneau: Denali Press, 1991), 187. Cleveland

Abbe, writing in the *Monthly Weather Review* (January 1899), 18, claimed to have found the word in the Vermillion *Dakota Republican* in 1867, but he did not cite the issue. Attempts to define *blizzard* in technical terms, e.g., *Glossary of Meteorology* (Boston: American Meteorological Society, 1959), 71, are often at variance with popular definitions, as the following indicates:

> A severe weather condition characterized by low temperatures and by strong winds bearing a great amount of snow (mostly fine, dry snow picked up from the ground). The U.S. Weather Bureau specifies, for *blizzard,* a wind of 32 mph or higher, low temperatures, and sufficient snow in the air to reduce visibility to less than 500 ft; and for *severe blizzard,* wind speeds exceeding 45 mph, temperatures near or below 10°F, and visibility reduced to near zero.

54. David L. Wheeler, "The Blizzard of 1886 and Its Effect on the Range Cattle Industry in the Southern Plains," *Southwestern Historical Quarterly* 94 (January 1991), 415–32; Edmund Morris, "Winter of the Blue Snow, 1886–1887," *Weatherwise* (December 1986), 304–7. For Russell's paintings, see Brian Dippie, *Looking at Russell* (Fort Worth, Tex.: Amon Carter Museum, 1987). "'Thirteen Were Saved,' or Nebraska's Fearless Maid, A Song of the Great Blizzard 1888," by William Vincent celebrates one rescue. The only full history I have found is William O'Gara, *In All Its Fury: A History of the Blizzard of January 12, 1888* (Lincoln: Nebraska Blizzard Club, 1947), which is composed of reminiscences.

    Earlier blizzards should not be neglected. The Disaster Relief Records for 1873 in the Minnesota Historical Society, St. Paul, are rich in detail. Seventy-three people died in the blizzard of January 7 through 9, 1873. Numerous others lost feet, toes, hands, and fingers as well as cattle. The state appropriated $5,000 for relief. One man who lost both feet and a hand in an earlier storm was awarded $40.

    Some languages distinguish among types of blizzards. The Mongolian word *dzud* is applied to storms that prevent cattle from grazing. A layer of snow only is called white dzud; icy dzud refers to a storm that leaves both snow and a layer of ice, whereas black dzud is accompanied by a drop in temperature that freezes the soil despite the insulation of the snow. More than 60 percent of Mongolia's livestock died in a dzud in 1945. *Mongol Messenger* (January 31, 1996), 4.
55. Paul Kocin, "Meteorological Analyses of the March 1888 "'Blizzard of '88,'" *EOS Transactions, American Meteorological Union* 69 (March 8, 1988), 137, 146–47; "An Analysis of the 'Blizzard of '88,'" *Bulletin of the American Meteorological Society* 64 (1983), 1258–72.
56. *New York Times* (March 13, 1888), 1–3.
57. Hugh Flick, "The Great Blizzard and the Blizzard Men of 1888," *New-York Historical Society Quarterly Bulletin* 19 (1935), 31. For other eyewitness accounts, see Napoleon Augustus Jennings, *New York in the Blizzard* (New York: Rogers and Sherwood, 1888); and Samuel M. Strong, *The Great Blizzard of 1888* (Brooklyn: By the author, 1938). Recent accounts include Judd Caplovitch, *Blizzard! The Great Storm of '88* (Vernon, Conn.: VeRo Publishing, 1988); Mary Cable, *The Blizzard of '88* (New York: Atheneum, 1988); Patrick Hughes, "The Blizzard of '88," *Weatherwise* (1981), 250; Irving Werstein, *The Blizzard*

of '88 (New York: Thomas Y. Crowell, 1960); Edward Oxford, "The Mighty Blizzard of March 1888," *American History Illustrated* 23 (March 1988), 11–19; Nat Brandt, "The Great Blizzard of '88," *American Heritage* 108 (1977), 32; and Mark L. Kramer and Gary Solomon, "Summary of the Blizzard of '88 Centennial Meeting," *Bulletin of the American Meteorological Society* 69 (August 1988), 981–83.

58. Meta Stern Lilienthal, *Dear Remembered World: Childhood Memories of an Old New Yorker* (New York: Richard Smith, 1947), 226–27. An excellent gendered reading of the storm is Marsha Ackermann, "Buried Alive! New York City in the Blizzard of 1888," *New York History* 74 (1993), 253–76.
59. For the maps, see issues of *Monthly Weather Review,* November 1888 through 1960, when the editors discontinued them without explanation. See also Patrick Hughes, *A Century of Weather Service, 1870–1970* (New York: Gordon and Breach, 1970), and Whitnah, *History of the U.S. Weather Bureau.*

CHAPTER 2. STALLED MAGNIFICENCE

1. Gerald M. Best, *Snowplow: Clearing Mountain Rails* (Berkeley, Calif.: Howell-North, 1966), 15–27. For an early description of the sheds on fire, see Mary McNair Mathews, *Ten Years in Nevada, or Life on the Pacific Coast* (Buffalo: Baker, Jones, 1880; reprint, Lincoln: University of Nebraska Press, 1985), 264.
2. Walter G. Berg, "Buildings and Structures of American Railroads: No. 6—Snow Sheds," *Railroad Gazette* (October 17, 1890), 717–18. Berg's articles appeared as a book, *Buildings and Structures of American Railroads* (New York: John Wiley, 1892), with two additional drawings of sheds for mountainsides, one on the Central Pacific, the other on the Oregon and Pacific. See also W. G. Curtis, "Snow Sheds on the Central Pacific," *Railroad Gazette* (October 18, 1889), 673; and "Southern Pacific Snow Sheds," *Railroad Gazette* (September 16, 1898), 669.
3. Robert Harold Brown, "Snow Fences: Then and Now," *Journal of Cultural Geography* 4 (Fall–Winter 1983), 87–88; "Snow Screens on Russian Railroads," *Railroad Gazette* (March 16, 1888), 176.
4. "Snow Fences, Stationary and Portable," *Railroad Gazette* (October 26, 1900), 697; "Snow upon Railways," *Scientific American Supplement,* no. 1317 (March 30, 1901), 21114–15; Brown, "Snow Fences: Then and Now," 93–97; George H. Billes, "Snow Removal from Country Roads by Successful Pennsylvania Plan," *Public Roads* 1 (January 1919), 38; "An Effective Snow Fence to Prevent Drifts," *American City* (October 1926), 605; R. A. Drought, "The Possibility of Natural Snow-Fences," *American City* (April 1928), 114; "A New Type of Snow-Fence," *American City* (August 1929), 189; "The Value of Snow-Fences in Keeping Highways Open," *American City* (October 1929), 209; "Snow-Fence in Wisconsin," *American City* (November 1930), 203; "Snow Fences," *Transactions of the American Geophysical Union,* Section on Hydrology (1943), 366–67.
5. "How to Use Snow Fence Effectively," *American City* (February 1974), 47–48; Ronald D. Tabler and Richard P. Furnish, "Benefits and Costs of Snow Fences on Wyoming Interstate 80," in *Transportation Research Record 860* (Washing-

ton, D.C.: Transportation Research Board, National Academy of Sciences, 1982), 13–20. For other uses of snow fencing, see Dora Gropp, *Experimental Snowfences at Prudhoe Bay* (Anchorage: Alaska Arctic Gar Study Co., March 1977); R. W. Verge and G. P. Williams, "Drift Control," in *Handbook of Snow,* ed. D. M. Gray and D. H. Male (Toronto: Pergamon Press, 1981), 636–46; "Snowfarming," Information on Trail Map, White Grass Ski Touring Center, Canaan Valley, W.V., 1994; and John O. Hibbs, "Snow Fence Design and Placement," paper presented at North American Snow Conference, Cleveland, Ohio, April 20, 1993. For additional information, see Lorne W. Gold, ed., *Annotated Bibliography of Snow Drifting and Its Control* (Ottawa: Division of Building Research, National Research Council of Canada, 1968).

6. William S. Huntington, "A Chapter on Snow-Plows," *Railroad Gazette* (November 12, 1870), 146–47; (October 31, 1891), 748–49. For examples of some unbuilt designs, see *Railroad Gazette* (September 24, 1870), 1, for Thomas L. Shaw's snowplow that was a flat blade that could be run into a snowbank, lifted up, then pulled back and dumped to the side; *Scientific American* (April 14, 1877), 226, for Robert G. Little's screw fan plow, a forerunner of the successful rotary plows a decade later; *Scientific American* (February 28, 1891), 130, for Arthur Gardiner's steam plow, which looks like lawn mower blades mounted in front of a fan; and *Railroad Gazette* (February 4, 1896), 85, for a snow-melting car built by the Boston and Maine Railroad that melted snow as fast as twenty-five men could shovel it.
7. Best, *Snowplow,* 38–46; John H. White, Jr., *The American Railroad Freight Car* (Baltimore: Johns Hopkins University Press, 1993), 97–98. The failure of Buckers is described in "The Snow Blockade on the Union Pacific," *Railroad Gazette* (February 24, 1872), 85. The Russell is illustrated in advertisements in the *Railroad Gazette* beginning in 1889, and in an article, "The Russell Snow Plow" (October 23, 1891), 742–43. All-steel versions of the Russell snowplow were manufactured in the 1930s. Russell Snow Plow catalog in the Northern Pacific Railroad Co. files, file 3653, no. 14, Minnesota Historical Society.
8. "A Snow Flanger," *Railroad Gazette* (February 18, 1887), 115–16; "Nevens' Flange Scraper," *Railroad Gazette* (February 8, 1888), 70–71; "A Snow Flanger Operated by Compressed Air," *Railroad Gazette* (November 19, 1897), 813; "Snow Plows and Flangers—Northwest Railroad Club," *Railroad Gazette* (January 11, 1889), 18; "Rail Flangers—Northwest Railroad Club," *Railroad Gazette* (February 15, 1889), 105–6. Later developments in flangers are described in *Railroad Gazette* (June 8 and August 29, 1894). "Train Accidents in 1886," *Railroad Gazette* (January 22, 1887), 62–63.
9. "Snow on the Northern Pacific," *Railroad Gazette* (March 3, 1876), 131; "Snow on the Central Pacific," *Railroad Gazette* (April 21, 1876), 177.
10. *Snowbound* (January 31, 1890), Reno, Nev. McCully printed his souvenir paper after he reached San Francisco. An item on page 4 says, "The two inside pages contain the matter which appeared in the original issue of 'The Snowbound' published on manilla paper and written in pencil. The outside pages were made up to complete a four page paper, which we take pleasure to issue at the request of passengers, as a souvenir of the blockade." McCully gave his address as the National Press Clipping Bureau, 502 Washington Street, San Francisco.

11. Milton F. Westheimer, "Memoirs," 18–19. These pages were sent to me in a letter of September 12, 1988, by Mr. Westheimer's daughter, Sue W. Ransohoff, in response to my author's query in the *New York Times.* I thank Mrs. Ransohoff for permission to quote from her father's manuscript.
12. Paul Swanson, "The Leslie Brothers and Their Giant Snowblower," *Trains* (January 1987), 27–39; W. H. Winterrowd, "The Development of Snow-Fighting Equipment," *Railway Maintenance Engineer* 16 (1920), 458–62; "The Rotary Steam Snow Shovel," *Railroad Gazette* (February 25, 1887), 128–29; advertisement, *Railroad Gazette* (February 10, 1888). See also Winterrowd, *Individual Paper on Snow Fighting Equipment,* circular no. S III-144 (Chicago: American Railroad Association, 1920).
13. Swanson, "Leslie Brothers," 32–33; "Competitive Trials of the Rotary and Other Machine Snow Plows," *Railroad Gazette* (May 23, 1890), 355–57; "Competitive Trials of Machine Snow Plows," *Railroad Gazette* (May 30, 1890), 371–72.
14. "The Rotary," ad, *Railroad Gazette* (December 25, 1891), xxxvi; Cy Warman, "The Battle of the Snow-Plows," *McClure's Magazine* 8 (November 1896), 92–96. Further improvements to the Jull and Leslie plows were reported in the *Railroad Gazette* (December 25, 1891), 912–13, and (January 15, 1892), 39. In 1902 the American Locomotive Company of New York City, which was building the rotary, issued pamphlet no. 10015, "Rotary Snow Plow," containing instructions on its operation. This pamphlet has been reprinted by Specialty Press, Ocean City, N.J.

    "An Electric Snow Sweeper," *Railroad Gazette* (October 9, 1891), 713; "Electric Snow Sweeper," *Scientific American* (November 12, 1892), 303; "The Rotary Snow Plow on Buffalo Street Railroads," *Railroad Gazette* (April 8, 1896), 233. The Buffalo machine had not a brush but scoops and Leslie-type fan blades at each end of the car. It was patented by George W. Ruggles of Rochester.
15. Edmund G. Love, *The Situation in Flushing* (New York: Harper & Row, 1965), 36–37. My thanks to David Jones of Okemos, Mich., for this reference. For later snow blockades, see Philip Earl, "Great Blizzard of '52," *Henderson Home News and Boulder City News* (December 27, 1984), 23–24. My thanks to Helen McBain of Henderson, Nev., for this reference. Theodore M. Oberlander and Robert A. Muller, "A Climate Hazard: Railroads against the Snowstorms of the Sierra Nevada in California," in *Essentials of Physical Geography Today,* 2d ed. (New York: Random House, 1987), 31–33. I thank Dr. Harold Klieforth of the Desert Research Institute for this reference.
16. Walt Whitman, "To a Locomotive in Winter," in *Leaves of Grass* (1892; reprint, New York: New American Library, 1958), 362. Jackson's photographs of the Leslie and Jull rotary plows are reproduced in Swanson, "Leslie Brothers," 32.
17. Don Cameron Shafer, *Punch* (January 18, 1911), 3; "A Winter Morning—Shoveling Out," *Every Saturday* (January 14, 1871). The excellent research on snow shovel patents was done by Pete Rothenhoefer while a Ph.D. candidate in American studies, George Washington University.
18. William P. Wentworth, patent no. 107,314, reissue no. 4,835; Eugene Campbell, patent no. 173,209, February 8, 1876; Carleton Jones, patent no. 173,964, February 22, 1876; William H. Hicks, patent no. 342,961, June 1, 1886; Lydia Fair-

weather, patent no. 399,394, March 12, 1889; Jessie A. Stauffer, patent no. 588,363, August 17, 1897; John A. Wiedersheim, patent no. 768,923, August 30, 1904; Frederick E. Kohler, patent no. 1,042,352, October 22, 1912.

The story of Duchamp's snow shovel is told in Calvin Tomkins, *The Bride and the Bachelors: Five Masters of the Avant Garde* (New York: Viking Press, 1965), 38–39, and by Herbert Molderings, "Objects of Modern Skepticism," in *The Definitively Unfinished Marcel Duchamp,* ed. Thierry de Duve (Cambridge, Mass.: MIT Press, 1992), 253, see also pp. 372 and 401. Molderings emphasizes the influence of the French scientist Henri Poincaré, who wrote that "the scientific fact is only the crude fact translated into a convenient language," and interprets the hanging snow shovel as an ironic statement about expected heaviness and unexpected lightness. I think it is more likely that Duchamp saw the shovel as an icon of the snow fighting arm(y), whose heroic but essentially futile efforts were chronicled each winter in newspapers and magazines. See note 35. Duchamp's original snow shovel apparently was lost and later models substituted in exhibitions and photographs. See *Marcel Duchamp,* ed. Anne D'Harnoncourt and Kynaston McShine (Greenwich, Conn.: New York Graphic Society for the Museum of Modern Art and the Philadelphia Museum of Art, 1975), 277–78. In 1986 the artist Hans Haacke created a work he calls *Broken R. M.,* which consists of a snow shovel with a broken handle suspended near a sign reading: "Art + Argent à tous les étages." Philadelphia Museum of Art.

19. Arthur Huberty, patent no. 1,200,186, October 3, 1916; Edgar C. Weaner, patent no. 1,214,397, January 30, 1917; Arthur Rishel, patent no. 1,352,384, September 7, 1920; Robert A. Smith, patent no. 2,183,976, December 19, 1939; George B. Nehls, patent no. 3,773,375, November 20, 1973; Eugene R. Jarvis, patent no. 4,265,475, May 5, 1981; Isaac Stewart, Jr., patent no. 4,547,011, October 15, 1985; Frank R. Moorefield, patent no. 4,848,819, July 18, 1989.
20. I thank Bradley H. Baltensperger and Jim Belote for their unpublished paper, "Snowdumps, Boardwalks, and Double Roofs: The Cultural Landscape for Coping with Heavy Snowfall in Michigan's Keweenaw Peninsula." See also Marsha Penti, "Cooper Country: Snow Country," in *1987 Festival of Michigan Folklife* (East Lansing: Michigan State University Museum, 1987), 54–55. Henry W. Staples, patent no. 258,260, May 23, 1882; Alvin Nelson, patent no. 3,154,336, October 27, 1964; Mack Holombo, patent no. 3,155,413, November 3, 1964; James Vonderacek, patent no. 4,193,626, March 18, 1980.
21. An early expression of the snow shoveler as mythic hero appears in a story by M. M. Raine and W. H. Eader, "How They Opened the Snow Road," *Outing* 49 (January 1907), 447–52, illustrated with four paintings by N. C. Wyeth. A group of miners trapped in the mountains without supplies dig their way out with "'No. 2' ore shovels" and trigger an avalanche with dynamite before it can destroy their road.

    For shoveling instructions, see, for example, Robert Kimber, "Moving Snow by Hand," *Country Journal* (January 1987), 10–13; Rochester, N.Y., *Democrat and Chronicle* (February 5, 1988), 5A; and *Washington Times* (January 8, 1996), C5. Shoveling from roofs has become a subject of special concern; see "Beware of Snowy Roofs!" *American Journal of Public Health* 78 (March 1988), 322; and

Dirk Thomas, "Snowbusting," *Country Journal* (November 1986), 60–65. David Huddle, "Shoveling," *New York Times Magazine* (January 10, 1988), 64.

22. For a review of more than twenty types of snow shovels available at present, see Matt Weiser, "When a Shovel Is Not Just a Shovel," *Tahoe Sunday Tribune* (February 23, 1992), C1.
23. Gregory R. Istre et al., "Surveillance for Injuries: Cluster of Finger Amputations from Snowblowers," *Public Health Reports* 104 (March–April 1989), 155–57; E. F. Lindsley, "I Built a Snowblower for $25," *Popular Science* (December 1956), 166–70; E. F. Lindsley, "Cold Facts about Snow Blowers," *Popular Science* (January 1962), 80–81, 204–6; "Snow Throwers," *Consumer Reports* (October 1989), 659–63; Elaine Underwood, "Every Time It Snows, It's Pennies from Heaven," *Adweek's Marketing Week* (January 21, 1991), 11; Willard S. Pratt, "Snow Throwers," *American City* (August 1952), 105.
24. John Rooney, Jr., "The Urban Snow Hazard in the United States: An Appraisal of Disruption," *Geographical Review* 57 (October 1967), 538–59; James W. Watson, "Beautiful Snow," in *The Humbler Poets: A Collection of Newspaper and Periodical Verse, 1870–1885*, ed. Slason Thompson (Chicago: McClurg, 1913), 371–73.
25. For examples of hysterical responses to the Buffalo storm of 1977, see Robert Bahr, *The Blizzard* (Englewood Cliffs, N.J.: Prentice-Hall, 1980).
26. Thomas Gutterbock, "The Effect of Snow on Urban Density Patterns in the United States," *Environment and Behavior* 22 (May 1990), 358–86; Rooney, "Urban Snow Hazard," 556–57.
27. Robert Davis, "Weather and Employee Absenteeism," unpublished paper, 1985. I thank Dr. Davis for sending me a copy of his paper. "Jackson Communiqué," *New Yorker* (February 18, 1985), 35; William J. Craig, "Seasonal Migration of the Elderly: Minnesota Snowbirds," *Southeastern Geographer* 32 (May 1992), 38–50.
28. Rooney, "Urban Snow Hazard," 544; Harold C. Cochrane and Brian A. Knowles, "Urban Snow Hazard in the United States: A Research Assessment," typescript (Washington, D.C.: National Science Foundation, 1975), 1–58.
29. Rooney, "Urban Snow Hazard," 556; Jacob Riis, "Midwinter in New York," *Century Magazine* 59 (February 1900), 520–33.
30. "Talk of the Town," *New Yorker* (January 23, 1989), 23–24.
31. Mary Douglas, *Purity and Danger: An Analysis of Concepts of Pollution and Taboo* (New York: Praeger, 1966; reprint, Baltimore: Pelican Books, 1970). *St. Paul & Minneapolis Pioneer Press* (February 14, 1886), 3.
32. *Daily Union and Advertiser* (February 13, 1871), 2; Blake McKelvey, "Snowstorms and Snow Fighting—The Rochester Experience," *Rochester History* 27 (January 1965), 1–24, and *Snow in the Cities: A History of America's Urban Response* (Rochester, N.Y.: Rochester University Press, 1995).
33. McKelvey, "Snowstorms and Snow Fighting," 11–12; R. S. DeBoer, "Winter Sports in Our City Parks," *American City* (January 1921), 55–58; Mary J. Breen, *Partners in Play: Recreation for Young Men and Women Together* (New York: A. S. Barnes, 1936), 134–40; "Report on the Problems of Snow Removal in the City of Rochester" (Rochester Bureau of Municipal Research, 1917), 22–24.
34. "50 and 100 Years Ago," *Scientific American* (January 1987), 8; "A Snow Melt-

ing Machine," *Scientific American* (February 27, 1897), 137; "Machine for Removing and Disposing of Snow," *Scientific American* (December 27, 1902), 461; H. L. Stidham, "The Removal of Snow," in George E. Waring, *Street-Cleaning and the Disposal of a City's Wastes . . .* (New York: Doubleday, 1898), 108; Rodney R. Fleming, "Snow-fighting's New Techniques," *American City* (November 1965), 83–85, 110, 112.

The earliest expressions of the concept *snow fighting* I have found occur in *Snow Removal: A Report of the Committee on Resolutions of the Snow Removal Conference Held in Philadelphia, April 16 and 17, 1914*, 5, and *Scientific American* (December 16, 1916), although Jacob Riis uses a military metaphor comparing "the warlike build of the electric sweeper" to a "modern man-of-war." Riis, "Midwinter in New York," 531.

35. Martin Melosi, *Garbage in the Cities: Refuse, Reform, and the Environment, 1880–1980* (Chicago: Dorsey Press, 1981), 68–69; Stidham, "Removal of Snow," 91, 95, 106, 108.
36. "Report on Snow Removal in Rochester," 4–7, 22; *Snow Removal, A Report of the Committee on Resolutions of the Snow Removal Conference, . . . 1914*, 1; advertisements in *American City* (January 1921), 64, 68, 70.
37. "Does It Pay to Clean City Streets in Winter?" *American City* (February 1921), 149; W. A. Van Duzer, "The Economy of Snow Removal," *American City* (January 1927), 58–60; "Snow Removal Cheaper Than Clogged Roads," *American City* (March 1927), 319; Charles M. Babcock, "Keeping a Billion-Dollar Investment Working," *American City* (January 1930), 133–34; V. R. Burton, "Cost Analysis of Snow Removal in Michigan," *American City* (April 1926), 386–93.
38. Samuel N. Baxter, "Street Trees Killed by Ice Cream Salt," *American City* (October 1921), 290; "Calcium Chloride Breaks Up Stubborn Street Ice," *American City* (January 1929), 160; Howard T. Barnes, "Removing—or Preventing—the Accumulation of Snow on City Streets?" *American City* (March 1930), 115–16. Many cities continued to use combinations of sand, crushed cinders, and salt. See P. E. Jarman, "Well-Organized Snow-Removal Plan for Sidewalks," *American City* (January 1929), 117–18.
39. A. L. H. Street, "The City's Legal Rights and Duties," *American City* (September 1927), 395.
40. Clarence J. Biladeau, "The Successful Campaign for Snow Removal in Berkshire County," *American City* (November 1923), 456–58; "Snow Removal," 1928 ed. (San Leandro, Calif.: Caterpillar Tractor Co., 1927), 4. The Caterpillar booklet claims that 75 percent of the population and 53 percent of the improved roads in the United States were located within the thirty-six states of the snowbelt, which also contained 66 percent of all motor vehicles registered in the world.
41. *Chicago Daily Tribune* (March 26, 27, and 28, 1930); Owen T. Lay, "Chicago's Greatest Snowstorm, March 25–26, 1930," *Monthly Weather Review* (April 1930), 146–48; *Washington Post* (February 8, 1936), 6.

Several aspects of the Chicago and Washington storms are interesting culturally. Lay notes that automobile traffic packed the wet snow so tightly into the streetcar tracks that it had to be removed by hand labor. A writer used the term *snow consciousness* to deride Chicagoans who badgered the Weather Bureau with calls asking how long the storm would last. The editors of the *Washington Post* commented on the "psychological change" in the capital as the city closed down and noted that "everything was deplorably inefficient; in fact, quite refreshingly so." These small voices of dissent in the chorus of deseasonalization gained strength in the 1970s and culminated in the livable winter cities movement, discussed later in this chapter.

42. "Don't Forget Snow Removal," *American City* (November 1932), 53–56; "A Forward Look to Winter!" (cartoon), *American City* (September 1939), 13; "Snow Removal in the Winter of 1933–34," *American City* (September 1933), 53–54.
43. Harry E. Wineberg, "The Hazard of the Wintery Crosswalk," *American City* (October 1930), 100–101.
44. John H. Nuttall, "What Shall We Do with Snow and Ice?" *American City* (February 1931), 102–3; Nuttall, "Warning—Ice and Snow Ahead, I," *American City* (November 1931), 90–91; Nuttall, "Warning—Ice and Snow Ahead, II," *American City* (December 1931), 84–85; Nuttall, "Warning—Ice and Snow Ahead, III," *American City* (January 1932), 90–92; "Don't Forget Snow Removal," 53–56; J. D. McVicar, "Snow-Types Met in Highway Snow-Removal," *Transactions of the American Geophysical Union* 20 (1940), 932–35.
45. *Boston Daily Globe* (February 15, 16, 17, 18, and 19, 1940).
46. *St. Paul Pioneer Press* (November 12, 1940); William Henry Hull, *All Hell Broke Loose* (Edina, Minn.: W. H. Hull, 1985); George R. Stewart, *Storm* (New York: Random House, 1941).
47. William H. Woodward, "Melting Ice with Salt Saves Men and Tires," *American City* (January 1943), 79; Harold S. Rand, "How Rochester Won the 'Battle of the Salt,'" *American City* (November 1948), 121–22; McKelvey, "Snowstorms and Snow Fighting," 18–19; "Salt without a Sting," *American City* (December 1948), 86; Eugene G. Moody, "Snow and Ice Removal Practices in Michigan Municipalities," Institute of Public Administration, University of Michigan, information bulletin no. 58 (1948); John Temmerman, "Inhibitor Blocks Salt Corrosion," *American City* (August 1949), 92.
48. "Salt Defended for Ice and Snow Control," *American City* (December 1949), 17; C. T. Roland and G. B. Hatch, "Snow Removal by Salt with Less Corrosion," *American City* (September 1949), 136–37; Edward J. Anderson, "The Snowless City," *American City* (January 1950), 100–101; Edward J. Neer, "Rochester Likes Its Rock Salt Straight," *American City* (September 1950), 118–19; "Cities Learn How to Store Rock Salt," *American City* (September 1951), 100–101; "How to 'Salt' Away a Snowstorm," *American City* (September 1954), 121; Warren J. Mann, "Salt Makes a Good Street," *American City* (January 1955), 102–3; Ray Blessing, "Colored Salt—Contented Cars," *American City* (October 1955), 155–56; "Snow and Ice Control with Chemicals and Abrasives," bulletin 152 (Washington, D.C.: Highway Research Board, 1960).

   On the pesticide campaigns, see Thomas R. Dunlap, *DDT: Scientists, Citizens, and Public Policy* (Princeton, N.J.: Princeton University Press, 1981); John H. Perkins, *Insects, Experts, and the Insecticide Crisis: The Quest for New Pest Management* (New York: Plenum Press, 1982); Pete Daniel, "A Rogue Bureaucracy: USDA Fire Ant Campaign of the Late 1950s," *Agricultural History* 64 (Spring 1990), 99–114; Edmund P. Russell III, "'Speaking of Annihilation': Mobilizing for War against Human and Insect Enemies, 1914–1945," *Journal of American History* 82 (March 1996), 1505–29, and, of course, Rachel Carson, *Silent Spring* (Boston: Houghton Mifflin, 1962). On the significance of the green lawn, see Virginia Scott Jenkins, *The Lawn: A History of an American Obsession* (Washington, D.C.: Smithsonian Institution Press, 1994).

By the 1970s, concern was growing over chemical pollution of water supplies. Canadian scientists led the way in attempting to minimize salt damage. John E. Fitzpatrick, an equipment operations engineer for the Ontario Department of Highways, designed the now familiar plywood beehive structures for storing sand and salt, thus reducing the polluting effects of these materials left in piles exposed to wind and rain. Laboratory studies of the melt rates of ice and snow at different temperatures and in different environments led to better understanding of the timing and amount of salt to be applied. Prewetting of salt and more efficient spreaders helped to reduce the amount of salt used, but not enough to satisfy many environmentalists. In 1972 Minnesota passed one of the first statutes restricting the use of salt for snow removal on highways. See John E. Fitzpatrick, "'Beehives' Protect Snow-Removal Salt and Prevent Water Pollution," *American City* (September 1970), 81–83; "Snow Removal Tailored to Reduce Salt Pollution," *American City* (January 1973), 16.

49. "How Many Snowplows?" *American City* (September 1955), 158–60; J. L. Galloway, "Commercial Weather Forecasting Pays Dividends," *American City* (October 1956), 125–27; "Snow Plowing Trends," *American City* (October 1957), 106–9; J. C. Thompson, "The Snowfall Probability Factor," *American City* (December 1959), 80–83; William S. Foster, "Some Elements of Good Snow Removal," *American City* (November 1959), 95–96, 165; Thomas Napier Adlam, *Snow Melting: Design, Installation, and Control of Systems for Melting Snow by Hot Water Coils Embedded beneath Walks, Roads . . .* (New York: Industrial Press, 1950); "Radiant Heat Snow Removal—What Does It Cost?" *American City* (February 1951), 13; "Snow Removal by Radiant Heat," *American City* (January 1952), 98–99; James B. Fullman, "Snow-Melting Systems," *American City* (June 1952), 92–94; "Effortless Snow Removal," *American City* (February 1957), 33.

By the 1980s the National Research Council of Canada concluded that "in situ melting systems . . . because of their large capital investment and high operating cost . . . are recommended only for sites where mechanical snow removal would be difficult, the use of chemicals would damage structures, traffic delays could not be tolerated or safety is an important consideration." G. P. Williams, "Thermal Methods of Control," in *Handbook of Snow,* 613.

50. Henry Morton Robinson, *The Great Snow* (New York: Simon & Schuster, 1947), 241, 1, 91, 170, and 277.

It is worth noting, I think, that the 1992 movie version of Edith Wharton's *Ethan Frome* credits a "Snow Wrangler," who presumably managed its many snow scenes. Several reviewers suggested that the snow, while not as bright as it ought to have been, was the best thing about the production. I thank Marjorie Baer for calling this to my attention.

51. *New York Times* (December 27, 28, and 29, 1947).

52. E. J. Kahn, Jr., "Our Snowbound Correspondents: Gilroy Was Here," *New Yorker* (January 17, 1948), 48–53.

53. Mari Sandoz, *White Thunder* (Philadelphia: Westminster Press, 1954; reprint, Lincoln: University of Nebraska Press, 1986). The story originally appeared as "The Lost School Bus," *Saturday Evening Post.* Mary Kay Roth, "White Hell," *Lincoln Journal-Star* (January 5, 1991), 4; "Air and Ground Forces Avert Nevada Winter Disaster," *Nevada Highways and Parks* (October 1949), 19–29; *Deseret News* [Salt Lake City] (January 22–30, 1949).

54. Patrick Hughes, "Francis W. Reichelderfer, Part II: Architect of Modern Meteorological Services," *Weatherwise* (August 1981), 148–57; Patrick Hughes, *A Century of Weather Service, 1870–1970* (New York: Gordon & Breach, 1970); Donald Whitnah, *A History of the United States Weather Bureau* (Urbana: University of Illinois Press, 1961), 201–40; Robert Marc Friedman, *Appropriating the Weather: Vilhelm Bjerknes and the Construction of Modern Meteorology* (Ithaca, N.Y.: Cornell University Press, 1989); Robert Henson, *Television Weathercasting: A History* (Jefferson, N.C.: McFarland, 1990); David Laskin, *Braving the Elements: The Stormy History of American Weather* (New York: Doubleday, 1996).

55. Hughes, *Century of Weather Service,* 121–24, 197–98.

56. J. B. Fulks, "The Early November Snowstorm of 1953," *Weatherwise* (February 1954), 12–16; A. K. Showalter to Francis Reichelderfer, November 9, 1953, "Unsatisfactory Forecasts of November 5–7"; Reichelderfer to Fulks, November 9, 1953, record group 27, Weather Bureau Correspondence, 1951–1955, box 30, file 614, National Archives and Records Administration, Washington, D.C.

57. L. E. Brotzman, memo for record, November 9, 1953; I. R. Tannehill to Francis Reichelderfer, November 10, 1953, with Brotzman's memo; Ernest J. Christie to Chief, U.S. Weather Bureau, November 16, 1953; R. C. Schmidt to Tannehill, November 18, 1953; B. Haurwitz, chairman, Committee on the Cyclone of 6–7 November 1953, to Reichelderfer, November 25, 1953. The long quotation is from Tannehill to Reichelderfer, November 19, 1953, Weather Bureau Correspondence, 1951–1955, box 30, file 614.

58. U.S. Department of Commerce, Weather Bureau, *Definitions of Hydrometeors and Other Atmospheric Phenomena,* adopted November 18, 1938, reprinted December 1, 1943, mimeograph, 3–6; I. R. Tannehill to Francis Reichelderfer, November 18, 1953; Reichelderfer to Tannehill, November 16, 1953, record group 27, Weather Bureau Correspondence, 1951–1955, box 30, file 614.

59. *Chicago Tribune* (January 25, 26, and 27, 1967). On January 14, 1978, Chicago received 52.6 centimeters (20.7 inches) of snow on top of 20.3 centimeters (8 inches) on the ground. So much for 200-year storms. According to columnist Mike Royko, Mayor Daley is supposed to have said, "Just keep the streets clean and the buses moving and you can steal anything you want," a lesson his successor Michael Bilandic forgot when he failed to get the streets plowed. This contributed to his defeat by Jane Byrne. Byrne's campaign manager shot her television commercials in the snow and was prepared to throw soap flakes over her if the weather improved. Bill and Lori Granger, *Fighting Jane: Mayor Jane and the Chicago Machine* (New York: Dial Press, 1980), 209–16. Douglas, *Purity and Danger.*

60. *New York Times* (February 10, 11, 12, and 13, 1969).
61. *Planning for Snow Emergencies* ([New York City]: Office of the Mayor, December 1969).
62. Marcy P. Cohen to author, September 1988. *Boston Globe* (February 8, 9, and 10, 1978).
63. *Washington Post* (February 19 and 20, 1979); Lance F. Bosart, "The Presidents' Day Snowstorm of 18–19 February 1979: A Subsynoptic-Scale Event," *Monthly Weather Review* (July 1981), 1542–66; *Washington Post* (February 12 and 13, 1983); *Washington Post* (January 23, 27, and 28, 1987); *Washington Post* (September 26, 1987).
64. Frederick Gutheim, "Livable Winter Cities," *Architectural Record* (February 1979), 111–16; William C. Rogers and Jeanne K. Hanson, *The Winter City Book: A Survival Guide for the Frost Belt* (Edina, Minn.: Dorn Books, 1980); John C. Royal, "How to Make Cities More Livable in Winter," *Canadian Geographic* (February–March 1984), 21–27.
65. Norman Pressman, "The Survival of Winter Cities: Problems and Prospects," in *The Future of Winter Cities,* ed. Gary Gappert (Newbury Park, Calif.: Sage, 1987), 49–70; Leo Zrudlo, *Psychological Problems and Environmental Design in the North* (Quebec: Université Leval, 1972), 126; William M. Smith and Tom Barchaky, "Habitability in Northern Housing Design," *Polar Record* 27 (1991), 39–42; June Calendar to author, October 2, 1988; Annie Dillard, *Pilgrim at Tinker Creek* (New York: Harper & Row, 1974), 43.
66. *Red Grooms: A Retrospective* (Philadelphia: Pennsylvania Academy of Fine Arts, 1985); Carter Ratcliff, *Red Grooms* (New York: Abbeville Press, 1964).
67. Andrew Ross, "The Work of Nature in the Age of Electronic Emission," *Social Text* 18 (1987–88), 116–28.
68. Don DeLillo, *White Noise* (New York: Viking Press, 1985), 55; Robert Henson, *Television Weathercasting* (Jefferson, N.C.: McFarland, 1990); David Hyatt, Kathy Riley, and Noel Sederstrom, "Recall of Television Weather Reports," *Journalism Quarterly* 55 (1978), 306–10; John Foster, "Will It Snow? How Much? When? How Cities Get the Word," *American City and County* (August 1983), 23–26; Cullen Murphy, "Under the Weather," *Atlantic* (February 1986), 16–18.
69. Al F. Wuori, "Snow Stabilization Studies," in *Ice and Snow: Properties, Processes and Application,* ed. W. D. Kingery (Cambridge, Mass.: MIT Press, 1963), 438–58; Jon Ecklund to author, August 6, 1991; Louise Erdrich, "Snow Houses," *Architectural Digest* (December 1994), 40; "Bob Levey's Washington," *Washington Post* (February 8, 1993), C12. Special thanks to Claudia for the snow turds story.

CHAPTER 3. SNOWMEN AND SNOWMANSHIP

1. Avon Neal and Ann Parker, *Ephemeral Folk Figures: Scarecrows, Harvest Figures, and Snowmen* (New York: Clarkson N. Potter, 1969); *Andy Goldsworthy: A Collaboration with Nature* (New York: Harry N. Abrams, 1990).
2. Bill Watterson, "Calvin and Hobbes," syndicated (February 21, 1990, and January 22, 1993).

3. Edmund S. Morgan, *The Birth of the Republic, 1763–1789* (Chicago: University of Chicago Press, 1956), 47.
4. Thomas Bailey Aldrich, *The Story of a Bad Boy* (Boston: Houghton Mifflin, 1870; reprint, Hanover, N.H.: University Press of New England, 1990), 141–54; Henry Adams, *The Education of Henry Adams* (Boston: Houghton Mifflin, 1918), 41–42.
5. Daniel Carter Beard, "Snow-Ball Warfare," *St. Nicholas* (January 1880), 263–66; Beard, "A Snow Battle," *St. Nicholas* (January 1881), 235–36; Beard, *The American Boys Handy Book* (New York: Charles Scribner's Sons, 1882; reprint, Boston: David R. Godine, 1983), 257–68. A hundred and ten years after Beard, a popular magazine published essentially the same instructions for building a snow fort and having a "peaceful" war, but with far more emphasis on planning and safety. See Malcolm Wells, "Snow Forts," *Country Journal* (January–February 1990), 76–79.

   For other descriptions of snowballing in the nineteenth century, see Thomas Miller, *The Boy's Winter Book: Descriptive of the Seasons, Scenery, Rural Life, and Country Amusements* (New York: Harper & Bros., 1847), 5–6; Henry Clarke Wright, *Growing Up in Cooper Country: Boyhood Recollections of the New York Frontier* (1849; reprint, Syracuse: Syracuse University Press, 1965); *The Cormany Diaries: A Northern Family in the Civil War,* ed. James C. Mohr (Pittsburgh: University of Pittsburgh Press, 1982), 59; "The Snow," *Child at Home* (February 1860), 1; John S. Jackman, *Diary of a Confederate Soldier,* ed. William C. Davis (Columbia: University of South Carolina Press, 1990), 110–11; James Russell Lowell, "A Good Word for Winter," in *Fireside Travels* (Boston: Houghton Mifflin, 1904), 366; Emory Pottle, "Memories of a Boy-Time Winter," *Outing* 43 (December 1903), 296; and Walter Brooks, *A Child and a Boy* (New York: Brentano's, 1915), 117–18. I thank Robert Lewis for the Cormany and *Child at Home* references.
6. The discovery of play is discussed in Bernard Mergen, *Play and Playthings* (Westport, Conn.: Greenwood Press, 1982), 57–80. Annie Dillard, *An American Childhood* (New York: Harper & Row), 1987, 45–49.

   In the future, children may have the option of snowball fighting in the comfort of their homes. Nintendo began marketing a video game called Snow Brothers in 1991, the object of which was to "hit a bunch of slippery bad guys with giant crusher snowballs. And bury them deep in the stuff, so they won't melt down." Snow Brothers, © 1991, Capcom USA, licensed by Nintendo.
7. Beard, *American Boys Handy Book,* 269–74; Samuel Van Brunt, "Snow-Sports for Girls and Boys," *St. Nicholas* (February 1880), 320–21; Neal and Parker, *Ephemeral Folk Figures,* 149–75; John Champlin and Arthur Bostwick, *The Young Folks' Cyclopedia of Games and Sports* (New York: Henry Holt, 1890), 660–61; Michael Gold, *Jews without Money* (New York: Horace Liveright, 1930), 242.
8. Neal and Parker, *Ephemeral Folk Figures,* 161; Thierry De Navacelle, *Woody Allen on Location* (New York: William Morrow, 1987), 343; Pete Cooke, "Clothes-Minded Neighbor Bugged by Naked Snowman," *Weekly World News* (April 11, 1989), 17.

9. Bill Watterson, *Attack of the Deranged Mutant Killer Monster Snow Goons* (Kansas City, Mo.: Andrews & McMeel), 1992.

A novelty item sold in airport gift shops and similar stores consists of a plastic dome similar to the familiar snow dome but filled with a liquid and bits of plastic representing lumps of coal, a hat, and a carrot. Called "Melted Snowmen," "California Snowmen," and similar names, they represent a further ironic mocking of the human/snowman condition in an age concerned with global warming and nuclear winter. (Patent pending, 1982, AE Enterprises, DeLand, Fla.).

10. P. K. Page, "The Snowman," in *The New Oxford Book of Canadian Verse in English* (Toronto: Oxford University Press, 1982), 182–83; Howard Nemerov, "Journey of the Snowmen," in *The Collected Poems of Howard Nemerov* (Chicago: University of Chicago Press, 1977), 241–42.
11. Fred Anders and Ann Agranoff, *Ice Palaces* (New York: Abbeville Press, 1983); Don Morrow, "The Knights of the Snowshoe: A Study of the Evolution of Sport in Nineteenth Century Montreal," *Journal of Sport History* 15 (Spring 1988), 5–40; Herman Pleij, "Urban Elites in Search of a Culture: The Brussels Snow Festival of 1511," *New Literary History* 21 (Spring 1990), 629–47.
12. C. C. Andrews, ed., *History of St. Paul, Minn.* (Syracuse, N.Y.: D. Mason, 1890). A useful history of the St. Paul carnivals based largely on the collections in the Minnesota Historical Society is Jean Spraker, "'Come to the Carnival at Old St. Paul': Souvenirs from a Civic Ritual Interpreted," in *Prospects: An Annual of American Cultural Studies* 11 (New York: Cambridge University Press, 1987), 233–46. My thanks to Ms. Spraker for her helpful comments on an earlier draft of this section of my book.
13. *St. Paul Pioneer Press* (January 15, 1886), 3.
14. Ibid. (January 21–February 5, 1886); Morrow, "Knights of the Snowshoe," 21–24; Spraker, "'Come to the Carnival,'" 240. See also Sylvie Dufresne, "Le Carnaval d'hiver de Montreal, 1833–1889," *Urban History Review* 11 (February 1983), 25–46.
15. *St. Paul Pioneer Press* (February 8, 1886), 6. Herbert Spencer is quoted in Daniel Rogers, *The Work Ethic in Industrial America, 1850–1920* (Chicago: University of Chicago Press, 1978), 94. The suspension of carnival activities on Sunday reflects a conservatism on the part of the organizers, since many Christian ministers were calling for less strict Sunday observance. "Sunday," wrote one reformer, "is the only day in the week when the over-tasked laborer can sleep over, when he has the sense of being master of his own household, when he can think leisurely of his relations to God and Man." Julius H. Ward, "The New Sunday," *Atlantic Monthly* (April 1881), 530.
16. Andrews, *History of St. Paul,* 201. For a useful theory of celebration employing the distinction between play and ritual, metaphor and metonymy, see Frank Manning, "Cosmos and Chaos: Celebration in the Modern World," in *Celebration of Society: Perspectives on Contemporary Cultural Performance* (Bowling Green, Ohio: Bowling Green University Popular Press, 1983), 3–30. On parades, see Susan Davis, *Parades and Power: Street Theater in Nineteenth Century Philadelphia* (Philadelphia: University of Pennsylvania Press, 1986).
17. *St. Paul Pioneer Press* (January 30, 1916), 1.

18. Ibid. (January 26–February 5, 1916).
19. Ibid. (January 26, 1916), 1 and 3; (February 4, 1916), 1.
20. F. Scott Fitzgerald, "The Ice Palace," in *Flappers and Philosophers* (New York: Charles Scribner's Sons, 1920), 47–71. For an interesting interpretation of gender in the 1990 carnival, see Robert Lavenda, "Festivals and the Creation of Public Culture: Whose Voice(s)?" in *Museums and Communities: The Politics of Public Culture,* ed. Ivan Karp, Christine Mullen Kreamer, and Steven D. Levine (Washington, D.C.: Smithsonian Institution Press, 1992), 99–100.
21. Spraker, "'Come to the Carnival,'" 245; Lavenda, "Festivals and the Creation of Public Culture," 98; *St. Paul Sunday Pioneer Press* (January 24, 1982), 1, 10, and supplement.
22. "Dartmouth Winter Carnival," *Literary Digest* 117 (February 10, 1934), 23; *Michigan Tech Lode Pictorial* 2 (February 8, 1964); Michigan Technological University, *Carnival Pictorial* (1989, 1991, 1992). I thank Dr. Ed Adams for copies of the carnival program. "Snowfest Official Program," *Tahoe World* (March 4–13, 1988); "Snow Sculpture Festivals," *New York Times* (January 6, 1992), sec. 5, 3. For the work of the St. Michael's students, I thank their instructor, Leslie Fry, for sending pictures of her project.
23. On Larkin Mead, see Arthur W. Peach, "The Snow Angel," [Vermont Historical Society] *News and Notes* 5 (January 1954), 34–35. The statue is now in the Brattleboro Public Library. Solon Hannibal Borglum, brother of the creator of the Mt. Rushmore portraits, was another sculptor inspired by snow. His depiction in marble of horses and riders caught in snowdrifts and blizzards brought him favorable notice in the 1920s. See Louise Eberle, "In Recognition of an American Sculptor," *Scribner's Magazine* 72 (September 1922), 379–84.
24. Harry L. Wells, "Coasting Down Some Great Mountains," *Cosmopolitan* (January 1896), 240.
25. E. John B. Allen, *From Skisport to Skiing: One Hundred Years of an American Sport, 1840–1940* (Amherst: University of Massachusetts Press, 1993). W. R. Rickmers, "The Elements of Ski-Running," in *Ski-running,* ed. E. C. Richardson (London: Horace Cox, 1904), 20. David Forkes, "Skiing: An English Language Bibliography, 1891–1971," typescript (Vancouver, 1975), 1, identifies this as the first book on skiing to be published originally in English.
26. Theodore A. Johnsen, *The Winter Sport of Skeeing* (Portland, Me.: Theo. A. Johnsen, 1905; reprint, New Hartford, Conn.: International Skiing History Association, 1994).
27. George Wharton James, *Winter Sports at Huntington Lake Lodge in the High Sierra: The Story of the First Annual Ice and Snow Carnival of the Commerce Club of Fresno, California* (Pasadena: Radiant Life Press, 1916), 35, 39.
28. *Winter Sports Verse,* ed. William Haynes and Joseph LeRoy Harrison (New York: Duffield, 1919), i, 17–18, 186.
29. The history of the development of skiing in Mt. Hood National Forest is contained in several boxes of record group 95, U.S. Forest Service, Division of Recreation and Lands, 1906–1951, National Archives and Record Administration, Washington, D.C. See Frank Waugh, "Recreation Uses in the Mt. Hood Area," 1920, box 127, and correspondence from Assistant Forester L. F. Kneipp to Homer A. Rogers in 1921; and W. B. Greeley to Sen. Charles L. McNary,

January 2, 1925, box 129. On the bureaucracy of the Forest Service, see Herbert Kaufman, *The Forest Ranger: A Study in Administrative Behavior* (Baltimore: Johns Hopkins University Press, 1960).

30. "List of Nominees for Membership on the Committee to Be Appointed by the Secretary of Agriculture to Study and Report upon the Mount Hood Area in Oregon"; Julius L. Meier to W. M. Jardine, August 29, 1928; Jardine to Meier, October 19, 1928, record group 95, U.S. Forest Service, box 128.

    Among the other members of the committee were the president of the Hood River Chamber of Commerce; the dean of forestry at Oregon Agricultural College (now Oregon State University, Corvallis); the president of the Farmers' Educational and Cooperative Union of America, Oregon & Southern Idaho Division; and the state forester.

31. Jardine to Meier, October 19, 1928, record group 95, U.S. Forest Service, box 128.

32. Jardine to John C. Merriam, February 2, 1929, record group 95, U.S. Forest Service, box 126. Frank A. Waugh and Aldo Leopold had both advocated the creation of wilderness areas before 1920. Jardine's position may reveal their influence. See Roderick Nash, *Wilderness and the American Mind,* 3d ed. (New Haven, Conn.: Yale University Press, 1982), 182–87.

33. "Public Values of the Mount Hood Area," p. 14, record group 95, U.S. Forest Service, box 125. The report was also published as Senate document 164, 71st Cong., 2d sess., 1930.

34. Ibid., 65, 92.

35. L. F. Kneipp to C. J. Buck, January 17, 1931; L. L. Tyler to Buck, March 16, 1931, record group 95, U.S. Forest Service, box 125.

36. James A. Mount to Sen. Charles McNary, February 6, 1935, record group 95, U.S. Forest Service, box 71. A similar letter from J. E. Carpenter for the Winter Sports Committee of the California Chamber of Commerce to F. A. Silcox, chief, U.S. Forest Service, and to Arno Cammerer, director, National Park Service, June 2, 1936, requested the erection of ski jumps on federal land and an end to the policy prohibiting private organizations from collecting fees for winter sports events. On June 16, C. E. Rachford, acting chief of the Forest Service, replied that "there is no reason . . . under established National Forest policy, why an association might not secure a permit for occupancy of an appropriate area of National Forest land and construct facilities where such events might be held subject to proper regulations with a charge to the public." Record group 95, U.S. Forest Service, box 71.

37. Allen, *From Skisport to Skiing,* 114–16, 141–42; Rachael Griffin and Sarah Munro, *Timberline Lodge* (Portland, Oreg.: Friends of Timberline, 1978), 5; Alan Gowans, *Styles and Types of North American Architecture: Social Function and Cultural Expression* (New York: HarperCollins, 1992), 257, 265.

38. Jean Weir, "The Arts and Furnishings," in Griffin and Munro, *Timberline Lodge,* 30–45; U.S. Forest Service, "Mount Hood Timberline Lodge," January 31, 1938, record group 95, U.S. Forest Service, box 129. An anonymous writer for the Federal Writers' Project rhapsodized:

    > Like the mountain upon which it is built, Timberline Lodge is symbolic of many things not seen in the timber and stone which make it.

As the winding road leading to it represented progress by laborers, not the least of whose rewards was the daily inspiration of the enlarged and expanding view of mountain tops, so the building itself exemplifies a progressive social program which has revived dormant arts and pointed the way for their perpetuation. It presents concretely the evidence that men still aspire to the dream, often secret but always universal, of becoming greater than themselves through association with others in a common purpose.

Art for the Millions: Essays from the 1930s by Artists and Administrators of the WPA Federal Art Project, ed. Francis V. O'Connor (Boston: New York Graphic Society, 1973), 189.

39. On Marshall, see Roderick Nash, "The Strenuous Life of Bob Marshall," *Forest History* 10 (October 1966), 18–25; and James Glover, "Romance, Recreation, and Wilderness," *Environmental Review* 14 (Winter 1990), 23–39. John Sieker, acting chief, Division of Recreation and Lands, to Mr. Clapp, July 22, 1938, record group 95, U.S. Forest Service, box 129.

40. Gilbert D. Brown, forest supervisor, press release, December 6, 1938; Robert Marshall to Regional Forester Buck, January 20, 1939; Marshall to Granger and Parkinson, May 10, 1939, record group 95, U.S. Forest Service, box 71. Marshall expressed his outrage over the granting of private concessions on national forests and parks in a letter to a regional forester, June 14, 1937, record group 95, U.S. Forest Service, box 71. He was also quick to praise foresters who showed initiative such as publishing guides to winter sports, see Marshall to Show, February 3, 1939, record group 95, U.S. Forest Service, box 71. For one of the few acknowledgments by private commercial interests of the Forest Service's contribution to the development of skiing, see "Ski Areas in the No. Pacific Region Developed by the U.S. Forest Service," *Ski Illustrated* (Spring 1937), 24–25, which illustrates plans for the warming hut at Leavenworth, Washington.

41. The thirteen archival boxes of material relating to Sun Valley in the Papers of W. Averell Harriman, Manuscript Division, Library of Congress, Washington, D.C., are alphabetized by the names of correspondents and some business topics and numbered consecutively 732–44. Subsequent footnotes will refer to box numbers. I thank Jacqueline McGlade for calling my attention to this material.

Rudy Abramson, *Spanning the Century: The Life of W. Averell Harriman, 1891–1986* (New York: William Morrow, 1992), chronicles Harriman's many activities and careers, devoting about 10 of 700 pages to Sun Valley. In an interview with Abramson shortly before his death, Harriman recalled his first glimpse of the valley: "There were all the mountains with gold in the background and the hills covered with snow. I fell in love with the place then and there" (224). The perception of snow as a valuable resource to be mined could hardly be stated more frankly.

A further indication of the Citizen Kane–like spirit in which Harriman began his Xanadu may be seen in his letter to Secretary of Commerce Ernest G. Draper, March 2, 1936: "I am a member of a group which has been contemplating the building and operating of a hotel in Idaho for winter sports enthusiasts. It is a new line of business for me and I am getting a great 'kick' out of it." Harriman Papers, box 736.

42. Abramson, *Spanning the Century,* 222–24; Dorice Taylor, *Sun Valley* (Bethany, Conn.: Ex Libris Sun Valley, 1980), 9–15; Felix Schaffgotsch telegram to Harriman, January 18, 1936, Harriman Papers, box 740; Carl R. Gray, president, UPRR, to Harriman, February 13, 1936, with Schaffgotsch's report, box 739.
43. Steve Hannagan to Harriman, March 28, 1936, Harriman Papers, box 733. This letter is actually a copy sent to Harriman by Victor H. Palmieri, president, Janss Corporation, November 18, 1965, after Sun Valley had been sold to Janss.
44. Abramson, *Spanning the Century,* 225–27; Joyce Zaitlin, *Gilbert Stanley Underwood: His Rustic, Art Deco, and Federal Architecture* (Malibu, Calif.: Pangloss Press, 1989), 149–57; "Sun Valley," brochure, November 1937, Harriman Papers, box 743; David O. Selznick to Harriman, January 5, 1937, box 742; Harriman, Memorandum on Sun Valley, March 15, 1939; C. T. Carey to Harriman, October 19, 1939; Harriman to W. P. Rogers, October 26, 1939, box 733.

    Selznick's nine-page, single-spaced letter is a delightful commentary on the opening of the lodge, remarking on everything from the hardness of the pillows to the "excess of butter used on everything." His version of his well-reported fistfight with the Chicago banker Charles F. Gore is that "Glore" [*sic*] pushed him to sit next to the French actress Lili Damita, who was married to Errol Flynn, whereupon Selznick "split open his nose." Selznick blamed John E. P. Morgan, whom Harriman had hired to organize a ski club of wealthy easterners. Morgan and Gore also made anti-Semitic remarks. A month later Morgan cheerfully reported on his extended trip to other western ski areas, including the Ahwahnee Hotel, which Underwood had designed for Yosemite Park in 1925. At the Ahwahnee, "now nicknamed the Marhihuana," Morgan wrote, "the guests were at least 75% jewish and gave us the same impression that Dexter Lindsay got of Sun Valley at New Years." Morgan to Harriman, January 28, 1937, Harriman Papers, box 738.
45. Harriman to Rogers, telegram, August 9, 1939; Rogers to Harriman, telegram, August 10, 1939, Harriman Papers, box 733; E. Bucher to Harriman, April 5, 1938, box 742; Harriman to Rogers, January 8, 1940, box 740; Don Fraser to Harriman, telegram, February 4, 1940, box 743; Christopher LaFarge to Harriman, May 13, 1940, box 738; Dr. Moritz, "Report on Injuries," April 15, 1940, box 738; R. P. Meiklejohn to Neil Regan, November 24, 1940, box 743.
46. Harriman to Alfred Biddle, March 27, 1937, Harriman Papers, box 732; Allen, *From Skisport to Skiing,* 109–14; John Jay, *Ski Down the Years* (New York: Award House, 1966), 104; Taylor, *Sun Valley,* 35–37; Harriman to H. C. Mann, April 15, 1936, box 742; Mann to William Jeffers, May 15, 1936, box 742; the names Ketchum and Harriman were penciled above the code words.
47. B. H. Prater to Harriman, October 4, 1937; Prater to Harriman, October 4, 1937, with L. Castagneto's report "Proposed Snow Sled Tractor Route—East Fork"; Prater to Harriman, October 6, 1937, with Castagneto's report "Bald Mountain Snow Sled Route"; Harriman Papers, box 736; Meiklejohn report "Aerial Tramway—Cannon Mountain, Franconia, New Hampshire," February 17, 1939; G. H. Trout to Harriman, telegram, June 30, 1939, box 733; J. Alden Wilson and Morlan W. Nelson, "A History of the Development of Oversnow

Vehicles," *Proceedings of the Western Snow Conference* (1968), 10–11; Prater to Mann, January 30, 1939; J. L. Haugh to Jeffers, August 22, 1939, box 733.

48. E. John B. Allen, *Teaching and Technique: A History of American Ski Instruction* (Latham, N.Y.: EPSIA Educational Foundation, 1987), 7–28; Allen, *From Skisport to Skiing,* 119–23; Schaffgotsch to John Ward, copy to Harriman, August 2, 1937; Otto Lang to Harriman, March 22, 1938; Harriman to Lang, March 26, 1938, Harriman Papers, box 738; Harriman, "Memorandum on Sun Valley," March 15, 1939, box 737.

    Harriman rejected an offer from Bertel N. Paaske, president of the Norden Ski Federation of America, to have the Norwegian "Holmekollen Technique" taught at Sun Valley. Harriman to Paaske, October 29, 1937, Harriman Papers, box 739.

    It should be remembered that during these crucial years Harriman was moving into a much larger world of international diplomacy and politics, as secretary of commerce, head of the Lend-Lease Program, ambassador to the Soviet Union, director of the Marshall Plan in Europe, and, in 1954, governor of New York.

49. *Skiing* (October 1989), 86–87, 106.

50. William A. Walsh, Jr., "The History and Technology of Man-Made Snow in Winter Recreation Areas," *Eastern Snow Conference Proceedings* (1974), 18; W. M. Pierce, Jr., "Method for Making and Distributing Snow," patent no. 2,676,471, April 27, 1954.

51. *Snow Country* (December 1991), 77; Steve Cohen, "High-Tech Snow," *Ski* (March 1992), 28–39.

52. Joe Tropeano, "Water," *Ski Area Management* (Winter 1963), 16–18; Walsh, "History and Technology of Man-Made Snow," 2; "The Bacteria at the Heart of a Good Snowfall," *New Scientist* (January 27, 1990), 38. Allen Best provides a vivid personal account of snowmaking at Silver Creek Ski Area in Colorado in "The Snow Also Rises," in *Western Water Made Simple* (Washington, D.C.: Island Press, 1987), 211–16.

53. Jill Wechsler, *Camelback: The Downs and Ups of a Banana Belt Ski Area* (Tannersville, Pa.: Coolmoor Publishing, 1989), 176; Cohen, "High-Tech Snow," 32–33; Joseph L. Sax, *Mountains without Handrails: Reflections on the National Parks* (Ann Arbor: University of Michigan Press, 1980), 70; *National Forest Landscape Management: Ski Areas,* agriculture handbook 671 (Washington, D.C.: GPO, 1994), vol. 2, chap. 7, 2.

    The complicated way demand for skiing is calculated is partially explored in Richard G. Walsh, Nicole P. Miller, and Lynde O. Gilliam, "Congestion and Willingness to Pay for Expansion of Skiing Capacity," *Land Economics* 59 (May 1983), 195–210.

54. U.S. Department of Agriculture, Forest Service, in cooperation with the National Ski Area Association, *Planning Considerations for Winter Sports Development* (Washington, D.C.: GPO, 1973); F. D. (Skip) Voorhees, "We're Losing Our Growing Room," *Ski Area Management* (Spring 1966), 20–22; Jim Spring, "Snowmess-at-Washington," *Ski Area Management* (Fall 1967), 52–56, 70; Bob Lochner, "Slim Davis: On Safeguarding the Sierras," *Ski Area Management* (Fall 1970), 32–33, 49; Peter Browning, "Mickey Mouse in the Mountains,"

Harper's (March 1972), 65–71, and (August 1972), 102–3; Jeanne Ora Nienaber, "Mineral King: Ideological Battleground for Land Use Disputes" (Ph.D. diss., University of California, Berkeley, 1973); Arthur B. Ferguson and William P. Bryson, "Mineral King: A Case Study in Forest Service Decision Making," *Ecology Law Quarterly* 2 (Summer 1972), 493–531; Susan R. Schrepfer, "Perspectives on Conservation: Sierra Club Strategies in Mineral King," *Journal of Forest History* 20 (October 1976), 176–90; "Commentary: Mineral King Goes Downhill," *Ecology Law Quarterly* 5 (1976), 555–74; U.S. Department of Agriculture, Forest Service Region 5, *Mineral King Final Environmental Statement* (San Francisco, February 26, 1976); John L. Harper, *Mineral King: Public Concern with Government Policy* (Arcata, Calif.: Pacifica Publishing, 1982).

55. Robert Cahnam, "The Water Crisis," *Ski Area Management* (Fall 1970), 30; Daniel Gibson, "Winning the Waste Disposal Battle," *Ski Area Management* (Fall 1970), 70–73, 95; "California Ski Resort to Make Snow from Treated Wastewater," *ENR* (February 1, 1993), 24; I. William Berry, "Vermont's Last Stand," *Ski Area Management* (Fall 1970), 37–39; John Hitchcock, "Vermont's Ecology Act One Year Later: The Meaning of Greening," *Ski Area Management* (October 1971), 22–24; Cindy Hill, "Loon Corp.'s Slippery Slope Proposal," *Wild Earth* (Summer 1991), 56–58; John Hitchcock, "Chemicals on Snow," *Ski Area Management* (Spring 1967), 14; Bill Tanner, "Erosion Control: Two Experiments," *Ski Area Management* (Spring 1967), 22–23; Ford Allen, "Conservation on the Ski Slopes," *Soil Conservation* (February 1972), 150–51; P. Walton Spear, "Preserving the Ecosystem," *Ski Area Management* (May 1988), 98–99, 120; Jim Buckingham, "Tree Planting Technique," *Ski Area Management* (July 1988), 66–67, 81; Wayne Victor Wilson, "The Decline in the Rate of Ski Facility Development: Changing American Attitudes toward the Environment and Economic Growth, 1968–1978" (Ph.D. diss., University of Massachusetts, 1981); G. Devarennes, "Composition physico-chimique de la neige artificielle et les impacts des eaux de fonte sur deux espèces végétales des écosystèmes montagneux au Québec," *Eastern Snow Conference Proceedings* (1993), 11–18.

The use of a melt-delaying chemical, PTX3, at the world championship races at Garmisch-Partenkirchen may have contributed to the death of the Austrian skier Ulrike Maier, according to Picabo Street of the U.S. ski team, who is quoted as saying that the chemical made the snow "grabby and weird" and turned "the course into a skating rink." *New York Times* (January 30, 1994), sec. 8, 1, 9.

An advertisement for a ground cover used by highway departments to "beautify their highways" suggests the extent to which ski areas have become extensions of interstates: "Attention Ski Area Owners! Protect and improve your slopes with Lofts Ski-Slope 901 Seed. Contains Crownvetch Ground Cover offers better skiing . . . better seeing . . . better snow-mobiling." *Ski Area Management* (June 1971), 11.

56. R. W. Hoham, A. E. Laursen, S. O. Clive, and B. Duval, "Snow Algae and Other Microbes in Several Alpine Areas in New England," *Eastern Snow Conference Proceedings* (1993), 165–73.

57. Sylvia Rodriguez, "Impact of the Ski Industry on the Rio Hondo Watershed,"

Annals of Tourism Research 14 (1987), 88–103; Perri Knize, "Water War," *Skiing* (December 1993), 90–97.

Similar disruptions of traditional life by ski area development, with varying results, have been studied in Japan and the Alps. See Okpyo Moon, *From Paddy Field to Ski Slope: The Revitalization of Tradition in Japanese Village Life* (New York: Manchester University Press, distributed by St. Martin's Press, 1990); and Hervé Gumuchian, *La Neige dans les alpes françaises du nord, une saison oubliée: L'Hiver* (Grenoble: Cahiers de l'Alpe, 1983).

58. William P. MacConnell and Robert R. Prescott, "Snow Barriers: Cure for Wind-Swept Slopes," *Ski Area Management* (Winter 1965), 34–37; Trail map showing snow fence drift lines for snow farming, White Grass Ski Touring Center, Davis, W.V., 1994; "They Could Steal Your Snow," *Ski Area Management* (Winter 1969), 44–45; Norman A. Wilson, "Planning for Weather Factors," *Ski Area Management* (April 1971), 26–27; Cohen, "High-Tech Snow," 36; Ronald I. Perla and B. Glenne, "Skiing," in *Handbook of Snow,* ed. D. M. Gray and D. H. Male (Toronto: Pergamon Press, 1981), 711.

59. Perla and Glenne, "Skiing," 719–20; Curtis Casewit, "The Wizard of Winter Park," *Ski Area Management* (Spring 1970), 42–43; William P. MacConnell and Lester F. Whitney, "A New Family of One-Man Snow Packers," *Ski Area Management* (Spring 1966), 18–19, 60; M. Martinelli, Jr., "Take the Plunge," *Ski Area Management* (February 1972), 26–28; Wilson and Nelson, "A History of the Development of Oversnow Vehicles," 9–18; Beartrac brochure, 1993; ad for Tucker Sno-Cat, *Ski Area Management* (Fall l962), 13; ad for Polaris Industries Sno-Traveller, *Ski Area Management* (Fall 1963), 13; "Area Products Being Talked About," *Ski Area Management* (Winter 1963), 24; Lew Cuyler, "The Future Is on the Fairways," *Ski Area Management* (September 1986), 87, 116.

60. See, for example, *Skiing Trauma and Safety: Fifth International Symposium,* technical publication 860 (Philadelphia: ASTM, 1985); *Skiing Trauma and Safety: Eighth International Symposium* (Philadelphia: ASTM, 1991); Stephen L. Fine, "Plug Your Legal Loopholes," *Ski Area Management* (Summer 1969), 19–21; Fine, "Ski Area Management in Court," *Ski Area Management* (Summer 1970), 20; Stanton E. Tefft, "Assignment of Responsibility," *Ski Area Management* (Winter 1966), 30–31.

61. Harrison Osborne, "Baby, It's Cold Out There," *Everyday Law* (January 1989), 33–37; Cleary, "Legal Scene," *Ski Area Management* (May 1988), 38–40; David Cleary, "Skier Responsibility Statutes Upheld," *Ski Area Management* (July 1988), 28–30, 74; Robert I. Rubin, "Ski Liability Law Cuts New Trails," *Trial* (October 1990), 108–13; John Underwood, "It's Pretty, It's Trendy, but Skiing Is Also Much Too Dangerous," *New York Times* (February 26, 1995), sec. 8, 9.

62. Peter Alford, "Snowmaking: Getting It All Together," *Ski Area Management* (Spring 1972), 31–33, 60; "Name That Snow," *Snow Country* 4 (December 1991), 32; Perri Knize, "Water War," *Skiing* (December 1993), 96; Peter Stark, "Let It Upsik . . . ," *Outside* (January 1995), 68; "Snow Conditions Terms and Definitions," *Mountain Messenger* (New England Ski Area Council, Woodstock, Vt., Winter 1994–95), 7.

63. Betsy R. Armstrong and Knox Williams, *The Avalanche Book* (Golden, Colo.: Fulcrum Publishing, 1992), 124–25; "Avalanches in Austria and Switzerland,"

Railroad Gazette (September 25, 1891), 671; Edward A. Beals, "Avalanches in the Cascades and Northern Rocky Mountains," *Monthly Weather Review* 38 (June 1910), 951–57.

64. Armstrong and Williams, *Avalanche Book,* 201–8; Anthony Will Bowman, "From Silver to Skis: The History of Alta, Utah, and Little Cottonwood Canyon, 1847–1966" (M.A. thesis, Utah State University, 1967), 66–83; Montgomery M. Atwater, *The Avalanche Hunters* (Philadelphia: Macrae Smith, 1968); Ronald I. Perla and M. Martinelli, Jr., *Avalanche Handbook,* agricultural handbook 489 (Washington, D.C.: GPO, 1976; rev. 1978); David McClung and Peter Schaerer, *The Avalanche Handbook* (Seattle: Mountaineers Books, 1993). My thanks to Bryan Dixon of Beartrac, the Snow Vehicle Division of LMC, for the Bowman reference.

65. Atwater, *Avalanche Hunters,* 18; Perla and Martinelli, *Avalanche Handbook,* 7–90; Armstrong and Williams, *Avalanche Book,* 68–69.

66. Muir quoted in Enos A. Mills, "Dangers of Snowslides," *Harper's Weekly* (December 24, 1904), 1999; Atwater, *Avalanche Hunters;* Mary Hallock Foote, *The Cup of Trembling and Other Stories* (1895; reprint, Freeport, N.Y.: Books for Libraries, 1970), 82. My thanks to Lynn Cothern for this reference.

67. Tom Lea, *The Primal Yoke* (Boston: Little, Brown, 1960), 335–36; Atwater, *Avalanche Hunters,* 9. Lea is also an artist whose best-known painting, *Two Thousand Yard Stare,* of a hollow-eyed soldier on a Pacific island during World War II, was published in *Life.* See Frederick Voss, *Reporting the War* (Washington, D.C.: National Portrait Gallery, 1994).

Ernest Hemingway noted the personalizing of mountains in Switzerland, where a priest told him that the mountain "is the great enemy of the mountain-dwelling people. . . . Yet we can never leave him. Perhaps in this, too, he shows he is our enemy." See "Swiss Avalanches," in *Ernest Hemingway Dateline: Toronto, The Complete Toronto Star Dispatches, 1920–1924,* ed. William White (New York: Charles Scribner's Sons, 1985), 455.

68. Armstrong and Williams, *Avalanche Book,* 44, 149–71, 189–97; Rogers to Harriman, January 25, 1952, Harriman Papers, box 732; J. E. Church, "Snow Perils and Avalanches," *Scientific Monthly* (April 1943), 309–31; Ernie Blake, "Avalanches under Control," *Ski Area Management* (Fall 1965), 25, 48; Edward LaChapelle, "The Control of Snow Avalanches," *Scientific American* (February 1966), 92–101; Perla and Martinelli, *Avalanche Handbook,* 111–75.

69. M. R. de Quervain, *Avalanche Classification* (Toronto: International Association of Scientific Hydrology, 1958); de Quervain, *On Avalanche Classification, A Further Contribution* (Davos: International Association of Scientific Hydrology, 1965); Malcolm Mellor, *Avalanches* (Hanover, N.H.: Cold Regions Research and Engineering Laboratory, 1968), 22–23; Perla and Martinelli, *Avalanche Handbook,* 223–24; McClung and Schaerer, *Avalanche Handbook,* 258; Fes de Scally and Jim Gardner, "The Hydrological Importance of Avalanche Snow, Kaghan Valley, Himalaya Mountains, Pakistan," *Proceedings of the International Snow Science Workshop* (Vancouver: Canadian Avalanche Association, 1988), 277–83; de Scally, "Can Avalanche Activity Alter the Pattern of Snowmelt Runoff from High-Mountain Basins?" *Eastern Snow Conference Proceedings* (1993), 213–21.

70. Cary J. Mock and Paul A. Kay, "Avalanche Climatology of the Western United States, with an Emphasis on Alta, Utah," *Professional Geographer* 44 (August 1992), 307–18; McClung and Schaerer, *Avalanche Handbook,* 256–57.

71. Charles Bradley, "Avalanches," *Cross Country Skier* (February 1982), 45; Bruce Tremper and David Ream, "Utah Avalanche Forecast Center User Survey and Formats for Avalanche Advisories," *Proceedings of the International Snow Science Workshop* (1988), 241–50; Tremper and Ream, "Utah Backcountry User Survey," *Avalanche Review* 7 (October 1, 1988), 1–2, 4–5, 8; Brad Meiklejohn, "'Safe Skiing': Backcountry Skiing in Avalanche Terrain," *Proceedings of the International Snow Science Workshop* (Bigfork, Mont., 1990), 97–110. Meiklejohn suggests using a red, yellow, and green light system to simplify warnings. For a good discussion of the problems of zoning, see Jack D. Ives and Misha Plam, "Avalanche-Hazard Mapping and Zoning Problems in the Rocky Mountains, with Examples from Colorado, U.S.A.," *Journal of Glaciology* 26 (1980), 363–75.

The history of snowmobiling, which dates back at least to 1898, remains to be written. The beginnings of such a history may be found in "Types of Motor Sleighs Produced during the Past Winter in the United States and Europe," *Popular Mechanics* (April 1912), 468–70; S. C. Whitlock, "The Story of the Michigan Snowmobile," *Michigan Conservation* 8 (June 1939); George D. Clyde, "The Utah Snow-Mobile," *Transactions of the American Geophysical Union,* pt. 1 (1944), 166–75; J. Alden Wilson and Morian W. Nelson, "A History of the Development of Oversnow Vehicles," *Proceedings of the Western Snow Conference* (1968), 9–18; Jim Burnett, *Adirondack Snow Flurries* (Cranberry Lake, N.Y.: Halstead Publishing, 1987), 14, 53–55; Bill Tanler, "Snowmobiles Are Here to Stay," *Ski Area Management* (Winter 1966), 16–20.

72. R. A. Young and R. Crandall, "Wilderness Use and Self-Actualization," *Journal of Leisure Research* 16 (1984), 149–60; Allan S. Mills, "Participation Motivations for Outdoor Recreation: A Test of Maslow's Theory," *Journal of Leisure Research* 17 (1985), 184–99; Celia Millward and Richard Millward, "Ski-Trail Names: A New Toponymic Category," *Names* 32 (September 1984), 191–217.

CHAPTER 4. OPENING THE SNOW FRONTIER

1. George Herbert, in *The Home Book of Proverbs, Maxims and Familiar Phrases,* ed. Burton Stevenson (New York: Macmillan, 1948), 2151.

2. Charles Dana Wilbur, *The Great Valleys and Prairies of Nebraska and the Northwest* (Omaha: Daily Republican Printer, 1881), 69–70. Both George Perkins Marsh and Grove Karl Gilbert recognized the importance of snow in providing water for cities and farms. In *Man and Nature* (1864; reprint, Cambridge, Mass.: Harvard University Press, 1965), Marsh comments on several schemes to tap underground water but favors "the creation of perennial springs by husbanding rain and snow water, storing it up in artificial reservoirs of earth, and filtering it through purifying strata, in analogy with the operations of nature" (378). Gilbert, noting the success of Mormon settlers in Utah in increasing streamflow for irrigation, concludes that "by removing foliage, that share of

the rain and snow which was formerly caught by it and thence evaporated, is now permitted to reach the ground, and some part of it is contributed to the streams. Snow beds that were once shaded are now exposed to the sun, and their melting is so accelerated that a comparatively small proportion of their contents is wasted by the wind. Moreover, that which is melted is melted more rapidly, and a larger share of it is formed into rills." Gilbert's study of water supply in the West was published as a chapter of John Wesley Powell's *Report on the Arid Region of the United States,* made to Congress in 1878 (reprint, Cambridge, Mass.: Harvard University Press, 1962), 89. Raymond M. Rice, "Snow Management Research in the High Sierra Range," *Journal of Range Management* 12 (January 1959), 15.

3. See, for example, Ralph E. Olson, *A Geography of Water* (Dubuque, Iowa: William C. Brown, 1970); Norman Smith, *Man and Water: A History of Hydro-Technology* (New York: Charles Scribner's Sons, 1975); Fred Powledge, *Water: The Nature, Use, and Future of Our Most Precious and Abused Resource* (New York: Hill & Wang, 1982); and Peter Rogers, *America's Water: Federal Roles and Responsibilities* (Cambridge, Mass.: MIT Press, 1993). For a comparison of urban water use, see Jean-Pierre Goubert, *The Conquest of Water: The Advent of Health in the Industrial Age* (French ed. 1986; Princeton, N.J.: Princeton University Press, 1989).
4. Luna Leopold, *Water: A Primer* (San Francisco: W. H. Freeman, 1974), 133–38; Sandra Postel, "Crisis on Tap: A World without Water," *Washington Post* (October 29, 1989), B3; Postel, *Last Oasis: Facing Water Scarcity* (New York: W. W. Norton, 1992); Peter H. Gleick, ed., *Water in Crisis: A Guide to the World's Fresh Water Resources* (New York: Oxford University Press, 1993).
5. Yi-Fu Tuan, *The Hydrologic Cycle and the Wisdom of God: A Theme in Geoteleology* (Toronto: University of Toronto Press, 1968), 3. For a charming description of the hydrologic cycle through snow, see Paul Gallico, *Snowflake* (Garden City, N.Y.: Doubleday, 1953).
6. Arnold Guyot, *Directions for Meteorological Observations and the Registry of Periodical Phenomena,* Smithsonian Miscellaneous Collections 148 (Washington, D.C.: GPO, 1870), 19–23. Earlier editions appeared in 1850, 1855, and 1860. The federal government uses the spelling *gage,* but I have opted for the more common form, *gauge,* except in direct quotation. For an illustration of Guyot's gauge, see W. E. Knowles Middleton, *Catalog of Meteorological Instruments in the Museum of History and Technology* (Washington, D.C.: Smithsonian Institution Press, 1969), 72. See also Middleton, *Invention of the Meteorological Instruments* (Baltimore: Johns Hopkins University Press, 1969); and *A History of the Theories of Rain and Other Forms of Precipitation* (New York: Franklin Watts, 1966).
7. B. E. Goodison, H. L. Ferguson, and G. A. McKay, "Measurement and Data Analysis," in *Handbook of Snow,* ed. D. M. Gray and D. H. Male (Toronto: Pergamon Press, 1981), 202; *Monthly Weather Review* (May 1901), 219. See also A. J. Henry, "The Density of Snow," *Monthly Weather Review* (March 1917), 102–13.
8. "Marvin, Charles Frederick," *Dictionary of American Biography,* supp. 3 (New York: Charles Scribner's Sons, 1973), 511–12.

9. *Instructions for Use of the Rain Gauge* (Washington, D.C.: Weather Bureau, 1892), 7–8. On the lack of self-registering meteorological instruments, see Robert Multhauf, "The Introduction of Self-Registering Meteorological Instruments," *Contributions from the Museum of History and Technology,* bulletin 228, U.S. National Museum, paper 23 (Washington, D.C.: Smithsonian Institution, 1961), 95–106.
10. *Instructions for Using Marvin's Weighing Rain and Snow Gauge* (Washington, D.C.: U.S. Weather Bureau, 1893), 3.
11. Cleveland Abbe, "Weight of Snow," *Monthly Weather Review* (April 1894), 175.
12. L. G. Carpenter and F. H. Brandenberg, "Melting Snow and River Floods," *Monthly Weather Review* 25 (May 1897), 209–10; Brandenberg, "The Water Supply for the Season of 1900 as Depending on Snowfall," *Monthly Weather Review* (November 1900), 493–99; Hiram Martin Chittenden, House Committee on Irrigation and Arid Lands, "Preliminary Examination of Reservoir Sites in Wyoming and Colorado," 55th Cong., 2d sess., 1897, H. Doc. 141.
13. Samuel Hays, *Conservation and the Gospel of Efficiency: The Progressive Conservation Movement, 1890–1920* (Cambridge, Mass.: Harvard University Press, 1959); Roderick Nash, *Wilderness and the American Mind* (New Haven, Conn.: Yale University Press, 1967; 3d ed., 1982); Marc Reisner, *Cadillac Desert: The American West and Its Disappearing Water* (New York: Viking Press, 1986); William Lilley and Lewis L. Gould, "The Western Irrigation Movement, 1878–1902: A Reappraisal," in *The American West: A Reorientation,* ed. Gene M. Grassley (Laramie: University of Wyoming Publications, 1966).
14. Donald Worster, *Rivers of Empire: Water, Aridity, and the Growth of the American West* (New York: Pantheon Books, 1985; reprint, New York: Oxford University Press, 1992); idem, *Nature's Economy: A History of Ecological Ideas* (San Francisco: Sierra Club Books, 1977; 2d ed., New York: Cambridge University Press, 1994).
15. James Edward Church, "The Human Side of Snow: The Saga of the Mount Rose Observatory," *Scientific Monthly* (February 1937), 141. Church published over a hundred scientific and popular papers and left ninety boxes of manuscripts and memorabilia to the University of Nevada, Reno. See Helen J. Poulton, *James Edward Church: Bibliography of a Snow Scientist,* bibliographical series 4 (Reno: University of Nevada Press, 1964); and Bernard Mergen, "Seeking Snow: James E. Church and the Beginnings of Snow Science," *Nevada Historical Society Quarterly* 35 (Summer 1992), 75–104. The claim that the first snow survey for streamflow forecasting was done by "Mr. Jarvis," chief engineer of the Chenango Canal in New York, in 1834, is not disputed, but nothing is known of his methods, and they did not lead to any further surveys. See Henry S. Santeford, "A Challenge in Snow and Ice," in *Advanced Concepts and Techniques in the Study of Snow and Ice Resources,* comp. Henry S. Santeford and James L. Smith (Washington, D.C.: National Academy of Sciences, 1974), 4.
16. James Edward Church, "A Direct Route from Susanville to Fall River Mills," *Sierra Club Bulletin* 2 (January 1898), 194–96.
17. James Edward Church, "Up from the 'Land of Little Rain' to the Land of Snows: Being the Journal of a Sledging Trip up Mount Whitney in Winter,"

Sierra Club Bulletin 7 (June 1909), 105–18; "The Mount Rose Weather Observatory, 1905–1907," *Sierra Club Bulletin* 6 (June 1907), 178; "The Mount Rose Weather Observatory," *Monthly Weather Review* (June 1906), 255–63; "A Midwinter Trip through Nevada's Mountain Park," *Sierra Club Bulletin* 8 (June 1912), 249–59.

18. Church, "Mount Rose Weather Observatory, 1905–1907," 181.
19. For an overview of the importance of Lake Tahoe, see Douglas H. Strong, *Tahoe: An Environmental History* (Lincoln: University of Nebraska Press, 1984); Donald J. Pisani, "Storm over the Sierras: A Study in Western Water Use" (Ph.D. diss., University of California, Davis, 1974); and Pisani and W. Turrentine Jackson, *Lake Tahoe Water: A Chronicle of Conflict Affecting the Environment, 1863–1939* (Davis: University of California, Institute of Governmental Affairs, 1972). The Pyramid Lake Paiutes have had to go to court numerous times to force the government to honor its treaty. See Martha C. Knack and Omer C. Stewart, *As Long As the River Shall Run: An Ethnohistory of Pyramid Lake Indian Reservation* (Berkeley: University of California Press, 1984); Senate Select Committee on Indian Affairs, *To Settle Certain Claims Affecting the Pyramid Lake Paiute Indian Tribe of Nevada: Hearing before the Select Committee on Indian Affairs* on S. 1558, 94th Cong., 1st sess., October 2, 1985; Senate Subcommittee on Water and Power, *Truckee-Carson–Pyramid Lake Water Rights Settlement Act: Hearing before the Subcommittee on Water and Power of the Committee on Energy and Natural Resources,* on S. 1554, 101st Cong., 2d sess., February 6, 1990.
20. Alexander McAdie to James Edward Church, February 20, 1909; Edwin Easterbrook to S. P. Ferguson [*sic*], September 28, 1910, Church Papers, NC 96, box 3. The earliest published description of the snow sampler is in James E. Church, "The Conservation of Snow: Its Dependence on Forests and Mountains," *Scientific American Supplement* 74 (September 7, 1912), 152–55, but a fuller description appeared in Church, "Recent Studies of Snow in the United States," *Quarterly Journal of the Royal Meteorological Society* 40 (January 1914), 43–52.

    That Church may have consciously seen this rivalry as a contest between those who studied snow for its own sake and saw in that study a unification of knowledge and those who merely wanted to understand how to control water supplies (arcadians v. imperialists) is suggested by what I am tempted to call a superfluous footnote in an essay recounting the development of snow surveys. After praising the growing cooperation among public and private, state and federal agencies concerned with irrigation, power, navigation, municipal water supply, and flood control, he writes: "The benefits have been both tangible and intangible. The latter are the greater, for rivalry (originally river contest) is ceasing and cooperation and confidence are taking its place." A footnote after "river contest" reads: "*Rivus,* Latin for river; hence rival is a river man." Church, "Organized Water," *Soil Conservation* 8 (November 1942), 111. Church's interpretation is slightly different from that of the *Oxford English Dictionary,* which suggests that *rival* refers to the bank of a river, hence rivals stand on opposite sides of the same river.
21. Charles A. Mixer, "The Water Equivalent of Snow on the Ground," *Monthly Weather Review* (April 1903), 173.

22. Robert E. Horton, "Snowfalls, Freshets, and the Winter Flow of Streams in the State of New York," *Monthly Weather Review* (May 1905), 196–202; "Horton, Robert E.," *Who Was Who in America,* vol. 2 (Chicago: Marquis, 1950), 263.

Although Horton maintained an interest in snow measurement, his job as a hydrologist took him into the expanding area of hydroelectric power. After serving as hydraulic expert for the Department of Public Works for the state of New York, he became an engineer with the Power Authority of the state, climaxing his career as a consultant to the Tennessee Valley Authority in 1942–43.

23. Charles Marvin to James Church, April 21, 1909, Church Papers, NC 96, box 3. I have not found the 1908 paper to which Marvin refers.

24. James E. Church, *Snow Surveying: Its Problems and Their Present Phases with Reference to Mount Rose, Nevada, and Vicinity* (Washington, D.C.: GPO, 1917).

25. Alexander McAdie may have named Church's device: "I hope the snow gage works. It ought to be called the Mount Rose Snow Gage, because it is the outcome of your own experience and our mutual discussion of the problem in Reno." McAdie to Church, November 10, 1909, Church Papers, NC 96, box 3; Church, "The Biography of Snow Surveying," *Proceedings of the Western Interstate Snow Survey Conference . . . February 18, 1933* (Carson City, Nev.: State Printing Office, 1934), 9.

26. R. A. Work et al., *Accuracy of Field Snow Surveys,* technical report 163 (Hanover, N.H.: U.S. Army Material Command, Cold Regions Research and Engineering Laboratory, 1965), 2.

27. Church, *Snow Surveying,* 1917.

28. Alfred H. Thiessen, "Measuring the Snow Layer in Maple Creek Canyon," *Monthly Weather Review* (April 1911), 601–3; Sylvester Q. Cannon, "Measurement of Snow in Big Cottonwood Canyon, Utah," *Monthly Weather Review* (April 1912), 609–11; B. F. Eliasson, "Snow Survey of Pole Creek Watershed, Sanpete County, Utah," *Monthly Weather Review* (May 1912), 770–71; J. Cecil Alter, "Predicting Water Supply for the Farmer," *Scientific American Supplement,* 1904 (June 29, 1912), 413–14; Thiessen to Chief, USWB, March 16, 1912, with clippings from the *Deseret News, Evening Telegram,* and *Salt Lake City News;* Thiessen to Chief, USWB, March 20, 1912, with Alter's report, record group 27, Weather Bureau Correspondence 1912–24, 532.3 SoDAK-WYO, box 2673, National Archives and Records Administration.

29. A. A. Justice to Official in Charge, Local Weather Bureau, Salt Lake City, Utah, April 3, 1914, record group 27, Weather Bureau Correspondence 1912–24, 532.3 UTAH, box 2623.

30. On Webster and Marsh, see Chapter 1. Bernhard Fernow, "The Influence of Forests on Water Supplies," U.S. Department of Agriculture, *Annual Report* (1889), 297–330, reissued with extensive alterations as *Relations of Forests to Water Supply,* U.S. Department of Agriculture, Forestry Division, bulletin 7 (Washington, D.C.: GPO, 1893), 123–70. Another early manifestation of the dispute is explored by William D. Rowley, "Forests and Water Supply: Robert L. Fulton, Science and U.S. Forest Policies," *Nevada Historical Society Quarterly* 37 (Fall 1994), 215–24. Rowley summarizes the arguments for clear-cutting advanced by Fulton, a railroad land agent, in *Science* in 1893, the rebuttal from Fernow in 1896, and an earlier criticism of Gilbert and Powell by the forester

Abbot Kinney in 1890. The fact that meteorologists are trained to deal with physical phenomena, while foresters work with biological systems, may have exacerbated the rivalry. Only the development of a holistic, ecological approach to climate could resolve the differences.

31. J. Cecil Alter, "Where the Snow Lies in Summer," *Monthly Weather Review* (May 1911), 758–61.
32. Alexander J. Jaenicke and Max H. Foerster, "The Influence of a Western Yellow Pine Forest on the Accumulation and Melting of Snow," *Monthly Weather Review* (March 1915), 115–24; "Remarks by the Weather Bureau," signed C.F.M. and B.C.K., ibid., 124–26; Alfred A. Griffin, "Influence of Forests upon the Melting of Snow in the Cascade Range," *Monthly Weather Review* (July 1918), 324–27.
33. Alfred J. Henry, "The Disappearance of Snow in the High Sierra Nevada of California," *Monthly Weather Review* (March 1916), 150–53; Henry F. Alciatore, "Snow Densities in the Sierra Nevada," *Monthly Weather Review* (September 1916), 523–27; Henry, "The Density of Snow," *Monthly Weather Review* (March 1917), 102–9; Alciatore, "Growth, Settling, and Final Disappearance of Snow Cover in the Sierra Nevada, 1915–16," ibid., 109–13; George D. Clyde, "Change in Density of Snow Cover with Melting," *Monthly Weather Review* (August 1929), 326–28; James E. Church, "The Conservation of Snow," *Scientific American Supplement* 74 (September 7, 1912), 155. Church made his case again in "Recent Studies of Snow," 43–52.
34. J. W. Pomeroy and R. A. Schmidt, "The Use of Fractal Geometry in Modelling Intercepted Snow Accumulation and Sublimation," *Proceedings of the Eastern and Western Snow Conference* (1993), 1–10. Fifteen years of experiment in Colorado showed that streamflow increased 16 percent in cut areas, but there was also greater soil erosion. C. G. Bates and A. J. Henry, "Forest and Streamflow Experiment at Wagon Wheel Gap, Colorado," *Monthly Weather Review,* supp. 30 (1928), 233. In the Sierra, Joseph Kittredge estimated that 13 to 27 percent of winter snowfall was intercepted by trees and that losses were greater during unusually snowy winters. Frequent storms, by contrast, meant little evaporation. *Forest Influences: The Effects of Woody Vegetation on Climate, Water, and Soil, with Applications to the Conservation of Water and the Control of Erosions and Floods* (New York: McGraw-Hill, 1948), 129–45.
35. Benjamin C. Kadel, "Mountain Snowfall Measurements," *Monthly Weather Review* (January 1913), 159–61; Andrew H. Palmer, "The Region of Greatest Snowfall in the United States," *Monthly Weather Review* (May 1915), 217–20; Henry, "Discussion," ibid., 220–21.
36. James E. Church, "Snow Surveying: Its Problems and Their Solution," *Monthly Weather Review* (December 1915), 607; Benjamin C. Kadel, "An Improved Form of Snow Sampler," *Monthly Weather Review* (October 1919), 697 and facing photo.
37. James Anderson, "Surveying America's Snows," *Scientific American* (December 11, 1920), 594, 603. It is interesting that Brown's cover shows a weather station with its instruments in a shelter perched precariously on a ledge above the surveyor. Surveyors always travel in pairs, both for safety and to take measurements properly.

Benjamin C. Kadel, *Measurement of Precipitation . . . ,* circular E, Instrument Division, 4th ed., rev. (Washington, D.C.: GPO, 1922); S. P. Fergusson, "Improved Gages of Precipitation," *Monthly Weather Review* (July 1921), 379–86; H. F. Alps, "Foot-Layer Densities of Snow," *Monthly Weather Review* (September 1922), 474–75; Charles E. Linney, "Snowfall and Run-Off of the Upper Rio-Grande," *Monthly Weather Review* (January 1923), 16–19; J. M. Sherier, "Mountain Snowfall and Flood Crests in the Colorado," *Monthly Weather Review* (December 1923), 639–41; James E. Church, "Wide Area Forecasting of Streamflow on the Columbia and Colorado," *Monthly Weather Review* (August 1925), 353–54; Church, "Sixteen Years of Snow-Surveying in the Central Sierra and Its Results," *Monthly Weather Review* (February 1926), 43–44; George D. Clyde, *Snow-Melting Characteristics* (Logan: Utah State Agricultural College, Utah Agricultural Experiment Station, August 1931); Harlowe M. Stafford, "California Snow Surveys," *Monthly Weather Review* (October 1929), 426–28.

38. Harlowe M. Stafford, "History of Snow Surveying in the West," *Proceedings of the Western Snow Conference* (1959), 2.
39. Stafford, "California Snow Surveys," 426–28.
40. Douglas Helms, *Readings in the History of the Soil Conservation Service,* NHQ, historical notes no. 1 (Washington, D.C.: U.S. Department of Agriculture, Soil Conservation Service, Economics and Social Sciences Division, 1992); A. Hunter Dupree, *Science in the Federal Government* (Cambridge, Mass.: Harvard University Press, 1957), 363; D. Harper Simms, *The Soil Conservation Service* (New York: Praeger, 1970), 11–13; H. H. Bennett, *Soil Conservation* (New York: McGraw-Hill, 1939), 819–25; Bennett and Lewis A. Jones, "The National Program of Soil and Water Conservation as Relating to Farm Lands," typescript (March 1932), record group 114, entry 23, Soil Conservation Service Research Information Files 1929–40, box 6, Bureau of Chemistry and Soils, mss. 1932–35, National Archives and Records Administration.
41. James E. Church, "On the Hydrology of Snow," *Transactions of the American Geophysical Union* 13 (1932), 277–78; Church, "Report of the Committee on Snow, 1933–34," ibid., 15 (1934, published 1935), 263. Church's first project was a questionnaire sent to 1,500 engineers, universities, power companies, and others asking about research on snowfall, snowcover, and runoff. He also included an extensive bibliography of published material in his annual reports.
42. James E. Church, "Plans and Ideals of the International Commission of Snow," International Union of Geodesy and Geophysics, *Association of Scientific Hydrology/Association internationale d'hydrologie scientifique,* bulletin 23 (6th General Assembly, Edinburgh, September 14–26, 1936), 3–4, and "Report of the Committee on Snow 1935–36," *Transactions of the American Geophysical Union* 17 (1936), 265–77. The familial metaphor seems to have been on Church's mind in these years. He wrote a long letter to Robert G. Stone of the Blue Hill Observatory, Milton, Mass., thanking him for a reprint of Stone's article on American weather observatories, which mentioned the Mt. Rose station, and concluded, "Blue Hill is at least the oldest sister of Mt. Rose and in some sense the latter's mother." Church to Stone, November 9, 1935, record

group 27, Weather Bureau Correspondence 1931–35, 532.4, Snowfall, box 3064, National Archives and Records Administration.

43. *Proceedings of the Western Interstate Snow Survey Conference . . . February 18, 1933.* On Lowdermilk, see Helms, *History of the SCS,* 35–43. For a lively description of the 1994 meeting, see Cullen Murphy, "In Praise of Snow," *Atlantic Monthly* (January 1995), 45–58.

44. James C. Marr, "Status of Coordination and Standardization of Snow-Surveying," *Transactions of the American Geophysical Union* 17 (1936), 530–33. A typescript of Marr's paper was attached to a memorandum from Henry M. Eakin, acting head, Sedimentation Studies, Soil Conservation Service, to W. C. Lowdermilk, Associate Chief SCS, February 3, 1936, reporting on the Pasadena meeting of the Western Interstate Snow-Survey Conference. The bureaucratic complexity and the extent of SCS involvement in snow surveying are indicated by the following:

> A main point of the committee meeting was the establishment of a policy of cooperation among all interested agencies. It seemed very gratifying to Mr. McLaughlin [chief of the Division of Irrigation] and the committee in general to have the Soil Conservation Service represented and to receive our assurances along the line indicated by Mr. Ramser's letter. [Ramser, of SCS Watershed and Hydrologic Studies, wrote to Lowdermilk on January 10 that his section would follow the established practices and recommendations of the Special Advisory Committee to the Water Resources Committee relating to snow surveys.] No detailed outline of our possible cooperation was developed, but in general it was deemed probable that the information we may develop regarding soil mantle in snow areas and such snow survey work as we may undertake in connection with any of our projects would be very helpful and welcome contributions. Mr. McLaughlin indicated they were communicating with you directly to work out detailed understanding and plans of cooperation in the near future as the general picture develops.

Record group 114, Records of the Soil Conservation Service, General Files of the Hydrologic Division, 1935–40, "Snow Survey," National Archives and Records Administration.

45. Church, "Report of the Committee on Snow 1935–36," 277; "Report of the Committee on Snow 1936–37," *Transactions of the American Geophysical Union* 18 (1937), 269–70.

46. Church anticipated Bernard's interdisciplinary approach. "The present world tendency toward the union of all forms of snow and ice in a single commission and also their humanization—an inevitable evolution irrespective of administrative relationship." Church, "Report of the Committee on Snow 1937–38," *Transactions of the American Geophysical Union* 19 (1938), 281; James C. Marr, "Status of Proposed Bulletin and Principles and Practice of Snow-Surveying," *Transactions of the American Geophysical Union* 20 (1939), 53–55; Merrill Bernard, "Progress toward a Rational Program of Snow-Melt Forecasting," *Transactions of the American Geophysical Union* 22 (1941), 176–77; W. W. McLaughlin, "Factors Affecting Run-Off Forecasts based on Snow Surveys," *Soil Conservation* 5 (December 1939), 148–51, 163.

47. Church, "Report of the Committee on Snow 1937–38," 289; Peter Adams and B. McArthur, "The Development and Roles of the Eastern Snow Conference," *Proceedings of the Eastern Snow Conference* 42 (1985), 1–24; Merrill Bernard and Ashton Codd, "Progress-Report on Mountain Snowfall-Programs of the Weather Bureau," *Transactions of the American Geophysical Union* 21 (1940), 123–31.
48. Cecil Alter, "Shielded Storage Precipitation Gages," *Monthly Weather Review* (July 1937), 262–65; J. A. Riley, Acting Chief, Station Operations Division, USWB, to Hydrologic Supervisors and Section Directors, July 29, 1941, record group 27, Weather Bureau Correspondence 1936–42, 532.3, box 132.
49. Ukichiro Nakaya, "Formation of Snow-Crystals in the Mountains and in the Laboratory in Japan (A Sound Film)," *Transactions of the American Geophysical Union* 21 (1940), 97–98; *Proceedings of the Central Snow Conference* (East Lansing: Michigan State College, Office of Short Courses, Special Courses, and Conferences, December 11–12, 1941); Fred Paget, "Report of Treasurer, Western Interstate Snow-Survey Conference, January 16, 1942," *Transactions of the American Geophysical Union* 23 (1942), 165.
50. *Transactions of the American Geophysical Union* 23 (1942), 393–424; Samuel Colbeck, "History of Snow-Cover Research," *Journal of Glaciology* (Special Issue, 1987), 61–63.
51. J. C. Stevens, "Hydrology's Part in the War Effort," James E. Church, "Snow-Study Program at Soda Springs near Donner Summit of Central Sierra Nevada," and "Western Snow-Conference Minutes of Business Meeting," *Transactions of the American Geophysical Union* 24 (1943), 5–6, 77–90; H. A. Van Norman, "Water Works Ski-Troopers," *American City* (October 1944), 67–69.
52. Western Snow Conference, Records of Sacramento Dinner Meeting, February 27, 1946, 37–38, Church Papers, NC 96, box 1, folder 4.
53. Thomas J. Henderson, "Western Snow Conference—The First Fifty Years," *Proceedings of the Western Snow Conference* (1982), 2.
54. James E. Church, "Science and Adventure," *Proceedings of the Western Snow Conference* (1949), i–xiii. Church died in 1959, on August 5, the Catholic feast day of Our Lady of the Snows. Fred A. Strauss, "Scientific Definitions of Snow Survey Terms," *Snow Surveyors' Forum* (1953), 27–28.
55. Stafford, "History of Snow Surveying in the West," 7, and A. A. Lang, "Discussion," 12; R. W. Gerdel, "Radioactive Snow Gage," *Weatherwise* (December 1952), 1271–129; Goodison, Ferguson, and McKay, "Measurement and Data Analysis," 217–20.
56. *CRREL's First Twenty-five Years, 1961–1986* (Hanover, N.H.: CRREL, 1986).
57. U.S. Army Corps of Engineers, *Snow Hydrology, Summary Report of the Snow Investigations* (Portland, Oreg.: U.S. Army Corps of Engineers, North Pacific Division, 1956); Walter T. Wilson, "Analysis of Winter Precipitation Observations in the Cooperative Snow Investigations," *Monthly Weather Review* (July 1954), 183–95; *Instructions for Climatological Observers,* circular B, 11th ed. (Washington, D.C.: GPO, January 1962); R. A. Work, *Stream-flow Forecasting from Snow Surveys,* Soil Conservation Service circular no. 914 (Washington, D.C.: GPO, 1953); C. E. P. Brooks and N. Carruthers, *Handbook of Statistical*

Methods in Meteorology (London: Her Majesty's Stationery Office, 1953); G. D. Rikhter, *Snow Cover, Its Formation and Properties,* trans. William Mandel (New York: Stefansson Library, 1950); Marcel R. de Quervain, *Snow and Ice Problems in Canada and the U.S.A.* (Ottawa: National Research Council of Canada, Division of Building Research, technical report 5, February 1950), 59–70; John Sherrod, Jr., "A Review of the First 10,000 Abstracts Prepared by SIPRE Bibliography," *Proceedings of the Western Snow Conference* (1955), 2–3. For a personal history of the Central Sierra Snow Lab, see David H. Miller, "An Evaluation of Research Programs at the Central Sierra Snow Laboratory, 1945–1964," *Proceedings of the Western Snow Conference* (1995), 116–23.

58. Henry W. Anderson, "Prospects for Affecting the Quality and Timing of Water Yield through Snowpack Management in California," *Proceedings of the Western Snow Conference* (1960), 44–50.
59. Howard W. Lull and Robert S. Pierce, "Prospects in the Northeast for Affecting the Quality and Timing of Water Yield through Snowpack Management," *Proceedings of the Western Snow Conference* (1960), 54–62; C. Anthony Federer, Robert S. Pierce, and James W. Hornbeck, "Snow Management Seems Unlikely in Northeastern Forests," *Proceedings of the Eastern Snow Conference* (1973), 102–13; Peter F. Ffolliott and David B. Thorud, "The Southwest's Frozen Assets: Snowpack Management," *Proceedings of the Western Snow Conference* (1977), 12–18; David M. Rockwood, "Human Attitudes in Snow Science," *Proceedings of the Western Snow Conference* (1963), 132–35.
60. Vincent J. Price, "Double-Barreled Snow Gage," *Soil Conservation* (February 1972), 147–49; Goodison, Ferguson, and McKay, "Measurement and Data Analysis," 217–19; R. C. Kattelmann, B. J. McGurk, N. H. Berg, J. A. Bergman, J. A. Baldwin, and M. A. Hannaford, "The Isotopic Profiling Snow Gage: Twenty Years of Experience," *Proceedings of the Western Snow Conference* (1983), 1–8; Paul A. Wheeler and Donald J. Huffman, "Evaluation of the Isotopic Snow Measurement Gage," *Proceedings of the Western Snow Conference* (1984), 48–52.
61. *Snow Surveys,* Soil Conservation Service, agricultural information bulletin 302 (Washington, D.C.: GPO, December 1965); Goodison, Ferguson, and McKay, "Measurement and Data Analysis," 215–17; Charles H. Howard, *Snow Sensor Evaluation in the Sierra Nevada, California* (Sacramento: Department of Water Resources, Division of Planning, 1976); R. T. Beaumont, "Mt. Hood Pressure Pillow Snow Gauge," *Journal of Applied Meteorology* 4 (October 1965), 626–31.
62. Richard C. Kattelmann, "Snowmelt Lysimeters: Design and Use," *Proceedings of the Western Snow Conference* (1984), 68–79.
63. Work et al., *Accuracy of Field Snow Surveys,* v–vi.
64. *Instructions for Climatological Observers;* Douglas M. A. Jones, "Effects of Housing Shape on the Catch of Recording Gages," *Monthly Weather Review* (August 1969), 604–8; R. M. Lhermitte and D. Atlas, "Doppler Fall Speed and Particle Growth in Stratiform Precipitation," in *Proceedings of the Tenth Weather Radar Conference* (Washington, D.C.: American Meteorological Society, 1963), 297–302; Louis J. Battan and Craig F. Bohren, "Radar Backscattering by Melting Snowflakes," *Journal of Applied Meteorology* 21 (December

1982), 1937–38; Olav Lillesaeter, "Parallel-Beam Attenuation of Light, Particularly by Falling Snow," *Journal of Applied Meteorology* 4 (October 1965), 607–13; Stanley E. Wasserman and Daniel J. Monte, *A Relationship between Snow Accumulation and Snow Intensity as Determined from Visibility,* NOAA technical memorandum NWS ER-41 (Garden City, N.Y.: U.S. Department of Commerce, National Oceanic and Atmospheric Administration, National Weather Service, Eastern Region, May 1971).

65. Lawrence C. Gibbons, Anthony J. Matthews, Vernon G. Plank, and Gobert O. Berthel, *Snow Characterization Instruments,* report AFGL-TR-83-0063 (March 1, 1983), 9, 13, 22.

66. Colbeck, "History of Snow-Cover Research," 64. Researchers at the U.S. Department of Agriculture's Scanning Electron Microscopy Labortory in Beltsville, Md., began analyzing individual snow crystals to determine water content in December 1993. By combining this information with microwave data from satellites on the number of snow grains in a snowpack, it may be possible to estimate the water content of the pack more accurately. Steven Vames, "Scientists' Sense of Snow," *Scientific American* 272 (March 1995), 32.

67. H. C. S. Thom, "Distribution of Maximum Annual Water Equivalent of Snow on the Ground," *Monthly Weather Review* (April 1966), 265–71; H. C. S. Thom and M. D. Thom, "Weekly Maps of Snow Cover Probability," *Synoptic Types Associated with Critical Fire Weather* (Berkeley: Pacific Southwest Forest and Range Experiment Station, 1964), 476–92; R. R. Dickson and Julian Posey, "Maps of Snow-Cover Probability for the Northern Hemisphere," *Monthly Weather Review* (June 1967), 347–53; Donald R. Wiesnet, "The Role of Satellites in Snow and Ice Measurement," in *Advanced Concepts and Techniques,* 447–56; G. A. McKay, "The Mapping of Snowfall and Snow Cover," *Proceedings of the Eastern Snow Conference* (1972), 103.

68. Manes Barton, "New Concepts in Snow-Surveying to Meet Expanding Needs," in *Advanced Concepts and Techniques,* 39–46; Barton and Michael Burke, "SNOTEL: An Operational Data Acquisition System Using Meteor Burst Technology," *Proceedings of the Western Snow Conference* (1977), 82–87.

"Billions of sand-sized meteorites enter the atmosphere daily. As each particle heats and burns in a region 50 to 75 miles above the Earth's surface, its disintegration creates a trail of ionized gases. The trails diffuse rapidly, usually disappearing within a second, but their short lifespan is adequate for SNOTEL communications to be completed." A radio signal is sent from the master station to the ionized layer, then relayed to the remote sites, which in turn send a signal back via the meteor burst. *Snow Surveys and Water Supply Forecasting,* bulletin 536 (Washington, D.C.: U.S. Department of Agriculture, Soil Conservation Service, June 1988).

69. R. M. Kogan, I. M. Nazarov, and Sh. D. Fridman, "Determination of Water Equivalent of Snow Cover by Method of Aerial Gamma-Survey," *Soviet Hydrology,* selected paper 2, American Geophysical Union, 1962, 183–87; L. K. Vershinina and A. M. Dimaksyan, eds., *Determination of the Water Equivalent of Snow Cover* (Jerusalem: Israel Program for Scientific Translations, 1971; originally published by the Main Administration of the Hydrometeorological Ser-

Notes to Pages 153–155

vice of the Council of Ministers of the USSR State Hydrological Institute, Leningrad, 1969); E. L. Peck, V. C. Bissell, E. B. Hones, and D. L. Burge, "Evaluation of Snow Water Equivalent by Airborne Measurement of Passive Terrestrial Gamma Radiation," *Water Resources Research* 7 (1971), 1151–59; F. M. Smith, C. F. Cooper, and E. G. Chapman, "Measuring Snow Depths by Aerial Photogrammetry," *Proceedings of the Western Snow Conference* (1967), 66–72.

70. Interview with Tom Carroll, August 17, 1989, Minneapolis; Carroll, "Volume and Rate of Streamflow Discharge from an Alpine Snowpack," paper presented at the Western Snow Conference, Anchorage, Alaska, 1974, but not published; Carroll, "An Estimate of Watershed Efficiency for a Colorado Alpine Basin," *Proceedings of the Western Snow Conference* (1976), 69–77.

71. Tom Carroll and Kenneth G. Vadnais, "Operational Airborne Measurement of Snow Water Equivalent Using Natural Terrestrial Gamma Radiation," *Proceedings of the Western Snow Conference* (1980), 97–106; Carroll and Lee W. Larson, "Application of Airborne Gamma Radiation Snow Survey Measurements and Snow Cover Modeling in River and Flood Forecasting," *Proceedings of the Western Snow Conference* (1981), 23–33; B. A. Shafer, D. T. Jensen, and K. C. Jones, "Analysis of 1983 Snowmelt Runoff Production in the Upper Colorado River Basin," *Proceedings of the Western Snow Conference* (1984), 1–11; Carroll and Russell D. Marshall, "Cost-Benefit Analysis of Airborne Gamma Radiation Snow Water Equivalent Measurements Made before the February 1985 Fort Wayne Flood," paper presented at the Sixth Conference of Hydrometeorology, American Meteorological Society, Indianapolis, 1985; Carroll, "Cost-Benefit Analysis of Airborne Gamma Radiation Snow Water Equivalent Data Used in Snowmelt Flood Forecasting," *Proceedings of the Western Snow Conference* (1986), 1–11.

72. Tom Carroll and Milan Allen, *National Remote Sensing Hydrology Program: Airborne and Satellite Snow Cover Measurements, A User's Guide* (Minneapolis: National Oceanic and Atmospheric Administration, National Weather Service, November 1, 1988), 14 (emphasis in original), 17; Work et al., *Accuracy of Field Snow Surveys,* v; H. S. Loijens, "Comparison of Water Equivalent of Snow Cover Determined from Airborne Measurements of Natural Gamma Radiation and from a Snow Course Network," *Proceedings of the Eastern Snow Conference* (1974), 112–22; Douglas R. Powell, "Observations on Consistency and Reliability of Field Data in Snow Survey Measurements," *Proceedings of the Western Snow Conference* (1987), 72.

73. "Provisions: Omnibus Western Water Law," *Congressional Quarterly Weekly Reports* (November 21, 1992), 3687–90; U.S. Congress, Office of Technology Assessment, *Alaskan Water for California: The Subsea Pipeline Option—Background Paper,* OTA-BP-O-92 (Washington, D.C.: GPO, 1992); Peter Rogers, *America's Water: Federal Roles and Responsibilities* (Cambridge, Mass.: MIT Press, 1993), 5. A good review of current water politics may be found in Leah J. Wilds, Danny A. Gonzalez, and Glen S. Krutz, "Reclamation and the Politics of Change: The Truckee-Carson–Pyramid Lake Water Rights Settlement Act of 1990," *Nevada Historical Society Quarterly* 37 (Fall 1994), 173–99.

1. *The Compact Edition of the Oxford English Dictionary* (Oxford: Oxford University Press, 1972), 2895–96; Mitford Matthews, ed., *A Dictionary of Americanisms on Historical Principles* (Chicago: Free Press, 1951), 133–34. For a review of the meanings of the French word *neige,* see Hervé Gumuchian, *La Neige dans les alpes françaises du nord, une saison oubliée: L'Hiver* (Grenoble: Cahiers de l'Alpe, 1983), 48–51.
2. Nelson H. H. Graburn and Stephen Strong, *Circumpolar Peoples: An Anthropological Perspective* (Pacific Palisades, Calif.: Goodyear Publishing, 1973), 137.
3. John Washington, *Eskimaux and English Vocabulary for the Use of the Arctic Expeditions* (London: J. Murray, 1850); Elliott Norse, *Ancient Forests of the Pacific Northwest* (Washington, D.C.: Island Press, 1990), 152; Tom Horton, *Bay Country: Reflections on the Chesapeake* (New York: Ticknor & Fields, 1987), 106; "Taking It to the Streets," *New Age Journal* 10 (January–February 1993), 58. My thanks to Andrew Mergen for the Horton reference and to James Deutsch for the *New Age Journal* article.
4. Roger Wells, *English-Eskimo and Eskimo-English Vocabularies* (Washington, D.C.: GPO, 1890); A. N. Formozov, *Snow Cover as an Integral Factor of the Environment and Its Importance in the Ecology of Mammals and Birds* (originally published in *Materials for Fauna and Flora in the USSR,* new ser., Zoology, 5, 20 (1946), 1–152, trans. William Prychodko and William O. Pruitt, Jr., for Boreal Institute, University of Alberta, Edmonton, occasional paper 1 (1965), 9; Yakov Malkiel and Pavel Sigalov, "Further Thoughts on Daco-Romance and Slavo-Rumanian Words for 'Snow,'" *General Linguistics* 28 (1988), 245–60; "Name That Snow," *Snow Country* (December 1991), 32; Bernard Mergen, "Names for Snow," *Snow Country* (May–June 1992), 87. My thanks to Joel Kuipers for the essay by Malkiel and Sigalov.
5. On Robert Flaherty's film *Nanook,* see the essays by Jay Ruby, Jo-Anne Birnie Danzker, and Paul Rotha and Basil Wright in *Studies in Visual Communication* 6 (Summer 1980), 2–76. Laura Martin, "'Eskimo Words for Snow': A Case Study in the Genesis and Decay of an Anthropological Example," *American Anthropologist* 88 (June 1986), 419. My thanks to Professor Martin for sharing her work with me. Martin is supported by Geoffrey K. Pullum, "The Great Eskimo Vocabulary Hoax," *Natural Language and Linguistic Theory* 7 (1989), 275–81. For a reply to Martin that challenges some of her interpretations, see Stephen O. Murray, "Snowing Canonical Texts," *American Anthropologist* 89 (1987), 443–44. An interesting meditation on Eskimo words for snow is Charles Fergus, "Upsik and Siqoq," in *The Wingless Crow* (Harrisburg: Pennsylvania Game Commission, 1984), 104–9. My thanks to Professor Thomas L. Althen of Metropolitan State College, Denver, for this reference. Another contribution to the discussion of Eskimo words for snow lists more than thirty: Margaret Visser, "It's Snowing What to Say," *Saturday Night* (February 1994), 30.
6. Quoted in Pullum, "Great Eskimo Vocabulary Hoax," 276. See also Benjamin Whorf, *Language, Thought and Reality* (Cambridge, Mass.: MIT Press, 1956).

7. "You Can't Snow an Eskimo . . . ," *Saturday Review* (February 4, 1961), 46–47. Another source of the belief in the acuity of Eskimos and the richness of their language is children's books.
8. *Monthly Weather Review* (July 1898), 311–12.
9. Cleveland Abbe, Jr., "American Definitions of 'Sleet,'" *Monthly Weather Review* (May 1916), 281.
10. Ibid., 283.
11. Ibid., 284–286; Gustav Hellmann, "Classification of the Hydrometeors, II," *Monthly Weather Review* (January 1917), 15.
12. Charles F. Brooks, "The Nature of Sleet and How It Is Formed," *Monthly Weather Review* (February 1920), 69–72; *The International Classification for Snow (With Special Reference to Snow on the Ground)* (Ottawa: Associate Committee on Soil and Snow Mechanics, National Research Council of Canada, 1954), 2. Fifteen years later, however, the UNESCO-supported *Illustrated Glossary of Snow and Ice,* edited by Terence Armstrong and others (Cambridge: Scott Polar Research Institute, 1969), still defined *sleet* as "precipitation of snow and rain together, or of snow melting as it falls" (35).

    N.B.: The International Commission of Snow and Ice became the International Commission on Snow and Ice sometime before 1950.
13. On the early years of snow science, see A. E. H. Tutton, *The High Alps: The Natural History of Ice and Snow* (London: K. Paul, Trench, Trubner, 1927); Samuel Colbeck, "History of Snow-Cover Research," *Journal of Glaciology* (Special Issue, 1987), 60–65; and Bernard Mergen, "Snow Sciences in North America: The Eastern and Western Snow Conferences in the Context of Environmental History," *Proceedings of the Eastern and Western Snow Conferences* (1993), 175–81. In 1910 I. B. Shchukevich published a study of 246 different forms of snow crystals that fell in the vicinity of St. Petersburg in the winter of 1907–8. See Formozov, *Snow Cover as an Integral Factor,* 5.
14. Gerald Seligman, *Snow Structure and Ski Fields* (New York: Macmillan, 1936), 25–26. Seligman may have misread *ganik* for *qanik.*
15. Ibid., 17–22. Seligman ignored Bentley's classification, which included hexagonal right pyramids, triangular plates, and twelve-sided plates, and simply renamed some of Bentley's photographs.
16. Ibid., 124–27.
17. Ibid., 128–31; G. C. Amstutz, "On the Formation of Snow Penitentes," *Journal of Glaciology* 3 (October 1958), 304–11; Armstrong et al., *Illustrated Glossary of Snow and Ice.*
18. Frederic Prokosch, *The Seven Who Fled* (New York: Harper & Bros., 1937), 80; "Twelve Kinds of Snow Recognized by Science," *Scientific Monthly* (February 1938), 174–75.
19. Henry Baldwin, "The Snow-Terminology of Ski-Runners," *Transactions of the American Geophysical Union* 19 (1938), 294–97. A copy of Park Carpenter's letter, "To Our Observers," may be found in the Papers of James E. Church, NC 96, box 5, Baldwin File, Special Collections, University of Nevada, Reno, Library. For a more recent classification, see Michael Brady, *Nordic Touring and Cross-Country Skiing* (Oslo: Dreyers Forlag, 1968), 48–49, offering waxing ad-

vice for fourteen kinds of falling and new snow, settled snow, and metamorphisized [*sic*] snow.

20. Henry Baldwin, "Classification of Snow Surfaces," *Transactions of the American Geophysical Union* 19 (1938), 726–28; Baldwin to Church, December 31, 1937, Church Papers, NC 96, box 5, Baldwin File.
21. Robert G. Stone, "Winter Sports," *Transactions of the American Geophysical Union* 19 (1938), 720–24, 722.
22. Ibid., 722, 724; Seligman, *Snoe Structure and Ski Fields,* 127.
23. Wayne Poulsen, "Snow-Surfaces in the High Sierras for Skiing Purposes," *Transactions of the American Geophysical Union* 20 (1939), 79–83. Later breakable crust was called *crud,* and some Americans adopted the Alpine term *piste* for hard-packed ski trails. *America's Ski Book* (New York: Charles Scribner's Sons, 1966), 439–40.
24. Merrill Bernard, "Outline of Weather Bureau Northeast States Winter-Sports Service," *Transactions of the American Geophysical Union* 20 (1939), 85–92.
25. Ad for "The Aspens," *Outside* (October 1989); "Name That Snow," 32; and "Snow Country Magazine Skier's Guide Supplement," insert, *Snow Country* (December 1991); Mergen, "Names for Snow," 87.
26. Marcel R. de Quervain, *Snow and Ice Problems in Canada and the U.S.A.,* technical report 5 (Ottawa: National Research Council of Canada, Division of Building Research, February 1950), and G. J. Klein, D. C. Pearce, and L. W. Gold, *Method of Measuring the Significant Characteristics of a Snow-Cover,* technical memorandum 18 (Ottawa: National Research Council of Canada, Associate Committee on Soil and Snow Mechanics, November 1950).
27. de Quervain, *Snow and Ice Problems,* 68–69; Klein, Pearce, and Gold, *Method of Measuring a Snow-Cover,* 2.
28. *International Classification for Snow,* 1.
29. Ibid., 4–6.
30. Ronald I. Perla and M. Martinelli, Jr., *Avalanche Handbook,* agricultural handbook 489 (Washington, D.C.: GPO, 1976; rev. 1978), 217; David McClung and Peter Schaerer, *The Avalanche Handbook* (Seattle: Mountaineers Books, 1993), 64–65.
31. *International Classification for Snow,* ii.
32. Ukichiro Nakaya, *Snow Crystals: Natural and Artificial* (Cambridge, Mass.: Harvard University Press, 1954), 78–88. Choji Magono and Hachiro Oguchi, "Classification of Snow Flakes and Their Structures," *Science Reports of the Yokohama University,* sec. 1, no. 4 (1955), 47–57; Magono and C. W. Lee, "Meteorological Classification of Natural Snow Crystals," *Journal of the Faculty of Science, Hokkaido University,* ser. 7 (Geophysics), 2, no. 4 (November 1966), 321–35.
33. Edward LaChapelle, *Field Guide to Snow Crystals* (Seattle: University of Washington Press, 1969, 1983), 14.
34. H. Bader, R. Haefeli, E. Bucher, J. Neher, O. Eckel, C. Tomas, and P. Niggli, "Der Schnee und seine Metamorphose," *Beiträge zur Geologie der Schweiz,* Geotechnische Serie Hydrologie 3 (1939); translated as *Snow and Its Metamorphism* (Wilmette, Ill.: U.S. Army Corps of Engineers, Snow, Ice, and

Permafrost Research Establishment, 1954); Edward LaChapelle, "Errors in Ablation Measurements from Settlement and Subsurface Melting," *Journal of Glaciology* 3 (1959), 458–67; J. Latham and J. Montagne, "The Possible Importance of Electrical Forces in the Development of Snow Cornices," *Journal of Glaciology* 9 (1970), 375–84; Richard Sommerfeld, "The Relationship between Density and Tensile Strength in Snow," *Journal of Glaciology* 10 (1971), 357–62; Charles Bradley, "The Location and Timing of Deep Slab Avalanches," *Journal of Glaciology* 9 (1970), 253–61; Sommerfeld and LaChapelle, "The Classification of Snow Metamorphism," ibid., 3–17; Sommerfeld, *Classification Outline for Snow on the Ground* (Fort Collins, Colo.: U.S. Department of Agriculture, Rocky Mountain Forest and Range Experiment Station, 1969).

35. Sommerfeld and LaChapelle, "Classification," 4. Italics in original.
36. Ibid., 4–6; Samuel C. Colbeck, "What Becomes of a Winter Snowflake?" *Weatherwise* (December 1985), 312–15.
37. As late as 1988 Ron Perla and Richard A. Sommerfeld offered a system for classifying snow crystals on the ground that recognized nine types by shape and size, analogous to the seven types of falling snow recognized by the 1954 international classification. To new snow crystals, initially metamorphosed crystals, rounded equant crystals, faceted solid crystals, skeletal crystals, and rounded crystals due to melting/freezing, they added cylindrical crystals, layered facets, and clusters due to melting/freezing. This system was meant to help fieldworkers using magnifications up to 20 times. "On the Morphology and Size of Snow Crystals," *Proceedings of the International Snow Science Workshop* (Vancouver: Canadian Avalanche Association, 1988), 34–36.
38. LaChapelle, *Field Guide,* 19 and 22.
39. Thomas S. Kuhn, *The Structure of Scientific Revolutions* (Chicago: University of Chicago Press, 1962; 2d ed., 1970).
40. On recent work in remote sensing, see Tom Carroll and Milan Allen, *National Remote Sensing Hydrology Program: Airborne and Satellite Snow Cover Measurements, A User's Guide* (Minneapolis: National Oceanic and Atmospheric Administration, National Weather Service, November 1, 1988). Roger F. Reinking, "Cloud Droplet Accretion on Snow Crystals," in *Advanced Concepts and Techniques in the Study of Snow and Ice Resources,* ed. Henry S. Santeford and James L. Smith (Washington, D.C.: National Academy of Sciences, 1974), 193–203.
41. Samuel C. Colbeck, "An Overview of Seasonal Snow Metamorphism," *Reviews of Geophysics and Space Physics* 20 (February 1982), 45. Also published in *Proceedings of a Workshop on the Properties of Snow,* special report 82–18, ed. R. L. Brown, S. C. Colbeck, and R. N. Yong (Hanover, N.H.: CRREL, 1982), 45–61.
42. Samuel C. Colbeck, "Classification of Seasonal Snow Cover Crystals," *Water Resources Research* 22 (August 1986), 59S.
43. Ibid., 64S. Some scientists feel that Colbeck's differences with Sommerfeld and LaChapelle are largely semantic, since the latter are well aware of the physical nature of the process. The term *equi-temperature* is unfortunate, since it is actually the magnitude of the temperature gradient that is important. McClung and Schaerer recommend against using either term. *Avalanche Handbook,* 54.
44. Colbeck, "Classification," 62S.

45. Working Group on Snow Classification: S. C. Colbeck (chair), E. Akitaya, R. Armstrong, H. Gubler, J. Lafeuille, K. Lied, D. McClung, and E. Morris, *The International Classification for Seasonal Snow on the Ground* (n.p.: International Commission on Snow and Ice of the International Association of Scientific Hydrology and International Glaciological Society, [1990]), 1.
46. Barry Lopez, "The Stone Horse," *Antaeus* (Autumn 1986), 226; reprinted in *Crossing Open Ground* (New York: Vintage Books, 1989), 11. My thanks to Andrew Mergen for this reference. On the nature of technical language in other fields, see Joseph Conrad, *Nigger of the Narcissus* (London: Nelson, 1919), 193: "to take liberty with technical language is a crime against the clearness, precision and beauty of perfected speech . . . technical language is an instrument wrought into perfection by ages of experience, a flawless thing for its purpose." Quoted in Joel C. Kuipers, "'Medical Discourse' in Anthropological Context: Views of Language and Power," *Medical Anthropology Quarterly,* new ser. 3 (June 1989), 99–123.
47. William Matthews, "Spring Snow," in *Rising and Falling* (Boston: Little, Brown, 1979), 3. Reprinted by permission of the author.

CHAPTER 6. ECOLOGY OF THE SNOW COMMONS

1. Garrett Hardin, "The Tragedy of the Commons," *Science* (December 16, 1968), 1243–48.
2. Wilson Bentley, "Twenty Years' Study of Snow Crystals," *Monthly Weather Review* (May 1901), 212–14; William Gibson, "A Winter Walk," *Harper's* (December 1885), 68–81; Frank Bolles, *Land of the Lingering Snow: Chronicles of a Stroller in New England from January to June* (Boston: Houghton Mifflin, 1891), 5; "Red Snow," *Scientific American* (March 25, 1882), 181; "Notes and News," *Science* (February 1, 1884), 139–40; "Colored Snow," *Monthly Weather Review* (October 1901), 465–66; Horace R. Byers, "Meteorological History of the Brown Snowfall of February 1936," *Monthly Weather Review* (March 1936), 86–87; William Thomas, "Tioga Pass Revisited: Interrelationships between Snow Algae and Bacteria," *Proceedings of the Western Snow Conference* (1994), 53–62; Harold E. Welch et al., "Brown Snow: A Long-Range Transport Event in the Canadian Arctic," *Environmental Science and Technology* (February 1991), 280–86.
3. Asa Fitch, "Winter Insects of Eastern New York," *American Journal of Science and Agriculture* 5 (1847), 283–84; O. Lugger, "The Snow-fly," University of Minnesota Agricultural Experiment Station, bulletin 48 (1896), 256–57; J. W. Folsom, "The Identity of the Snow-flea (*Achorutes nivicola* Fitch)," *Psyche: A Journal of Entomology* 9 (1902), 315–21. For general background, see Ray F. Smith, Thomas E. Mittler, and Carroll N. Smith, eds., *History of Entomology* (Palo Alto, Calif.: Annual Reviews, in cooperation with the Entomological Society of America, 1973).
4. Ernest Thompson Seton, *Life-Histories of Northern Animals: An Account of the Mammals of Manitoba,* vol. 1, Grass-Eaters; vol. 2, Flesh-Eaters (New York: Charles Scribner's Sons, 1909); "Seton, Ernest Thompson," *Dictionary of*

American Biography, supp. 4 (New York: Charles Scribner's Sons, 1974), 735–37. See also *Trail of an Artist-Naturalist: The Autobiography of Ernest Thompson Seton* (New York: Charles Scribner's Sons, 1940; reprint, New York: Arno Press, 1978); John Henry Wadland, *Ernest Thompson Seton: Man in Nature and the Progressive Era, 1880–1915* (New York: Arno Press, 1978); Betty Keller, *Black Wolf: The Life of Ernest Thompson Seton* (Vancouver: Douglas & McIntyre, 1984); and H. Allen Anderson, *The Chief: Ernest Thompson Seton and the Changing West* (College Station: Texas A&M University Press, 1986).

5. Seton, *Life-Histories,* vol. 2, 1101–2; vol. 1, 329–30.
6. Ibid., vol. 1, plate 9; Abbott H. Thayer, *Concealing-Coloration in the Animal Kingdom* (New York: Macmillan, 1909), 12. For an evaluation of Thayer's theories in the context of biological science, see Sharon Kingsland, "Abbott Thayer and the Protective Coloration Debate," *Journal of the History of Biology* 11 (Fall 1978), 223–44.
7. Thayer, *Concealing-Coloration,* 117.
8. John Harshberger, "Preliminary Notes on American Snow Patches and Their Plants," *Ecology* 10 (July 1929), 275–81; Gregg Mitman, *The State of Nature: Ecology, Community, and American Social Thought, 1900–1950* (Chicago: University of Chicago Press, 1992).
9. Harshberger, "Preliminary Notes," 275–76.
10. Gwen Schultz, *Ice Age Lost* (Garden City, N.Y.: Doubleday, Anchor Books, 1974); Hugh M. Raup, "Botanical Problems in Boreal America," *Botanical Review* 7 (March 1941), 148–208, (April 1941), 209–48. Confusing and contradictory evidence on climate changes in glacial and interglacial periods continues to mount, making it difficult to give firm answers to questions about plant behavior. See William K. Stevens, "Data Give Tangled Picture of World Climate between Glaciers," *New York Times* (November 1, 1994), C4, reporting on the work of the British paleoecologists Michael H. Field and Brian Huntley, and a German colleague, Helmut Müller, who analyzed pollen from lake bottoms in France and Germany.
11. Raup, "Botanical Problems," 148–98.
12. Robert Griggs, "The Edge of the Forest in Alaska," *Ecology* 15 (1934), 80–96; Griggs, "Timberlines in the Northern Rocky Mountains," *Ecology* 19 (1938), 560, 562; E. A. Johnson, "The Relative Importance of Snow Avalanche Disturbance and Thinning on Canopy Plant Populations," *Ecology* 68 (1987), 43.
13. Raup, "Botanical Problems," 198–203, 210–31, 233; Carol Kaesuk Yoon, "Warming Moves Plants up Peaks, Threatening Extinction," *New York Times* (June 21, 1994), C4.
14. E. Laurence Palmer, "The Cornell Nature Study Philosophy," Cornell Rural School leaflet 38 (September 1944), 3–80; Liberty Hyde Bailey, *The Nature-Study Idea; An Interpretation of the New School–Movement to Put the Young into Relation and Sympathy with Nature* (New York: Macmillan, 1903), 4–5.
15. Anna Botsford Comstock, *Handbook of Nature Study* (Ithaca, N.Y.: Comstock Publishing Associates, 1911); E. W. Jenkins, "Science, Sentimentalism or Social Control? The Nature Study Movement in England and Wales, 1899–1914," *History of Education* 10 (1981), 33–43; Muriel Blaisdell, "Morgan, Ann Haven," in *Notable American Women: The Modern Period* (Cambridge, Mass.: Harvard

University Press, 1980), 497–98; Ann Haven Morgan, *Field Book of Animals in Winter* (New York: G. P. Putnam's Sons, 1939); Pamela M. Henson, "'Through Books to Nature': Anna Botsford Comstock and the Nature Study Movement," in *Using Nature's Languages: Women Engendering Science, 1690–1990,* ed. Barbara T. Gates and Ann B. Shteir (forthcoming). My thanks to Pam Henson for these and other references.

16. Corydon Bell, *The Wonder of Snow* (New York: Hill & Wang, 1957), 196–99, 234–54; Vincent Schaefer, "The Production of Ice Crystals in a Cloud of Super-Cooled Water Droplets," *Science* 104 (1946), 457; Schaefer, "The Inadvertent Modification of the Atmosphere by Air Pollution," *Bulletin of the American Meteorological Society* 50 (1969), 199; Horace R. Byers, "History of Weather Modification," in *Weather and Climate Modification,* ed. W. N. Hess (New York: John Wiley & Sons, 1974), 3–44; Theodore Steinberg, *Slide Mountain, or, The Folly of Owning Nature* (Berkeley: University of California Press, 1995), 106–34.
17. On ersatz snow dances and snow-calling contests, see *Tahoe Daily Tribune* (September 22, November 25, and December 6, 1993). Thanks to Alexa Mergen and Matt Weiser for these references. Clark Spence, *The Rainmakers: American "Pluviculture" to World War II* (Lincoln: University of Nebraska Press, 1980), 136.
18. Byers, "History of Weather Modification," 3–44; B. J. Mason, *Clouds, Rain, and Rainmaking,* 2d ed. (New York: Cambridge University Press, 1975); W. E. Knowles Middleton, *A History of the Theories of Rain and Other Forms of Precipitation* (New York: Franklin Watts, 1965).
19. George Breuer, *Weather Modification: Prospects and Problems* (Cambridge: Cambridge University Press, 1979); Barrington S. Havens, *History of Project Cirrus* (Schenectady, N.Y.: General Electric Research Laboratory, Research Publications Services, 1952). A clear indication of the linkage of atomic science and weather modification in the public mind was a Weather Control Bill introduced by Sen. Clinton Anderson of New Mexico in 1951. He wrote: "To some persons this may be considered somewhat presumptuous, if not sacrilegious, as there are those who say that they are still willing to leave the weather to Divine Providence. To this I can only reply that in my judgment the sphere of Divine Providence is no more invaded by weather control than by the splitting of the atom." His bill, modeled on the Atomic Energy Act of 1946, never became law.
20. Walter McDougall, . . . *the Heavens and the Earth: A Political History of the Space Age* (New York: Basic Books, 1985), 4; Albert Rosenfeld, *The Quintessence of Irving Langmuir* (New York: Pergamon Press, 1966).
21. Vincent Schaefer, "Artificially Induced Precipitation and Its Potentialities," in *Man's Role in Changing the Face of the Earth,* vol. 2, ed. William L. Thomas (Chicago: University of Chicago Press, 1956), 607–18.
22. *Final Report of the Advisory Committee on Weather Control* (Washington, D.C.: GPO, December 31, 1957), vol. 1, vi; vol. 2, 201.
23. U.S. House Committee on Science and Astronautics, *Weather Modification,* 86th Cong., 1st sess., February 16, 1959, no. 36, 10–11, 32–33; Vincent Schaefer, "After Sixteen Years," *Proceedings of the Western Snow Conference* (1963), 96.

24. Ralph W. Johnson, "Federal Organization for Control of Weather Modification," in *Controlling the Weather,* ed. Howard J. Taubenfeld (New York: Dunnellen, 1970), 132–33.
25. J. Eugene Haas, "Sociological Aspects of Weather Modification," in *Weather and Climate Modification,* ed. W. N. Hess (New York: John Wiley, 1974), 787–811.
26. J. Eugene Haas, "Social Impact of Induced Climate Change," in R. C. d'Arge et al., eds., *Economic and Social Measures of Biologic and Climatic Change,* Climate Impact Assessment Program (CIAP), monograph 6 (Washington, D.C.: U.S. Department of Transportation, 1975).
27. Leo Weisbecker, *The Impacts of Snow Enhancement: Technology Assessment of Winter Orographic Snowpack Augmentation in the Upper Colorado River Basin* (Norman: University of Oklahoma Press, 1974), xxxii. In a brief digest of his findings, *Snowpack, Cloud-Seeding, and the Colorado River: A Technology Assessment of Weather Modification* (Norman: University of Oklahoma Press, 1974), Weisbecker concedes that "we have the capability to modify the weather on a regular basis over widespread areas long before we could have a detailed understanding of its possible environmental and ecological effects" (43).
28. Weisbecker, *Impacts of Snow Enhancement,* xvii. The farmers' belief that snow from seeded clouds was slipperier than natural snow was confirmed by research. I thank Judy Horton for relating this story. Letter from Horton to author, October 3, 1988.
29. Weisbecker, *Impacts of Snow Enhancement,* xxix–xxx; Harold Klieforth, "On the Potential of Weather Modification for Enhancing Water Supply in Nevada" (Reno: Desert Research Institute, University of Nevada System, 1969), n.p.
30. Barbara Farhar, "A Societal Assessment of the Proposed Sierra Snowpack Augmentation Project" (Boulder, Colo.: Human Ecology Research Services, 1976), 73–74, 105–9.
31. Subcommittee on the Environment and the Atmosphere of the House Committee on Science and Technology, *Weather Modification: Hearing before the Subcommittee on the Environment and the Atmosphere,* 95th Cong., 1st sess., October 26, 1977, no. 32, testimony of Harlan Cleveland, 22–23; Klieforth, "Potential of Weather Modification," 26; Richard A. Kerr, "Cloud Seeding: One Success in 35 Years," *Science* (August 6, 1982), 519–21; *The World of Weather Resources Management,* proceedings of the 9th annual meeting of the North American Interstate Weather Modification Council, November 12–14, 1982 (Sinaloa, Mexico, and Mesilla Park, N.M.: Office of the Executive Director, June 1983), 88–94.

    At the same time Cleveland was making his claims for weather modification, others were beginning to question its environmental impact. See Eric I. Hemel and Clifford G. Holderness, *An Environmentalist's Primer on Weather Modification* (Stanford, Calif.: Stanford Environmental Law Society, September 1977).
32. David W. Reynolds, "A Report on Winter Snowpack-Augmentation," *American Meteorological Society Bulletin* 69 (November 1988), 1294.
33. William O. Pruitt, Jr., "Animals in the Snow," *Scientific American* (January 1960), 61–68.

34. William O. Pruitt, Jr., *Wild Harmony: Animals of the North* (New York: Nick Lyon Books, 1988), 6–15.

35. Ibid.

36. Ibid., 165–80; Pruitt, "Some Ecological Aspects of Snow," *Ecology and Conservation* 1 (1970), 91. See also Donald W. Stokes, "Weathering the World of Snow," *National Wildlife* 21 (December–January 1983), 4–10, on snow as both burden and blessing to plants and animals.

37. Pruitt, "Some Ecological Aspects," 88.

38. William O. Pruitt, Jr., *Boreal Ecology* (London: E. Arnold, 1978), 58–59. Among those Pruitt implicitly criticized, I suspect, was Laurence Irving, whose *Arctic Life of Birds and Mammals, Including Man* (Berlin: Springer Verlag, 1972), emphasizes biological adaptation.

39. Edmund Telfer and John Kelsall, "Adaptation of Some Large North American Mammals for Survival in Snow," *Ecology* 65 (1984), 1828–34.

40. Paul R. Ehrlich, Dennis E. Breedlove, Peter F. Brussard, and Margaret A. Sharp, "Weather and the 'Regulation' of Subalpine Populations," *Ecology* 53 (Early spring 1972), 243.

41. Ann Zwinger and Beatrice Willard, *Land above the Trees: A Guide to American Alpine Tundra* (New York: Harper & Row, 1972; reprint, Tucson: University of Arizona Press, 1989), 53–54. On mosses and lichens, see R. E. Longton, *Biology of Polar Bryophytes and Lichens* (New York: Cambridge University Press, 1988).

42. G. A. Bird, D. B. Rachar, and L. Chatapaul, "Increased Skeletonization of Leaf Litter under Snow Following Timber Harvest," *Ecology* 68 (February 1987), 221–23; C. W. Aitchison, "A Possible Subnivean Food Chain," in *Winter Ecology of Small Mammals,* special publication 10, ed. Joseph F. Merritt (Pittsburgh: Carnegie Museum of Natural History, 1984), 363–72.

43. Alan D. Maccarone, "Effect of Snow Cover on Starling Activity and Foraging Patterns," *Wilson Bulletin* 99 (March 1987), 94–97; Nicholas A. M. Verbeek, "Breeding Ecology of the Water Pipit," *Auk* 87 (July 1970), 425–51; Paul Hendricks, "Habitat Use by Nesting Water Pipits (*Anthus spinoletta*): A Test of the Snowfield Hypothesis," *Arctic and Alpine Research* 19 (1987), 313–20; S. Wijk, "Performance of *Salix herbacea* in an Alpine Snow-Bed Gradient," *Journal of Ecology* 77 (1989), 853–69.

44. Nike J. Goodson, David R. Stevens, and James A. Bailey, "Effects of Snow on Foraging Ecology and Nutrition of Bighorn Sheep," *Journal of Wildlife Management* 55 (1991), 214–22; W. Kent Brown and John B. Theberge, "The Effect of Extreme Snowcover on Feeding-Site Selection by Woodland Caribou," *Journal of Wildlife Management* 54 (1990), 161–68; Richard D. Taber and Thomas A. Hanley, "The Black-Tailed Deer and Forest Succession in the Pacific Northwest," in *Sitka Black-Tailed Deer: Proceedings of a Conference in Juneau, Alaska,* series R10-48 (Juneau: U.S. Department of Agriculture Forest Service, Alaska Region, 1979), 33–52, cited in Elliott Norse, *Ancient Forests of the Pacific Northwest* (Washington, D.C.: Island Press, 1990), 92.

45. Glenn D. Delgiudice, L. David Mech, and Ulysses S. Seal, "Physiological Assessment of Deer Populations by Analysis of Urine in Snow," *Journal of Wildlife Management* 53 (1989), 284–91.

46. Norse, *Ancient Forests,* 90; Mary Meagher, "Range Expansion by Bison of Yellowstone National Park," *Journal of Mammalogy* 70 (August 1989), 670–75.
47. D. A. Walker, James C. Halfpenny, Marilyn D. Walker, and Carol A. Wessman, "Long-Term Studies of Snow-Vegetation Interactions," *BioScience* 43 (May 1993), 287–301; D. A. Walker et al., "Hierarchic Studies of Snow-Ecosystem Interactions: A 100-Year Snow-Alternation Experiment," *Proceedings of the Eastern and Western Snow Conference* (1993), 407–14.
48. The paintings of Marris, Woods, Ostergaard, Ruesch, and others may be seen in any issue of *Wildlife Art News,* published bimonthly by Pothole Publications, St. Louis Park, Minn.
49. *The Art of Robert Bateman* (New York: Viking Press, 1981); Ramsay Derry, *The World of Robert Bateman* (Toronto: Madison Press Books, 1985); Stanwyn G. Shelter, *Portraits of Nature: Paintings of Robert Bateman* (Washington, D.C.: Smithsonian Institution Press, 1987), 131–32.
50. Craig Bohren, "Strange Footprints in Snow," *Weatherwise* (June 1989), 168–70.

CHAPTER 7. THE MODERN MINDS OF WINTER

1. On Kent, see Eliot Clark, "American Painters of Winter Landscape," *Scribner's* (December 1922), 768. Richard V. West, *"An enkindled eye": The Paintings of Rockwell Kent, A Retrospective Exhibition* (Santa Barbara, Calif.: Santa Barbara Museum of Art, 1985); Rockwell Kent, *N by E* (New York: Brewer & Warren, 1930; reprint, Middletown, Conn.: Wesleyan University Press, 1978); Kent, *It's Me O Lord: The Autobiography of Rockwell Kent* (New York: Dodd, Mead, 1955; facsimile reprint, New York: Da Capo Press, 1977); Kent, *Wilderness: A Journal of Quiet Adventure in Alaska* (New York: G. P. Putnam's Sons, 1920; reprint, Los Angeles: Wilderness Press, distributed by Ward Ritchie Press, 1970), xiv–xv.
2. George Santayana, *Character and Opinion in the United States* (New York: George Braziller, 1920; reprint, Doubleday, Anchor Books, 1956), 131; Stevens quoted in George S. Lensing, *Wallace Stevens: A Poet's Growth* (Baton Rouge: Louisiana State University Press, 1986), 273.
3. Napier Shaw, *Manual of Meteorology* (Cambridge: Cambridge University Press, 1926), vol. 1, 123. For the most useful interpretations of "The Snow Man," in roughly chronological order, see Samuel French Morse, *Wallace Stevens: Poetry as Life* (New York: Pegasus, 1970); Harold Bloom, *Wallace Stevens: The Poems of Our Climate* (Ithaca: Cornell University Press, 1977), 53–63; Armin Paul Frank, "Emerson and Stevens: The Poem as Hibernal Architecture," in *Poetic Knowledge: Circumference and Centre, Papers from the Wuppertal Symposium 1978* (Bonn: Bouvier Verlag, 1980), 141–51; Helen Vendler, *Wallace Stevens: Words Chosen out of Desire* (Knoxville: University of Tennessee Press, 1984), 46–50; David H. Hesla, "Singing in Chaos: Wallace Stevens and Three or Four Ideas," *American Literature* 57 (May 1985), 240–62; Joan Richardson, *Wallace Stevens: The Early Years, 1879–1923* (New York: William Morrow, 1986); Jerome Griswold, "Zen Poetry, American Critics; American Poetry, Zen Criticism: Robert Aitken, Basho, and Wallace Stevens," in *Zen in American Life and*

Letters, ed. Robert S. Ellwood (Malibu: Undena, 1987), 1–15; Robert Randolph, "'The Snow Man': Nausea or Numin?" *ANQ* 3, new ser. (July 1990), 119–21; Judith Butler, "'The Nothing That Is': Hegelian Affinities," in *Theorizing American Literature: Hegel, the Sign, and History* (Baton Rouge: Louisiana State University Press, 1991), 269–87; B. J. Leggett, *Early Stevens: The Nietzschean Intertext* (Durham, N.C.: Duke University Press, 1992), 187–94.

4. Glen MacLeod, *Wallace Stevens and Modern Art: From the Armory Show to Abstract Expressionism* (New Haven, Conn.: Yale University Press, 1993).

   In 1985 the artist Jasper Johns illustrated an edition of Stevens's poems edited by Helen Vendler, and MacLeod reprints an encaustic on canvas titled *Winter,* which consists of a shadow on a brick wall, falling snow, and a childlike drawing of a snowman.

5. Frank Townsend Hutchens, "Snow Scenes in Oil Painting," *Art Amateur* (March 1901), 100; Letter to J. Alden Weir, December 16, 1891, quoted in Kathleen A. Pyne, "John Twachtman and the Therapeutic Landscape," in *John Twachtman Connecticut Landscapes* (New York: Harry N. Abrams, 1989), 53.

6. Robert Frost, "Afterflakes," in *Collected Poems, Prose and Plays* (New York: Library of America, 1995), 276; Rachel Hadas, *Form, Cycle, Infinity: Landscape Imagery in the Poetry of Robert Frost and George Seferis* (Lewisburg, Pa.: Bucknell University Press, 1985), 166; James Louis Pean, "Snow in Frost" (Ph.D. diss., St. John's University, 1985).

7. Deborah Chotner, "Twachtman and the American Winter Landscape," in *John Twachtman Connecticut Landscapes,* 80. See examples of the work of Palmer, Harrison, and Carlson in William H. Gerdts, *Art across America: Two Centuries of Regional Painting, 1710–1920,* vol. 1 (New York: Abbeville Press, 1990), 169, 170, and 178; Birge Harrison, "The Appeal of the Winter Landscape," *Scribner's* (March 1914), 403–7; G. Frank Muller, "Koeniger, Painter of Snow," *International Studio* (June 1925), 210–15.

8. Harris quoted in Brian Osborne, "The Iconography of Nationhood in Canadian Art," in *The Iconography of Landscape,* ed. Denis Cosgrove and Stephen Daniel (London: Cambridge University Press, 1989), 172. See also Ronald Nasgaard, *The Mystic North: Symbolistic Landscape Painting in Northern Europe and North America, 1890–1940* (Toronto: University of Toronto Press, 1984); and Roger Boulet, *The Canadian Earth: Landscape Painting in the Group of Seven* (Markham and Scarborough, Ont.: Prentice Hall, Cerebrus, 1982). The other members of the Group were J. E. H. MacDonald, Franklin Carmichael, Frank Johnson, Arthur Lisner, and Frederick Varley. Tom Thomson was associated with them in style, but his accidental drowning in 1917 precluded his participation in their group exhibitions.

9. D. G. Carmichael, *The McMichael Canadian Collection* (Kleinburg, Ont.: McMichael Collection, 1979), 65–77; Robert Stacey, "The Myth—and Truth—of the True North," and Esther Trepanier, "The Expression of a Difference: The Milieu of Quebec Art and the Group of Seven," in *The True North: Canadian Landscape Painting, 1896–1939,* ed. Michael Tooby (London: Lund Humphries in association with Barbican Art Gallery, 1991), 36–63 and 98–116.

10. Eliot Clark, "American Painters of Winter Landscape," *Scribner's* (December 1922), 765–68; Thomas Folk, *The Pennsylvania School of Landscape Painting:*

An Original American Impressionism (Allentown, Pa.: Allentown Art Museum, 1984); Gerdts, *Art across America,* 278.

11. Benjamin O. Flower, "Edward W. Redfield: An Artist of Winter Locked Nature," *Arena* 36 (July 1906), 20–26; Folk, *Pennsylvania School;* Gerdts, *Art across America,* 46–47, 278; Patricia Hills, *Turn-of-the-Century America: Paintings, Graphics, Photographs, 1890–1910* (New York: Whitney Museum of American Art, 1977), 27, 79; Martha Hutson-Saxton, "Walter Emerson Baum, Pennsylvania Artist," *American Art Review* 6 (December 1994–January 1995), 150–57.

    A special case needs to be made for the Norwegian-born artist Jonas Lie, best known for his depiction of engineering works and the Panama Canal. In 1930 he did a series of paintings near Lake Kora in the Adirondacks. His snow lies heavy, crusts the branches of the pines, and spots the trunks of birches, but the wind seems to have stopped. Anthony N. B. Garvan, *Adirondack Winter—1930: Paintings by Jonas Lie* (Blue Mountain Lake, N.Y.: Adirondack Museum, 1971).

12. Bruce Robertson, *Reckoning with Winslow Homer: His Late Paintings and Their Influence* (Bloomington: Cleveland Museum of Art in cooperation with Indiana University Press, 1990), 95–97; Marianne Doezema, *George Bellows and Urban America* (New Haven, Conn.: Yale University Press, 1992).
13. J. A. Blaikie, "The Mask of Silence," *Art Journal* (January 1885), 1; H. P. Robinson, "Winter Photography for the Artist," *Art Journal* (January 1890), 1–6; F. C. Lambert, "Snow Pictures," *Photo-American* 4 (March 1893), 128; Frank Townsend Hutchens, "Winter in Watercolors," *Art Amateur* (March 1901), 99. My thanks to Julie K. Brown for the last reference.
14. Geraldine Wojno Kiefer, "Alfred Stieglitz and *The Steerage:* An Empirio-Critical Correlation," *Word & Image* 7 (January–March 1991), 58–60.
15. Marianne Fulton Margolis, ed., *Alfred Stieglitz Camera Work: A Pictorial Guide* (New York: Dover, 1978), 30; Kiefer, "Alfred Stieglitz," 60.
16. Rudolf Eickemeyer, *Winter* (New York: R. H. Russell, 1903), 30–31, iii–iv; Mary Panzer, *In My Studio: Rudolf Eickemeyer, Jr., and the Art of the Camera, 1885–1930* (Yonkers, N.Y.: Hudson River Museum, 1986).
17. For good examples of Adams's snow photos, see Andrea Gray, *Ansel Adams: An American Place, 1936* (Tucson: University of Arizona, Center for Creative Photography, 1982), and Jonathan Spaulding, *Ansel Adams and the American Landscape: A Biography* (Berkeley: University of California Press, 1995).
18. Ray Atkeson, *Ski and Snow* (New York: U.S. Camera, 1960), 6; "Portfolio in Black and White" [Atkeson], *Snow Country* (December 1991), 78–83; "Alpine Abstracts" [Berko], *Snow Country* (September 1993), 86–89. Compare the work of Josef Muench, "Winter through the Camera's Eye," *Natural History* (January 1950), 24–27.
19. F. Jack Hurley, *Marion Post Wolcott: A Photographic Journey* (Albuquerque: University of New Mexico Press, 1989). Her photographs are in the Farm Security Administration files in the Prints and Photographs Division of the Library of Congress.
20. Huebler's photographs in *Winter* (Hanover, N.H.: Hood Museum of Art, Dartmouth College, and University Press of New England, 1986), 131.
21. Charlie Chaplin, *My Autobiography* (New York: Simon & Schuster, 1964),

303–4; C. F. McGlashan, *History of the Donner Party* (Truckee, Calif.: Crowley & McGlashan, 1879; 2d ed., rev., San Francisco: Bancroft, 1880), 135.

22. Robin Bates with Scott Bates, "Fiery Speech in a World of Shadows: Rosebud's Impact on Early Audiences," *Cinema Journal* 26 (Winter 1987), 3–26; Frank Brady, *Citizen Welles: A Biography of Orson Welles* (New York: Charles Scribner's Sons, 1989), 267; Charles Maland, *Frank Capra* (New York: Twayne, 1980), 133; Raymond Carney, *American Vision: The Films of Frank Capra* (New York: Cambridge University Press, 1986), 279; Daniel M. Kimmel, "Weather at the Movies," *Weatherwise* (June 1990), 128–34.
23. Andrew Wyeth to William E. Phelps, February 2, 1949, Phelps Papers, Archives of American Art, Smithsonian Institution. My thanks to Liza Kirwin for this reference. A slightly different version is quoted in Wanda Corn, *The Art of Andrew Wyeth* (Greenwich, Conn.: New York Graphic Society for the Fine Arts Museum of San Francisco, 1973), 26.
24. Andrew Wyeth, *Wyeth at Kuerners* (Boston: Houghton Mifflin, 1976), 143, 314, 315.
25. John Arthur, *Spirit of Place: Contemporary Landscape Painting and the American Tradition* (Boston: Little, Brown, Bulfinch Press, 1898); Richard Meryman, *Andrew Wyeth* (New York: HarperCollins, 1996).
26. Frank H. Goodyear, Jr., *Welliver* (New York: Rizzoli, 1985), 14, 139, 143.
27. John Ashbery in Goodyear, *Welliver,* 14.
28. T. V. F. Brogan, "Poetry," and John Neubauer, "Science and Poetry," in *The New Princeton Encyclopedia of Poetry and Poetics,* ed. Alex Preminger et al. (Princeton, N.J.: Princeton University Press, 1993), 938–42, 1120–27; Robert Wallace, *Writing Poems,* 3d ed. (New York: HarperCollins, 1991), 391; Octavio Paz, *The Bow and the Lyre: The Poem, the Poetic Revelation, Poetry and History,* trans. Ruth L. C. Simms (Austin: University of Texas Press, 1973), 11; Haward Nemerov, "Because You Asked about the Line between Poetry and Prose," in *The Vintage Book of Contemporary Poetry,* ed. J. D. McClatchy (New York: Vintage, 1990), 130.

    There are 125 snow and snowman poems in the subject index of *The Columbia Granger's Index to Poetry* (New York: Columbia University Press, 1994), which indexes only anthologies. The annual number published in magazines, big and little, may be estimated by consulting *Index of American Periodical Verse: 1992,* edited by Rafael Catala and James D. Anderson (Metuchen, N.J.: Scarecrow Press, 1994), 601–2, 628, which has twenty-eight poems with "snow" in the title and another fifty with "winter." This index has been published yearly since 1971, when there were only ten poems with "snow" in the title. A conservative calculation would put the total number of American poems about snow published since 1776 somewhere over 3,000. My sample is small, and I do not claim to have exhausted the possibilities of snow as a symbol, but since I am treating each poem as an expression of an insight about snow, completeness is not necessary.
29. Elinor Wylie's "Velvet Shoes" was first published in *Nets to Catch the Wind* (1921) and later in *Collected Poems of Elinor Wylie* (New York: Alfred A. Knopf, 1932), 40. I am aware that the poem can be read as a self-portrait, and as a meditation on perfection and on pure sensuousness, as well as on silence.

Snow works well as a metaphor of all these meanings. See Judith Farr, *The Life and Art of Elinor Wylie* (Baton Rouge: Louisiana State University Press, 1983), 17, 75–79.

30. Anthony Sobin, "Problems in Painting: Spring Landscape with Melting Snow," *Literary Review* 36 (Winter 1993), 190.
31. James Joyce, "The Dead," *Dubliners* (1916; reprint, New York: Viking Press, 1964), 166. For a good discussion of the meanings of snow in the story, see Gerald Doherty, "Shades of Difference: Tropic Transformation in James Joyce's 'The Dead,'" *Style* 23 (Summer 1989), 225–37. Joyce's contemporary W. B. Yeats found snow an appropriate metaphor for madness, another form of negation. See Yeats, "Mad as the Mist and the Snow," in *The Collected Poems of W. B. Yeats,* ed. Richard J. Finneran (New York: Collier Books, 1989), 265–66. Edna St. Vincent Millay, "The Buck in the Snow," in *Collected Poems* (New York: Harper & Bros., 1956), 228; Howard Nemerov, "First Snow" and "The View from an Attic Window," in *The Collected Poems of Howard Nemerov* (Chicago: University of Chicago Press, 1977), 480, 225–27; Loren Eiseley, "The Snowstorm," in *The Innocent Assassins* (New York: Charles Scribner's Sons, 1973), 111–12. For Donald Hall, see "The Snow," in *Old and New Poems* (New York: Ticknor & Fields, 1990), 59–61.
32. Michael McFee, "Snow Goat," in *Vanishing Acts* (Frankfort, Ky.: Gnomon Press, 1989), 31; Robert Penn Warren, "Function of Blizzard," in *Being Here: Poetry 1977–1980* (New York: Random House, 1980), 45.
33. May Swenson, "Snow by Morning," in *The New Yorker Book of Poems* (New York: William Morrow, 1974), 652; Richard Hugo, "Snow Poem" and "Bear Paw," in *Making Certain It Goes On: The Collected Poems of Richard Hugo* (New York: W. W. Norton, 1984), 216, 362–63.
34. Robert Bly, "Three Kinds of Pleasure," in *Silence in the Snowy Fields* (Middletown, Conn.: Wesleyan University Press, 1962), 11; Merle Brown, quoted in Victoria Frankel Harris, "Criticism and the Incorporative Consciousness," *Centennial Review* 25 (Fall 1981), 418–21; James Engell, "Imagination," in *New Princeton Encyclopedia of Poetry and Poetics,* 566–74; T. V. F. Brogan, "Intuition in Poetry," ibid., 624–25.

    As with other themes, memory and intuition received attention from the early European modernists as well as Americans. Thomas Mann's *Magic Mountain* [*Der Zauberberg,* 1924] (New York: Alfred A. Knopf, 1927) is an obvious example. In the chapter titled "Snow," Hans Castorp dreams of a better humanity, then forgets the plan to achieve it. He skis down the mountain after his dream to embrace life, not death or the unattainable. Snow is a symbol of transcendence and of deception for Mann, who calls snowflakes "insignia," "orders," and "agraffes" (480).
35. Jane Flanders, "The Students of Snow," in *The Students of Snow* (Amherst: University of Massachusetts Press, 1982), 3, copyright © 1982 by Jane Flanders.
36. A. R. Ammons, "Snow Log," in *The Selected Poems, 1951–1977* (New York: W. W. Norton, 1977), 78.
37. A. R. Ammons, *The Snow Poems* (New York: W. W. Norton, 1977); Michael McFee, "A. R. Ammons and *The Snow Poems* Reconsidered," *Chicago Review*

33 (Summer 1985), 33; Ammons, "A Poem Is a Walk," *Epoch* 18 (Fall 1968), 116, quoted in McFee, "A. R.Ammons and *The Snow Poems,*" 37; Ammons, *Garbage* (New York: W. W. Norton, 1993).

38. Ammons, *Snow Poems,* 81, 237, 265–69, 140, 290.
39. Cathleen Calbert, "Lunatic Snow," *Literature and Psychology* 38 (1992), 65.
40. John Frederick Nims, "The Six-Cornered Snowflake," in *The Six-Cornered Snowflake and Other Poems* (New York: New Directions, 1990), 12. In the notes to his poems, Nims explains his allusions and the use of stanzas whose shapes imitate the objects they describe. The practice was called *technopaignia* in ancient Greece. He also notes that "the three equal axes between opposite points intersect in 60-degree angles at their center" (50).
41. Bern Oldsey, "The Snows of Ernest Hemingway," *Wisconsin Studies in Contemporary Literature* 4 (Spring–Summer 1963), 172–98; Robert E. Fleming, "Wallace Stevens' 'The Snow Man' and Hemingway's 'A Clean, Well-Lighted Place,'" *ANQ,* new ser. 2 (April 1989), 61–62.
42. Conrad Aiken, "Silent Snow, Secret Snow," in *Among the Lost People* (New York: Charles Scribner's Sons, 1934), 128–57.
43. Jeanne Schinto, "The Disappearance," *Yale Review* (Spring 1988), 441–62, also published in *Shadow Bands* (Princeton, N.J.: Ontario Review Books), 1988, 127–47. My thanks to Jeanne Schinto for sharing her work with me. See also Sharon Dilworth, *The Long White* (New York: W. W. Norton, 1988), 3–15, for snow as a symbol of confinement and mental depression.
44. Robert Stone, "Helping," in *The Best American Short Stories, 1988,* ed. Mark Halprin (Boston: Houghton Mifflin, 1988), 285–313.

    Snow in American fiction rages in the prairie blizzards of Ole Rölvaag, Willa Cather, and Laura Ingalls Wilder, symbolizing the harsh realities of the pioneer, and muffles the screams of dying New Hampshire villages in the novels of Russell Banks; it falls in the cities and towns of Richard Wright, Ralph Ellison, and Toni Morrison, where it confines, obscures, and finally erases all hope. See Rölvaag, *Giants in the Earth* (New York: Harper & Bros., 1927); Cather, *My Ántonia* (Boston: Houghton Mifflin, 1926); Wilder, *The Long Winter* (New York: Harper & Bros., 1940); Banks, *Affliction* (New York: Harper & Row, 1989); Wright, *Native Son* (New York: Harper & Bros., 1940); Ellison, *Invisible Man* (New York: Random House, 1952); Morrison, *Beloved* (New York: Alfred A. Knopf, 1987).
45. Henry Beston, *Northern Farm: A Chronicle of Maine* (New York: Rinehart, 1948; reprint, Henry Holt, 1994), 4; Edwin Way Teale, *Wandering through Winter: A Naturalist's Record of a 20,000 Mile Journey through the North American Winter* (New York: Dodd, Mead, 1965; reprint, New York: St. Martin's Press, 1990); Diane Kappel-Smith, *Wintering* (Boston: Little, Brown, 1984; reprint, New York: McGraw-Hill, 1986).

    Mikhail Prishvin's *Nature's Diary* (Moscow: Foreign Language Publishing House, 1958; reprint, New York: Penguin, 1987), makes an interesting contrast with Beston. Prishvin was an agronomist, rural schoolteacher, journalist, and novelist. This book, published in Russian in 1925, records a year spent at a research station near Lake Pleshcheyevo, about 160 kilometers (100 miles) northeast of Moscow. Although he has more to say about hunting than Beston,

Prishvin also sees the motion of shadows on snow as an expression of cosmic order.

46. Sigurd Olson, *The Singing Wilderness* (New York: Alfred A. Knopf, 1957), 193; Aldo Leopold, *A Sand County Almanac* (New York: Oxford University Press, 1949; reprint, Oxford University Press, 1968), 4.
47. Annie Dillard, *Pilgrim at Tinker Creek* (New York: Harper & Bros., 1974; reprint, Harper Perennial Library, 1985), 43; Abbott Thayer, *Concealing-Coloration in the Animal Kingdom* (New York: Macmillan, 1909), 12.
48. Letters from Juliana C. Nash, Virginia K. Anderson, Janice K. White, and Morgan Richards. My thanks to them for their good stories.
49. Judyth Powers to author. My thanks to Ms. Powers for her story.
50. Letters from Jerre F. Conder, Pat Winter, and Susan M. Watkins. My thanks to them. Dan Graf, "The Smell of Snow, Q & A," *New York Times* (December 29, 1992), C8.
51. Susan Stewart, *On Longing: Narratives of the Miniature, the Gigantic, the Souvenir, the Collection* (Baltimore: Johns Hopkins University Press, 1984; reprint, Durham, N.C.: Duke University Press, 1993).
52. Letter from Marcy P. Cohen; Liam Callanan, "The Hit List, J Street," *Washington Post Magazine* (March 14, 1993), 11.
53. William K. Stevens, "Deeper Look at Cold, Snowy Winter Reveals Balmier Future," *New York Times* (February 8, 1994), C4. But compare A. Frei and D. A. Robinson, "North American Snow Cover Variability from Satellite Data (1972–1993) and Comparison with Model Output," *Proceedings of the Eastern and Western Snow Conference* (1993), 43–50, which is more cautious in drawing conclusions. Oscar Suris, "There's No Business Like Snow Business in a Period of Thaw," *Wall Street Journal* (April 23, 1993), A1, A4. On the world's water scarcity, see Sandra Postel, "Crisis on Tap: A World without Water," *Washington Post* (October 29, 1989), B3.
54. James H. Cragin, Alan D. Hewitt, and Samuel Colbeck, "Elution of Ions from Melting Snow: Chromatographic Versus Metamorphic Mechanisms," CRREL Report 93-8 (Hanover, N.H.: Cold Regions Research and Engineering Laboratory, 1993), 1.
55. "Snow Is a Living Thing" (Tokyo: Snow Research Center, 1990); Suzuki Bokushi, *Snow Country Tales: Life in the Other Japan,* trans. Rose Lesser and Jeffrey Hunter (New York: Weatherhill, 1986), 8. This book should not be confused with Yasunari Kawabata's *Snow Country,* trans. Edward G. Seidensticker (New York: Alfred A. Knopf, 1969), a novella about a tragic love affair in a mountain resort, published in Japanese in the 1930s. For two of Hiroshige's woodblock prints of snow scenes, see *Winter* (1986), 58–59. I thank Barbara Sandrisser for stimulating my interest in Japanese snow.
56. Some examples of Chinese snow painting can be found in *Winter.* For the meanings of *xue,* see Herbert A. Giles, *A Chinese-English Dictionary,* 2d ed., vol. 1 (New York: Paragon, 1964), 602, and *A Modern Chinese-English Dictionary* (Hong Kong: Hai Feng, 1968), 1008. I thank my former student Juan Liu for her help.
57. Hervé Gumuchian, *La Neige dans les alpes françaises du nord, une saison oubliée:*

L'Hiver (Grenoble: Cahiers de l'Alpe, 1983), 79–86, 148. Gumuchian followed this book with *Les Territoires de l'hiver, ou la montagne française au quotidien* (Grenoble: Cahiers de l'Alpe, 1984), a brief essay on mountain life and culture.

58. Gilles Vigneault, "Mon Pays," in *Poems of a Snow-Eyed Country,* ed. Richard Woollatt and Raymond Souster (Don Mills, Ont.: Academic Press Canada, 1980), 66–67. I thank whoever told me about "Mon Pays" at the 1988 meeting of the International Snow Science Workshop in Whistler, B.C. I think it may have been Doug Andruseik, Bruce Allen, or Darro Stinson. They were all helpful in introducing me to snow science and Canadian culture.
59. Frederick Philip Grove, "Snow," in *Tales from the Margin: The Selected Short Stories of Frederick Philip Grove,* ed. Desmond Pacey (Toronto: Ryerson Press–McGraw-Hill, 1971), 261–70. See also Helmut Bonheim, "F. P. Grove's 'Snow' and Sinclair Ross' 'The Painted Door'—The Rhetoric of the Prairie," in *Encounters and Explorations: Canadian Writers and European Critics,* ed. Franz K. Stanzel and Waldemar Zacharasiewicz (Würzburg: Konigshausen & Neumann, 1986), 58–72. Grove's book *Over Prairie Trails* (Toronto: McClelland & Stewart, 1922), contains some beautiful descriptions of snow. According to his biographer, Grove was a shadowy character, born Felix Paul Greve in Radomno, Prussia, in 1879, who came to North America in 1909, settling in Manitoba about 1912. D. O. Spettigue, "Grove, Frederick Philip," in *The Oxford Companion to Canadian Literature,* ed. William Toye (Toronto: Oxford University Press, 1983), 324–27.
60. Grove, "Snow," 261, 262, 264, 267, 268. Rosalie Murphy Baum, in "Snow as Reality and Trope in Canadian Literature," *American Review of Canadian Studies* 17 (1987), 323–33, sees the snow in Grove's story as "treacherous," but I think this anthropomorphizes snow more than Grove intended.
61. Margaret Atwood, "Small Poems for the Winter Solstice," in *True Stories* (New York: Simon & Schuster, © 1981), 27–39. Reprinted by permission of the author.
62. Margaret Atwood, *Survival: A Thematic Guide to Canadian Literature* (Toronto: Anansi, 1972), 49–66. More recently Atwood has expanded her thesis in *Strange Things: The Malevolent North in Canadian Literature* (Toronto: Oxford University Press, 1995).
63. Farley Mowat, *The Snow Walker* (Toronto: McClelland & Stewart, 1975; reprint, New York: Bantam Books, 1977), 2. Some ideas about Canadian snow are further developed in Rudy Wiebe, *Playing Dead: A Contemplation concerning the Arctic* (Edmonton, Alta.: NeWest, 1989); Glen Norcliffe and Paul Simpson-Housley, "No Vacant Eden," in *A Few Acres of Snow: Literary and Artistic Images of Canada* (Toronto: Dundurn, 1992), 1–15; and Michael Dorland, "A Thoroughly Hidden Country: *Ressentiment,* Canadian Nationalism, Canadian Culture," *Canadian Journal of Political and Social Theory/Revue canadienne de théorie politique et sociale* 12 (1988), 130–64.
64. Morley Callaghan, *Our Lady of the Snows* (New York: St. Martin's Press, 1985), 1, 40, 81, 84, 92, 110, 160, 186. For the Catholic legend of the appearance of the Virgin Mary and a miraculous August snowfall near Rome in the fourth century, see M. S. Conlan, "Our Lady of the Snows," in *New Catholic Encyclopedia* (New York: McGraw-Hill, 1967), vol. 10, 837.

Note to Page

249

1. Samuel Colbeck et al., *The International Classification for Seasonal Snow on the Ground* (International Commission on Snow and Ice of the International Association of Scientific Hydrology, [1991]); Colbeck, "What Becomes of a Winter Snowflake?" *Weatherwise* (December 1985), 312–15; Frank Riley, "A Snowball's Chance," *New Scientist* (January 14, 1988), 45–48; Colbeck, "An Overview of Seasonal Snow Metamorphism," in *Proceedings of a Workshop on the Properties of Snow,* ed. R. L. Brown, S. C. Colbeck, and R. N. Yong (Hanover, N.H.: U.S. Army Cold Regions Research and Engineering Laboratory, 1981), 45–61; Colbeck, "Classification of Seasonal Snow Cover Crystals," *Water Resources Research* 22 (August 1986), 59S–70S; Ronald Perla and M. Martinelli, Jr., *Avalanche Handbook,* rev. ed. (Washington, D.C.: U.S. Department of Agriculture, Forest Service, 1978).

Source Credits

Grateful acknowledgment is made for permission to reprint the following copyrighted works.

"Snow Log," from *The Selected Poems, Expanded Edition* by A. R. Ammons. Copyright © 1987, 1977, 1975, 1974, 1972, 1971, 1970, 1966, 1965, 1964, 1955 by A. R. Ammons. Reprinted by permission of W. W. Norton & Company, Inc.

The Snow Poems by A. R. Ammons. Copyright © 1977 by A. R. Ammons. Reprinted by permission of W. W. Norton & Company, Inc.

"Small Poems for the Winter Solstice," from the poetry volume *True Stories* by Margaret Atwood © 1981 Simon & Schuster. Reprinted with permission of the author.

"The Snowstorm," from *The Innocent Assassins* by Loren Eiseley. Copyright © 1973 by Loren Eiseley. Reprinted with the permission of Scribner, a Division of Simon & Schuster.

"The Students of Snow," reprinted from *The Students of Snow* by Jane Flanders. Amherst: University of Massachusetts Press, 1982. Copyright © 1982 by Jane Flanders.

"Afterflakes," from *The Poetry of Robert Frost,* edited by Edward Connery Lathem. Copyright 1936 by Robert Frost. © 1964 by Leslie Frost Ballantine. © 1969 by

Holt, Rinehart & Winston, Inc. Reprinted by permission of Henry Holt & Co., Inc.

"Poem of Falling Snow," by Tom Hansen, was published by *The Literary Review,* 1988. Reprinted by permission of the author.

"Snow Poem," from *Making Certain It Goes On: The Collected Poems of Richard Hugo* by Richard Hugo. Copyright © 1984 by Richard Hugo. Reprinted by permission of W. W. Norton & Company, Inc.

"Snow Goat," by Michael McFee, is reprinted from his book *Vanishing Acts,* by permission of Gnomon Press.

"March Snow," from *Birding, or Desire* by Don McKay. McClelland & Stewart, 1983. Reprinted by permission of the author.

"Spring Snow," from *Rising and Falling* by William Matthews. Little, Brown, 1979. Reprinted by permission of the author.

"First Snow" and "Snowflakes," from *The Collected Poems of Howard Nemerov* by Howard Nemerov. University of Chicago Press, 1977. Reprinted by permission of Margaret Nemerov.

"The Six-Cornered Snowflake," from *The Six-Cornered Snowflake and Other Poems* by John Frederick Nims. Copyright © 1990 by John Frederick Nims. Reprinted by permission of New Directions Publishing Corp.

"The Snowman," by P. K. Page. Copyright © 1967 by P. K. Page. Reprinted by permission of the author.

"Snow Songs, iv," from *Each in His Season,* by W. D. Snodgrass. Copyright © 1993 by W. D. Snodgrass. Reprinted with the permission of BOA Editions, Ltd., 260 East Ave., Rochester, N.Y. 14604.

"Problems in Painting: Spring Landscape with Melting Snow," by Anthony Sobin, first published in *The Literary Review,* Winter 1993. Reprinted by permission of the author.

"The Snow Man," from *Collected Poems* by Wallace Stevens. Copyright 1923 and renewed 1951 by Wallace Stevens. Reprinted by permission of Alfred A. Knopf, Inc.

"Function of Blizzard," from *Being Here: Poetry 1977–1980* by Robert Penn Warren. Copyright © 1978, 1979, 1980 by Robert Penn Warren. Reprinted by permission of Random House, Inc.

"Velvet Shoes," from *Collected Poems* by Elinor Wylie. Copyright 1932 by Alfred A. Knopf, Inc., and renewed 1960 by Edwina C. Rubenstein. Reprinted by permission of the publisher.

Index

A

Abbe, Cleveland, 30, 38–39, 124, 162–64
Adams, Ansel, 219–20, 221
Adams, Henry, 82–83
Adelson, Fred B., 14
Agassiz, Louis, 22–23
Aiken, Conrad, 236–37
airborne snow survey, 155–56
Aldrich, Thomas Bailey, 82
Allen, E. John B., 94, 107
Allen, Woody, 85
Alter, J. Cecil, 133–35, 144
American Geophysical Union, 140, 145–47, 168–69. *See also* Committee on Snow
Ammons, A. R., 232–34
animals in snow: bighorn sheep, 202; buffalo, 200, 202; caribou, 199–200, 202; elk, 200, 202–4; hare, 32, 199; lynx, 199; moose, 185–86, 199–200; shrew, 185; squirrel, 32, 185, 200; wolf, 199. *See also* birds; ecology of snow
Arctic, xvi, 188
Armstrong, Tim, 18
Army Signal Corps, 12
art: snow as art, 55–56, 81; snow scenes, 14–17, 27–29, 76–77, 84–85, 185–87, 203–5, 208–9, 210–26, 256n26, 258n46. *See also names of individual artists*
Ashbery, John, 224–25
Atkeson, Ray, 101, 115, 219–20
Atwater, Montgomery "Monty," 116–17, 175
Atwood, Margaret, 245
avalanche: mountain, 114–19; roof, 15

B

Bailey, Liberty Hyde, 190
Baldwin, Henry, 166–68
Balls, Matthew, 140
Bara, Jana, 26
Barnes, Howard T., 61, 140
Barry, Marion, 75
Barton, Manes, 154
Bateman, Robert, 187, 203–5

Bates, Robin, 222
Baxter, Samuel N., 60
Beard, Daniel Carter, 83–84
Beaumont, R. T., 151
"Beautiful Snow" (Watson), 18
Bell, Corydon, xv
Bellows, George, 216–17
Belt Reader, 153
Bennett, Hugh H., 140, 143
Bentley, Wilson Alwyn, 30–31, 168, 183
Bergeron, Tor, 191
Berko, Ferenc, 219–20
Bernard, Merrill, 143–44
Beston, Henry, 238
Bickman, Martin, 20
birds: camouflage in snow, 185–86, 203–4; survival in snow, 9, 33, 36, 185, 201–4. *See also* ecology of snow
Bjerknes, Vilhelm, 70
black snow, 184
Blaikie, J. A., 216
blizzards, 33–38, 65–69, 260nn53, 54
Blodgett, Lorin, 11
Bly, Robert, 230–31
Boas, Franz, 161
Bohren, Craig, 204
Bokushi, Suzuki, 242
Bolles, Frank, 32–33, 183
Bordman, H. P., 130, 133
Borglum, Solon Hannibal, 273n23
Boston, xiv, 65, 73–74
Boston Massacre, 82
Bostwick, Arthur, 84
Bradford, William, 15
Bradley, Charles, 119, 175
Bradley, Steve, 113
Brandenberg, F. H., 125
Braun-Blanquet, Josias, 187
Brode, Robert, 193
Brooks, C. E. P., 149
Brooks, C. T., 25
Brooks, Charles F., 164
Brown, Howard, 138–39
Brown, Robert Harold, 43
brown snow, 184
Bryant, William Cullen, 22, 27
Buck, C. J., 99
Buffalo, N.Y., xiv
Bulosan, Carlos, xv
Bureau of Land Management, 150
Bureau of Reclamation, 143, 196
Burke, Michael, 154
Burroughs, John, 31–32, 184, 238
Burton, V. R., 60

C

Cahill, Holger, 101
Calbert, Cathleen, 234
California, 122–23, 138–39, 149, 150
"California Snowmen," 272n9
Callaghan, Morley, 246
"Calvin and Hobbs," 81–82, 236
Camelback Ski Corporation, 110
Canada: attitudes toward snow, 243–46; cities and snow, 76; snow scenes, 16–17, 26–27, 203–4, 212–14; snow science, 24, 171–72, 198, 201; winter carnival (Montreal), 87
Capra, Frank, 222–23
Carlson, John F., 212
Carney, Raymond, 223
Carpenter, L. G., 125
Carpenter, Park, 166–68
Carroll, Tom, 155–56
Carruthers, N., 149
Caulfeild, Vivian, 166
Central Sierra Snow Laboratory, 146, 148–49, 151
Central Snow Conference, 144–45
Champlin, John, 84
Chaplin, Charlie, 222
Chicago, 63, 72, 269n59
Chickering, Frances E. Knowlton, 23–25, 30, 257n37
China, 243
Chittenden, Hiram Martin, 125
Chloromonas skiensis, 111
Chotner, Deborah, 212
Church, James E., xiv, 12, 126–47, 149–51, 157, 168
Church, Frederic, 15
cities: effects of snow on, 54, 57–58; snow management, 58–60, 63–67; urban snow hazard, 53, 55–56. *See also names of cities;* livable winter city; snow removal; weather reports
Civilian Conservation Corps, 100, 109
Clark, Eliot, 214
Cleveland, Harland, 197

climate, 1–2, 241, 253n2
cloud seeding. *See* weather modification
Clyde, George Dewey, 140–42
Codd, Ashton, 144
Colbeck, Samuel, xv, 177–80, 242
Cold Regions Research and Engineering Laboratory (CRREL), 149, 153, 177, 241
Colorado River Basin Project Act, 196
Commission on Snow and Ice (International Association of Scientific Hydrology), 140–41, 144, 171–73, 180
Committee on Snow (American Geophysical Union), 140–41, 143, 145, 149
Comstock, Anna Botsford, 190
Conder, Jerre, 240
Cragin, James H., 242
Crèvecoeur, St. John de, 4–5
Cullen, Maurice, 26
Cullings, E. S., 140
Curran, Jim, 105

D

Daley, Richard, 72, 269n59
Davis, Robert, 54
de Gautier, Felisa Rincón, xv
DeLillo, Don, 77–78
DeLorean, John, 113
de Quervain, Marcel, 118, 171–72
Dickinson, Emily, 19–22, 24, 208, 225
Dillard, Annie, 76, 83, 238
Donner, Tamsen, 7–8
Doten, S. B., 133
Douglas, Mary, 72
Duchamp, Marcel, 52, 210, 222
Duchin, Marjorie Oelrichs, 104
Dukakis, Michael, 73–74
Durrie, George Henry, 9, 15–17
Dyhlen, C. G., 166

E

Eastern Snow Conference, 140, 144, 147
Eaton, Walter Prichard, 95–96
ecology of snow: alga, 111, 184, 198; birds, 185, 201–2; food chain, 201; insects, 184, 190; mammals, 185–86, 198–202; plants, 187–90, 198, 201, 203. *See also* animals in snow; birds
Ehrlich, Paul, 200–201
Eickemeyer, Rudolf, 218–19
Eiseley, Loren, 228–29
Emerson, Ralph Waldo, 1, 10–11, 25, 207–8, 225
Eskimo words for snow, 159–62, 199
Espy, James Pollard, 11–12

F

Fairweather, Lydia, 51–52
Fall Velocity Indicator, 153
Farhar, Barbara, 196
Federal Weather Modification Board, 196–97
Fergusson, S. P., 130
Fernald, M. L., 188
Fiddler, Jimmie, 78
Findeisen, Walter, 191
Finney, E. A., 43
Fisher, Alvan, 14
Fitch, Asa, 184
Fitzgerald, F. Scott, 93
Flanders, Jane, 231–32
Fleming, J. A., 146
Fleming, James R., 11–12
Flick, Hugh, 37
Flower, Benjamin O., 215
flurry, 70–72
Folsom, Justus Watson, 184
Folsom, L. Edwin, 19
Foote, Mary Hallock, 116
forests and snow, 1–2, 133–36, 188–89
Formozov, A. N., 160
Forrest, Linn, 100
Forry, Samuel, 11
Fraser, Don, 105
Fredkovsky, Reuben, 92
Frost, A. B., 203
Frost, Robert, 211–12, 223, 228
"Frosty the Snowman," 86

G

Gallen-Kallela, Akseli, 212
Gerdel, R. W., 148
Gibson, William, 32, 183
Gilbert, Grove Karl, 281n2
Glaisher, James, 17, 22
global warming. *See* climate

Gold, Lorne, 174
Gold, Michael, 84
Goldsworthy, Andy, 81
Goodyear, Frank H., 224
Gould, Helen F., 22
Gowans, Alan, 100, 256n29
Graburn, Nelson, 160
Graf, Dan, 240
Great Snow, The (Robinson), 67–68
Greeley, W. B., 96
green snow, 184
Griggs, Robert, 188–89
Grooms, Red, 76–77
Grove, Philip, 244–45
Gumuchian, Hervé, 243
Gutheim, Frederick, 75
Gutterbock, Thomas, 54
Guyot, Arnold H., 13, 123–24, 130, 163

H

Haas, J. Eugene, 194–95
Haines, John, xvi, 41
Hall, Donald, 228
Hannagan, Steve, 103
Hansen, B. L., 148
Hardin, Garrett, 183
Harriman, W. Averell, 96, 103–7, 117, 275n41
Harris, Lawren, 212–14, 221
Harris, T. W., 190
Harrison, Birge, 212
Harshberger, John, 187–88
Hartmann, Sadakichi, 219
Hauser, Hans, 107
Hawthorne, Nathaniel, 9
Hayakawa, S. I., 162
Hays, Samuel, 125
Hellmann, Gustav, 163–64
Helms, Douglas, xv
Hemingway, Ernest, 236
Henry, A. J., 137
Henry, Joseph, 12
Henson, Robert, 78
Herbert, George, 121
Hertz, Heinrich, 217
Hewitt, Alan D., 242
Hibbard, Aldro Thompson, 215
Hill, James J., 57, 91–92
Hill, Louis W., 91–92
Hiroshige, Utagawa, 243
Hobbs, William Herbert, 140
Hoham, Ron, 111
Holmes, Oliver Wendell, 26
Homer, Winslow, 50, 215–16
Hooker, Joseph Dalton, 188
Hooker, William Jackson, 188
Hope, Quentin, 3
Horton, Floyd V., 100
Horton, Robert E., 130–32, 145
Horton, Tom, 160
Houghton, Mich., xiv, 149
Huberty, Arthur, 52
Huddle, David, 52–53
Huebler, Douglas, 221–22
Hugo, Richard, xiii, 230
Huitfeld, Fritz, 166
Hulten, Eric, 188–89
Hunt, William Morris, 6–7
Hutchens, Frank Townsend, 216
Hyde, Arthur M., 98
hydrology, 141, 149–52
hydrologic cycle, 123

I

Ickes, Harold, 107, 140
Inness, George, 15
Institute of Snow Research, 241

J

Jackson, A. Y., 212–13
Jackson, William Henry, 27–28, 49–50
Jackson, Miss., 55
James, George Wharton, 95
Japan, 174, 193, 242
Jardine, W. M., 96–97
Jeffers, William, 106
Jefferson, Thomas, 1–2
Johnsen, Theodore A., 95
Johnson, Eastman, 15
Johnson, W. E., 145
Joyce, James, 228
Jull, Orange, 49–49
Justice, A. A., 134

K

Kadel, Benjamin C., 135–38
Kahn, E. J., 68–69
Kappel-Smith, Diane, 238
Kasson, Joy, 8

Kelsall, John, 200
Kent, Rockwell, 208–9, 215–16, 224
Kepler, Johannes, 3–4, 235
Keweenaw Field Station, 149
Kiefer, Geraldine Wojno, 218
Kilmer, Joyce, 230
Kinsey, Joni Louise, 28
Kirk, Ruth, xv
Kittredge, Joseph, 141
Klein, G. J., 171–72
Klieforth, Harold, 196
Kneipp, L. F., 99
Kocin, Paul, 35
Koeniger, Walter, 212
Koziol, Felix, 116
Krieghoff, Cornelius, 16–17
Kupperman, Karen Ordahl, 4

L

LaChapelle, Edward, 116, 118, 174–78
LaFarge, Christopher, 105
LaFarge, John, 28
Lake Tahoe, 129, 131–33, 169
Langmuir, Irving, 190, 192–93
law and snow, 61, 64, 113–14, 117, 196
Lawson, Ernest, 215
Lea, Tom, 117
Lee, C. W., 174–75, 181
Leopold, Aldo, 238
Leopold, Luna, 122
Leslie, Edward, 48–49
Leslie, John, 48–49
Lilienthal, Meta Stern, 37–38
Linde, Carl L., 99
Lindsay, John, 73
livable winter cities, xvii, 75–76
Longfellow, Henry Wadsworth, 22, 27
Loomis, Elias, 13
Lopez, Barry, 182
Lowdermilk, W. C., 140–41
Lowell, James Russell, 26–27
Love, Edmund G., 50
Lugger, O., 190
Lunn, Arnold, 166, 170

M

MacDougall, Walter, 192
Mach, Ernst, 217
Magono, Choji, 174–75, 181
Markham, Edwin, 72
Marr, James C., 142
Marris, Bonnie, 203
Marsh, George Perkins, 135, 281n2
Marshall, Robert, 102–3
Martin, Laura, 161
Marvin, Charles Frederick, 124, 131, 135–38
Mason, B. J., 174
Matisse, Henri, 56
Matthews, Mitford, 34
Matthews, William, 159, 182
McAdie, Alexander, 127
McCarthy, D. F., 25
McClung, David, 119
McCully, George T., 46–47
McFee, Michael, 229, 233
McGlashan, C. F., 222
McKay, Don, 247
McKay, G. A., 154
Mead, Larkin G., 94
Meier, Jack, 100, 103
Meier, Julius L., 97
Melville, Herman, 257n35
Merchant, Carolyn, xv
Merriam, C. Hart, 189
Merriam, John C., 97–99
Metcalf, Willard, 215
Millay, Edna St. Vincent, 228
Mineral King Valley, 110–11
Mixer, Charles A., 130–32
Montagne, John, 175
Monte, Daniel, 152
Monthly Weather Review, 30, 38–39, 124, 130, 153, 162
Moore, Charles Herbert, 15, 28–29
Moore, Henry, 56
Moore, Willis L., 135
Moran, Thomas, 27
Morgan, Ann Haven, 190
Morrice, James Wilson, 26, 221
Mount, James A., 99
Mountain of the Holy Cross, 27–28
Mount Hood, 96–103
Mount Hood Pressure Snow Pillow, 151
Mount Rose Snow Sampler, 132, 285n25
Mowat, Farley, 246
Muir, John, 33, 116, 238
Munch, Edvard, 212
Murphy, Cullen, 78
Museum of Modern Art, 56

N

Nakaya, Ukichiro, 144, 174, 176, 191
Nash, Roderick, 125
National Academy of Sciences, 194
National Environmental Policy Act, 110, 112, 194
National Oceanic and Atmospheric Administration (NOAA), 155, 194
National Park Service, 100
National Science Foundation, 193–95
Neal, Avon, 84–85
Nelson, E. W., 185
Nemerov, Howard, xix, 87, 225, 228
Nevada, 46–47
New England, 4–5, 11, 16, 22
Newlands Reclamation Act, 125, 129
New York City, 35–38, 56–59, 67–69, 71–72, 73, 76–77
Nims, John Frederick, 235–36
niphometrology, 144
nivean frontier. *See* snow frontier
Niwot Long-Term Ecology Research, 202–3
Norboe, Paul M., 138
Norse, Elliott, 160, 202
North American Snow Conference, 78
Notman, William, 26–27
Nuttall, John H., 64–65

O

Oldenburg, Claes, 56
Oldsey, Bern, 236
Olmsted, Frederick Law, Jr., 97–99
Olson, Sigurd, 238
Operation Haylift, 69
Operation Snowbound, 69
Orton, Clark, 33
Orville, Howard T., 193
Ostergaard, Clark, 203
Our Lady of the Snows, 246, 289n54, 309n64

P

Paepcke, Walter, 220
Page, P. K., 81, 87
Paget, Fred, 145–46
Palmer, Walter Launt, 212, 219
Parker, Ann, 84–85
Parshall, Ralph L., 142
Paz, Octavio, 225
Perla, Ron, 116
Pierce, Wayne M., 108–9
Poe, Edgar Allan, 257n35
Pollack, Vincenz, 115
Pomeroy, J. W., 136
Poulsen, Wayne, 169–70
Powers, Judyth, 239–40
Pressman, Norman, 76
Priestly, R. E., 166
Prishvin, Mikhail, 307n45
Project Cirrus, 192
Prokosch, Frederic, 166
Pruitt, William O., 198–200
Pseudomonas syringae, 110
Pyne, Stephen, xv–xvi

Q

qali, 199, 208, 220

R

radioisotope snow gauge, 148–51
railroads, 41–50
rammsonde, 113
Rango, Albert, 180–81
Raunkiaer, Christen, 189
Raup, Hugh, 188
Redfield, Edward Willis, 214–15
red snow, 182, 184
Reed, Arden, 9
Reichelderfer, Francis W., 69–72
Reiffel, Charles, 50–51
Reisner, Marc, 125
Richardson, John, 188
Rickmers, W. R., 94–95, 166
Riis, Jacob, 56
Robinson, Bruce, 216
Robinson, David A., 241
Robinson, Henry Morton, 67–68
Rochester, N.Y., 57–59
Rodin, Auguste, 55–56
Rodriguez, Sylvia, 111–12
Rogers, W. P. "Pat," 105, 117
Roker, Al, 77
Rooney, John, 53–56
Roosevelt, Franklin D., 99–100, 139, 236
Roosevelt, Theodore, 184–85
Ross, Andrew, 77

Rossby, C. G., 70
Ruesch, Pat, 203
Ruskin, John, 14–15
Russian words for snow, 160–61

S

Santayana, George, 208
Sax, Joseph, 110
Schaefer, Vincent, 171–72, 176, 190–93
Schaerer, Peter, 119
Schaffgotsch, Felix, 103, 107
Schinto, Jeanne, 237
Schmidt, R. A., 136
Schneider, Hannes, 107
Scott, Julia H., 22
Scott, Willard, 77
Seligman, Gerald, 164–66, 170–71, 176
Selznick, David O, 104, 276n44
Seton, Ernest Thompson, 184–85, 203
Shafer, Bernard, 156
Shaw, Napier, 210
Simmons, Herman Georg, 188
SIPRE (Snow, Ice, and Permafrost Research Establishment), 149
skiers' words for snow, 94–95, 105, 114, 166–71
skiing, 94–108
ski safety, 113, 117
ski trails, 113, 119, 278n55
sleet, 162–63
Sloan, John, 84–85, 216
slush, 65, 77, 162, 167, 170, 180
Smallwood, Charles, 24
Smith, C. Alphonso, 163
Smith, Margery Hoffman, 100
Smith, Robert A., 52
Smith, Seba, 6
Smith, Tony, 56
Smithsonian Institution, 12, 123
Snodgrass, W. D., 183, 203
SNOTEL (Snow Telemetry), 154–55
snow
—classifications: crystals, xx, 164, 172–75, 181; snowcover, xiv, 78–79, 165–66, 168–69, 171–82; storms, 70–79, 162–63; weight, 65. *See also* blizzards; Eskimo words for snow; flurry; skiers' words for snow; sleet; slush; and snowflakes
—definitions, 2, 159, 164–65, 178
—metaphors, xiii, 2–3, 9–11, 13–14, 67–68, 93, 96, 116–17, 119, 166, 207–46
snowballs, 82–84
snowbirds, 55
snowblower, 53
"Snow-Bound" (Whittier), 18–19
snowcover maps, 38–39, 154
snowfall, xiv, xix–xxi, 162, 165, 172, 177, 251n2
snow fences, 43–45
snowflakes, xx, 3–4, 23–24, 30–31, 164, 166, 174, 179, 181
snow flea, 32, 184, 190
snow frontier, 38–39, 183, 241
snow grooming, 112–14
snow humor, 38, 52, 71, 76–79, 81–82, 95, 111, 114, 119, 147–48, 222, 253n5
snowmaking, 108–14. *See also* weather modification
"Snow Man, The" (Stevens), xvii, 207, 209–10
snowmen, 84–87, 208
snow metamorphism, 165, 172–73, 175–81, 247–49
snowpack management, 149–57
snowsheds, 41–43
Snow Rate Meter, 153
snow removal
—chemical: Banox, 66; calcium chloride, 61, 64; Nalco 8181-C, 66; salt, 60–61, 66, 268n8
—plows, 45–46, 48–49, 60–64, 67
—shoveling, 42, 50, 52–53, 57, 63
—snow fighting, 59–60. *See also* cities
snow sampler, xiv, 129–34, 137–39, 142–44, 148, 152
snow shovels, 50–53
"Snow-Storm, The" (Emerson), 1, 10–11, 18, 208
snow survey, 125, 130, 132–33, 138–44, 147–57
snow vehicles, 106–7, 113, 281n71
Sobin, Anthony, 227
Soil Conservation Service, 137, 140, 142–43, 146, 149, 151–52, 154–55, 157
Sommerfeld, R. A., 174–76, 178
Spence, Clark, 191
Stafford, Harlowe, 139
Staples, Henry W., 52
Stefansson, Vilhjalmur, xvii, 236

Stevens, Wallace, 11, 20, 207–10, 225–27, 236, 244
Stewart, George R., 65, 68
Stewart, Susan, 240
Stidham, H. L., 58–59
Stieglitz, Alfred, 217–18
Stoddard, George Allen, 45
Stone, Robert, 237
Stone, Robert G., 168–69
Strong, Stephen, 160
Sturgis, Russell, 29
Sun Valley Lodge, 103–9
Suzor-Côté, Marc-Aurèle de Foy, 26
Swenson, May, 229

T

Tabler, Ronald D., 44
Tait, Arthur Fitzwilliam, 15
Tannehill, I. R., 70–72
Taubenfeld, Howard, 193
Taylor, Bayard, 219
Teale, Edwin Way, 238
Telfer, Edmund, 200
Tey Manufacturing Company, 108
Thayer, Abbott H., 186–87, 203, 238
Thiessen, Alfred H., 133–34
Thom, H. C. S., 153
Thomas, William, 184
Thoreau, Henry David, 9–10, 25, 238
Tiffany, Louis, 17
Timberline Lodge, 100–103
Tropeano, Joseph, 109–10
Tuan, Yi-Fu, 123
Turner, Frederick Jackson, 38, 183
Twachtman, John, 210–12, 214–15, 219
Tyler, L. L., 96–97, 99
Tyndall, John, 25

U

Underwood, Gilbert Stanley, 100, 104
U.S. Army Corps of Engineers, 143, 148–49, 151
U.S. Department of Agriculture, 13, 71, 96, 123–24, 140, 194
U.S. Forest Service, 96–103, 110–11, 115–16, 135, 137, 143, 148, 150, 168, 170, 175
U.S. Geological Survey, 196
Utah, 42, 115–16, 133–34, 138, 175

V

Van Brunt, Samuel, 84
Van Slyke, W. A., 91
Verbeek, Nicholas A. M., 201
Vermont, 100, 108, 111–12
Vigneault, Gilles, 243–44
Villon, François, 3
Vonnegut, Bernard, 190, 193
Von Neumann, John, 70

W

Walker, D. A. "Skip," 202
Wallace, Henry, 140, 143
Ward, Robert DeC., 163
Warhol, Andy, 53
Waring, George E., 58
Warman, Cy, 49
Warren, Israel Perkins, 22
Warren, Robert Penn, 229
Washington, John, 160
Washington, D.C., 62, 74–75, 241
Wasserman, Stanley, 152
water: consumption, 122–23; content of snow, 123–24, 129–30, 133–34, 137, 150; pollution from contaminated snow, 66, 111–12; snow as source, 121–26, 149–50; used in snowmaking, 109
Watson, James W., 18, 53
Watterson, Bill, 81–82, 86
Waugh, Frank A., 97–99
Weaner, Edgar C., 52
Weather Bureau, 69–72, 115, 123–25, 127–28, 130–39, 143–44, 146, 152–56, 163, 170, 196
weather modification, 190–98
weather reports, 70–71, 78
Webster, Noah, 1–2, 135
Wegener, Alfred, 191
Weisbecker, Leo, 195–96
Welles, Orson, 222
Welliver, Neil, 214, 223–26
Wells, Roger, 160
Welty, Eudora, 55
Wentworth, William P., 51
Western Snow Conference, 132, 140–42, 144–49, 167, 193
Westheimer, Milton F., 47–48
Wharton, Edith, 68, 268n50
Wheeler, David I., 34

White, Janice, 239
Whitman, Walt, 50
Whittier, John Greenleaf, 18–19, 53, 62, 237
Whorf, Benjamin, 161–62
Wilbur, Charles Dana, 122
Willard, Beatrice, 201
Wilson, Wilford M., 163
Winter, Pat, 240
winter carnivals: Dartmouth, 94; Michigan Technological University, 94; Montreal, 87; St. Michael's College, 94; St. Paul, 57, 87–93, 103
Wolcott, Marion Post, 220–21
Woods, Sarah, 203
Work, R. A. "Arch," 142, 146, 152, 156
Worster, Donald, xv–xvi, 125–26, 135
Wright, C. S., 166
Wyeth, Andrew, 223
Wylie, Elinor, 226–27

Y

yellow snow, 184, 198, 201

Z

Zwinger, Ann, 201